Cyberwarfare: Information Operations in a Connected World

SECOND EDITION

Mike Chapple | David Seidl

JONES & BARTLETT
LEARNING

World Headquarters
Jones & Bartlett Learning
25 Mall Road, Suite 600
Burlington, MA 01803
978-443-5000
info@jblearning.com
www.jblearning.com

Jones & Bartlett Learning books and products are available through most bookstores and online booksellers. To contact Jones & Bartlett Learning directly, call 800-832-0034, fax 978-443-8000, or visit our website, www.jblearning.com.

22831-1

Production Credits

VP, Content Strategy and Implementation: Christine Emerton
Director of Product Management: Laura Carney
Product Manager: Ned Hinman
Content Strategist: Melissa Duffy
Content Coordinator: Mark Restuccia
Project Manager: Kristen Rogers
Senior Digital Project Specialist: Angela Dooley
Director of Marketing: Andrea DeFronzo
Marketing Manager: Suzy Balk
Content Services Manager: Colleen Lamy
VP, Manufacturing and Inventory Control: Therese Connell
Product Fulfillment Manager: Wendy Kilborn
Composition: Straive
Project Management: Straive
Cover Design: Briana Yates
Media Development Editor: Faith Brosnan
Rights & Permissions Manager: John Rusk
Rights Specialist: James Fortney
Cover Image (Title Page, Part Opener, Chapter Opener):
© Yurii Andreichyn/Shutterstock
Printing and Binding: McNaughton & Gunn

Library of Congress Cataloging-in-Publication Data

Names: Chapple, Mike, 1975– author. | Seidl, David, author.
Title: Cyberwarfare : information operations in a connected world / Mike Chapple, David Seidl.
Description: Second edition. | Burlington, Massachusetts : Jones & Bartlett Learning, [2023] |
Identifiers: LCCN 2021013019 | ISBN 9781284225440 (paperback)
Subjects: LCSH: Cyberspace operations (Military science) | Information warfare. | Computer security.
Classification: LCC U167.5.C92 C43 2023 | DDC 355.4/1–dc23
LC record available at https://lccn.loc.gov/2021013019

6048

Printed in the United States of America
25 24 23 22 21   10 9 8 7 6 5 4 3 2 1

Contents

PART TWO   Offensive and Defensive Cyberwarfare   101

CHAPTER 5   **The Evolving Threat: From Script Kiddies to Advanced Attackers**   103

CHAPTER 9 Defense-in-Depth Strategies   193

Contents

This book is dedicated to our friend and colleague, Dewitt Latimer.
Rest in peace, dear friend.

Preface

Purpose of This Book

This book is part of the Information Systems Security & Assurance Series from Jones & Bartlett Learning (www.jblearning.com). Designed for courses and curriculums in IT Security, Cybersecurity, Information Assurance, and Information Systems Security, this series features a comprehensive, consistent treatment of the most current thinking and trends in this critical subject area. These titles deliver fundamental information security principles packed with real-world applications and examples. Authored by Certified Information Systems Security Professionals (CISSPs), they deliver comprehensive information on all aspects of information security. Reviewed word for word by leading technical experts in the field, these books are not just current, but forward-thinking— putting you in the position to solve the cybersecurity challenges not just of today but of tomorrow as well.

Consider this scenario: On a quiet June morning, sometime in the future, military commanders at U.S. Strategic Command in Nebraska gather around a computer screen, pointing at a series of social media posts that have just appeared on their screens. They are somewhat bewildered, because the posts seemed to come out of the blue, and they haven't heard anything from higher levels of command yet. Here's what they see:

International Cable News

Breaking News: President Jones killed in attack on White House. More to come.

Unified Press Agencies

Military troops ordered on high alert after DC attacks. Retaliatory strikes expected.

The commanding general picks up the hotline phone to place a call to the National Military Command Center to obtain further direction. She gets a puzzled look on her face when she hears a rapid busy signal. Normally, the watch officer in Washington answers the phone immediately. She quickly flips on the television and finds static instead of the normal cable news broadcast.

A young soldier in the command center turns to a computer connected to the public Internet and finds that he is unable to connect to any websites. The command center has no contact with the outside world and has received information that an attack against the nation's capital has resulted in the death of the commander in chief.

An alarmed airman approaches the general and reports: "Ma'am, we've lost control of one of our Predator drones. Station 6 is no longer able to control the flight, and the drone appears to be following orders from someone else." The alarmed command staff turns its attention to a monitor that is still streaming live video from the drone over a secure network and watches in horror as the drone begins to land at an airstrip in the Middle East. It is surrounded by inquisitive foreign military officers before the feed goes dead.

Did the attack against Washington actually take place? Is this the beginning of a major armed conflict? Were the social media posts legitimate or the results of cyberattacks against the press's social media accounts? Are communication circuits dead because of a bomb dropped on a communications complex or a cyberwarfare attack against the command center? How did the enemy gain control of that drone? What is going to happen next?

Of course, this is a fictional scenario. But each of the attacks described here has occurred in one form or another over the past decade. In this book, you will learn about the role that cyberwarfare plays in modern military operations. In today's connected world, it has become almost impossible to separate cyberwarfare from traditional warfare. The tools and techniques of cyberattacks have become part of the modern military arsenal, and cyberattacks can be expected before, during, and after armed conflict.

This book is divided into three parts. In Part One, you will learn about the history of cyberwarfare. Information is a military asset and has played an important role in armed conflict from the days of Sun Tzu and Julius Caesar to the present. With the emergence of the cyber domain, electronic battles have joined the ranks of air, land, sea, and space warfare. This cyberwarfare leads to a variety of new concerns. Military planners must learn how to select and attack cybertargets. Military ethicists must apply long-standing principles of ethics and the law of armed conflict to domains that were never previously envisioned.

In Part Two of this book, you will learn how offensive cyberwarfare has become an important part of the modern military arsenal. The rise of the advanced persistent threat has changed the face of cyberwarfare, and military planners must now be conscious of the Cyber Kill Chain: the series of cyberwarfare actions that include reconnaissance, weaponization, delivery, exploitation, installation, command and control, and acting on objectives. You will read about the history of cyberwarfare and how it evolved from a novelty in the 1990s to a powerful integrated weapon in recent years. You will learn about the various types of malware that plagued the Internet in the 1990s and early 2000s and how malware evolved into military weapons used to destroy nuclear facilities in recent years. You will also learn how nonstate actors have appeared on the cyberwarfare stage armed with more power than ever before.

You will also learn about the defensive strategies that militaries have developed to protect themselves against cyberattacks. The concept of defense in depth is critical to building a well-rounded defense that will stand up to cyberwarfare events. Military defenses have evolved to include technological defenses such as cryptography, endpoint protection, firewalls, and data loss prevention systems.

In Part Three, you will learn how cyberwarfare may evolve in the future. You will read about military doctrine's evolution to include this new domain of warfare and how military planners use threat modeling and deterrence to plan strategic and tactical cyberwarfare operations. You will also learn how recent events have opened a Pandora's box, setting the stage for future cyberwarfare attacks.

New to This Edition

This edition is revised and expanded to include new developments in U.S. military cyberwarfare doctrine as well as changes in the U.S. military's organizational structure for cyberwarfare. The text incorporates recent examples of cyberwarfare operations waged against the United States and its allies as well as examples of offensive cyberwarfare operations undertaken by those nations. Events like the 2020 SolarWinds Orion breach, large-scale cyberwarfare attacks, and defensive concepts like zero trust are all part of the updates. Cyberwarfare changes quickly, and new attacks, threat actors, and methodologies appear constantly. This book will help prepare you to think about cyberwarfare by combining historical context, current operational strategies, and insights into the future.

Learning Features

The writing style of this book is practical and conversational. Step-by-step examples of information security concepts and procedures are presented throughout the text. Each chapter begins with a statement of learning objectives. Illustrations are used both to clarify the material and to vary the presentation. The text is sprinkled with Notes, Tips, FYIs, Warnings, and sidebars to alert the reader to additional helpful information related to the subject under discussion. Assessments appear at the end of each chapter, with solutions provided in the back of the book. Chapter summaries are included in the text to provide a rapid review or preview of the material and to help students understand the relative importance of the concepts presented.

Audience

The material is suitable for undergraduate or graduate computer science majors or information science majors, students at a 2-year technical college or community college who have a basic technical background, or readers who have a basic understanding of IT security and want to expand their knowledge.

Cloud Labs

This text is accompanied by Cybersecurity Cloud Labs. These hands-on virtual labs provide immersive mock IT infrastructures where students can learn and practice foundational cybersecurity skills as an extension of the lessons in this textbook. For more information or to purchase the labs, visit go.jblearning.com/Cyberwarfare2e

Acknowledgments

The authors would like to thank the many people who contributed to the successful publication of this book.

First and foremost, we had a tremendous team of subject matter experts assisting us in the preparation of our manuscript. Justin Hensley, director of information security and infrastructure at the University of the Cumberlands, served as our technical editor and gave us valuable input during the writing process.

We would like to thank Melissa Duffy and Ned Hinman of Jones & Bartlett Learning for their continued friendship and support on this project. They helped bring together an incredible team of professionals who ensured its success. This manuscript benefited from the skills of a top-notch editing team. Sheryl Nelson corrected our typos, pointed out areas of ambiguity in the text, and served as a wonderful advisor throughout the project. Thank you all for your guidance and support.

We also extend our thanks to Carole Jelen of Waterside Productions, our literary agent. Carole's decades of experience and wonderful network of contacts proved themselves invaluable once again on this project.

Finally, we would like to thank the many people we never met who contributed to this book. Artists, layout specialists, and technical staff at Jones & Bartlett Learning helped this book make the leap from our minds to the printed page or electronic text that you read today. Thank you all for your help.

About the Authors

MIKE CHAPPLE, PhD, is teaching professor of information technology at the University of Notre Dame's Mendoza College of Business. He previously served as Notre Dame's senior director for IT service delivery, where he oversaw the information security function. In past positions, he served as both a consultant and an active duty Air Force officer. He is a technical editor for *Information Security* magazine and has written 25 other books, including the *Security+ Study Guide, Information Security Illuminated,* and the *Official CISSP Study Guide.* He earned his undergraduate and PhD degrees from Notre Dame in computer science and engineering. He also holds a master's degree in computer science from the University of Idaho and an MBA from Auburn University.

DAVID SEIDL is the vice president for information technology and CIO at Miami University of Ohio. During his career, he has served in a variety of technical and information security roles, including serving as the senior director for campus technology services at the University of Notre Dame where he oversaw cloud operations, ERP, databases, identity management, and a broad range of other technologies and services. He also served as Notre Dame's director of information security, led Notre Dame's information security program, and previously taught a popular course on networking and security for Notre Dame's Mendoza College of Business. David has written books on security certification and cyberwarfare including the *Security+ Study Guide, The CYSA+ Study Guide, The Official (ISC)2 Practice Tests,* and numerous other books. David holds a bachelor's degree in communication technology and a master's degree in information security from Eastern Michigan University, as well as CISSP, GPEN, GCIH, CSA+, and Pentest+ certifications.

PART ONE

The Cyberwarfare Landscape

Information as a Military Asset

WHAT DO YOU THINK OF when you imagine military assets? Chances are that your thoughts immediately go to the traditional implements of armed conflict—tanks, naval ships, rifles, fighters, and bombers. These are, after all, the traditional, highly visible ways that armies engage each other on the battlefield. However, throughout the history of warfare, information has played a crucial role in shaping the ways that nations engage each other in warfare.

From the earliest wars waged between ancient forces to the complex maneuvering of today's tremendous military forces, military leaders have sought to use information as a weapon. They seek information about opposing forces and attempt to spread misinformation about their own plans and objectives in the hope of skewing the outcome of battles. Information is, and always has been, a critical military asset.

Over the past century, society has made dramatic improvements in the ways people store, process, and transmit information. During the Second World War, the few computers that were available to the military were highly complex devices that took up tremendous amounts of space and had primitive computing capacity. Today, it's hard to find someone who doesn't have a smartphone in his or her pocket that possesses literally millions of times the computing power of those old "supercomputers."

It only makes sense that the advances in information technology (IT) that have changed the way people do business also change the way nations fight wars. The first natural extension of this technology is to perform the same tasks in the military sector that·IT performs in the private sector. After all, armies need human resources systems, spreadsheets, and electronic mail. It's easier to present an intelligence briefing using PowerPoint slides than transparencies on an overhead projector. Militaries can also use these technologies for military-specific purposes, such as maintaining target databases, calculating missile trajectories, and similar tasks. Information is a powerful tool in warfare, and information technology is a way to magnify the impact that superior information can have on the battlefield.

Once you understand information as a critical military asset, it is not a significant leap to imagine information as a military target. If an information system provides a military with a battlefield advantage, the opposing force would surely want to deny their enemy the use of that weapon. Actions they can take to destroy enemy information technology are, therefore, now high on the priority lists of modern military forces.

How do militaries attack the information infrastructure of their adversaries? It is certainly possible to engage them using traditional weapons. Dropping a bomb on a data center is a very effective way of destroying the computer systems it contains. But what if the enemy has a backup data center? In addition, sending a bomber to a data center deep within enemy territory puts friendly airmen in harm's way and requires an overt hostile action that may trigger an undesired escalation in the conflict between two nations.

This text explores the concept of *cyberwarfare*. This involves taking the information war to a new level—not only seeing information as a military asset and a potential target, but also using information technology as a potential weapon. Attacks that take place in the cyber domain use information technology resources to wage war against the technology infrastructure of an opposing force. This may include attacks designed to cripple enemy information systems but, as you will learn, may also include the use of electronic weapons to destroy traditional targets.

Chapter 1 Topics

This chapter covers the following topics and concepts:

- What cyberwarfare is
- How warfare has evolved over the course of history
- What the role of information in warfare is
- What role cyber plays in the domains of warfare
- What categories of information operations the cyber domain entails
- What the techniques of information operations include

Chapter 1 Goals

When you complete this chapter, you will be able to:

- Describe the relationship between cyberwarfare, information warfare, and information operations
- Explain the role that information has played in armed conflict over the course of military history
- Describe the concept of cyberwarfare and how it relates to the traditional domains of armed conflict
- Describe the techniques used to effectively fight in the cyber domain

What Is Cyberwarfare?

Cyberwarfare includes a wide range of activities that use information systems as weapons against an opposing force. The strategy outlined by the United States Director of National Intelligence (DNI) reflects the fact that the history of cyberwarfare is at a turning point. This domain of fighting is emerging, and the actions that countries take over the next several decades will shape doctrine, tactics, techniques, and procedures for years to come.

> **NOTE**
> Although this text offers a concise definition of cyberwar, it is important to point out that there is no agreed-upon definition of cyberwar among military planners.

The DNI's threat assessment in this area, relied upon by U.S. government officials, considers the cyberthreat to be a major threat to national security over the coming years. The risks come from two major activities of cyberwar:

- **Cyberattacks** are nonkinetic, offensive operations that are intended to cause some form of physical or electronic damage. These cyberattacks are what most people envision when they hear the term *cyberwar*. Cyberattacks may range from a computer virus designed to disrupt the control systems of unmanned aerial vehicles (UAVs or *drones*) to stealthy invasions of an adversary's information systems to alter information used to make military decisions.
- **Cyberespionage** involves intrusions onto computer systems and networks designed to steal sensitive information that may be used for military, political, or economic gain. Cyberespionage is akin to traditional intelligence-gathering operations that seek to gain access to protected information.

Cyberwarfare is the combination of activities designed to participate in cyberattacks and cyberespionage, on either side of the attack. Militaries certainly seek to attack other forces and will use the weapons of cyberwarfare to gain advantage when possible. At the same time, militaries must defend themselves against the cyberwarfare activities of other nations and nonstate actors. The combination of these offensive and defensive activities is cyberwarfare.

Likelihood of Cyberwar

Is all-out cyberwarfare likely to occur? Military and technology experts around the world hotly debate this question. Although there is no broad agreement on this question, the most common thought, echoed by the U.S. government's threat assessment, is that large-scale catastrophic cyberattacks are unlikely in the short term. Very few groups possess the ability to wage sophisticated, sustained cyberwarfare. Outside of the governments of the United States, China, Iran, North Korea, Israel, and Russia, only a handful of countries are known to have significant cyberwarfare programs. It is unlikely that any of these nations would launch a significant cyberattack against an adversary unless it was part of a larger war that crossed traditional domains.

Although it may be unlikely that the world will see a massive cyberattack in the next few years, that does not mean that cyberwarfare won't take place. Remember, cyberwarfare has two major activities: cyberattack and cyberespionage. It is extremely likely that each of the nations just identified has extremely sophisticated cyberespionage capabilities and is currently using them against many different adversaries.

As an example, the U.S. National Security Agency (NSA) appeared in the global spotlight after a defense contractor, Edward Snowden, released information that he alleged provided the world with a glimpse into the inner workings of an agency dedicated to cyberespionage activities. News reports alleged a massive worldwide electronic spying operation that caught billions of people in its dragnet. According to the reports, the NSA cooperated with technology companies to systematically undermine the security of products and collect information about system users and, when the companies would not cooperate, conducted cyberespionage operations to gain surreptitious access to those information systems.

It would be naïve to think that the United States is the only nation that sees the value of information that may be gained through cyberespionage. There is clear and convincing evidence that other technologically advanced nations have military units conducting similar activities designed to retrieve sensitive information from potential future adversaries and use it for military, political, and economic advantage. Cyberespionage is not only likely, it is happening on a large scale every day.

Finally, although it is unlikely that a nation will launch a massive cyberattack against another nation, it is very likely that cyberattacks will occur. Throughout this text, you will read examples of such attacks that have taken place over the past decade, and these examples will only continue to multiply. What, then, is the difference between these attacks and all-out cyberwarfare?

Many cyberattacks that take place today are not traceable back to a national government. They are the work of **nonstate actors**: individuals or groups who seek to participate in cyberwarfare but do so independently, without the endorsement of a national government. These individuals and groups may be extremely motivated to conduct hostile actions to advance their agendas but lack the sophistication and technical capability to conduct a sustained cyberwar. They do, however, pose the threat of causing significant damage against a limited scope of targets.

Consider, as an example, the hacker group known as Anonymous. This loosely organized collective of activist hackers has waged a collective cyberwar against organizations that it has found distasteful. The targets of Anonymous have included the Church of Scientology, government agencies, financial institutions, and proponents of defending intellectual property. Using Internet message boards, the members of Anonymous vote to select targets for their attacks and then wage cyberwarfare against their victims. These types of attacks can have crippling effects on their targets, but organizations of this type simply do not possess the scale to wage a massive cyberwar against an organized opponent.

Cyberwarfare Terminology

The terminology of cyberwarfare is not agreed upon and may often seem confusing and overlapping. This text covers the major activities of cyberwarfare, including both offensive and defensive activities. Some would argue that defensive activities are not truly cyberwar, but the authors disagree and choose to include them. Students of cyberwar must understand both offensive and defensive capabilities, tactics, and procedures. They are inseparable.

Also, many people make the distinction between cyberwarfare and **information operations**, a broader term used to describe the many ways that information affects military operations. The military defines information operations as actions taken to affect an adversary's information and information systems while defending your own information and information systems.

Finally, the military services also talk about **information warfare** as information operations conducted during a time of crisis or conflict to achieve specific objectives.

What is commonly agreed upon is that cyberwarfare activities (including cyberattack and cyberespionage) are part of information operations and that information operations includes activities (such as psychological operations and military deception) that are not included in cyberwarfare. This text uses the terms *information operations* and *information warfare* interchangeably.

The Evolving Nature of War

In 1775, a ragtag colonial military consisting of American patriots engaged the British military, the strongest armed force in the world, in battles at the towns of Lexington and Concord in Massachusetts. The colonists had very little going for them. They were outgunned, ill trained, and unprepared for a major military assault. They fought a conventional military battle using the linear formations of the time, and despite the odds, they prevailed, sending the British into a retreat toward Boston. That's when things began to get interesting.

Rather than allowing the British to retreat, the colonists set up ambush positions along the route and began to fight a type of guerrilla warfare that the British army had never before seen. Colonists hid in the woods, behind rocks, and in ditches and relentlessly attacked the retreating forces. British General Hugh Percy, sent to rescue the retreating forces, said, "The rebels attacked us in a very scattered, irregular manner, but with perseverance and resolution, nor did they ever dare to form into any regular body. Indeed, they knew too well what was proper, to do so. Whoever looks upon them as an irregular mob, will find himself very much mistaken."

Fast-forward 235 years. In 2010, a nuclear enrichment facility located in Natanz, Iran, suffered critical technical problems that caused significant damages to centrifuges critical to the uranium-enrichment process. News reports quickly linked this to a computer worm known as Stuxnet that seemed specifically targeted at damaging the Natanz facility. Although no nation has officially claimed credit, both the United States and Israeli governments have openly hinted at their involvement in the attack.

What do these two military actions have in common? They both mark major turning points in the evolving history of warfare. Before the American Revolution, it was common (in Europe) to fight in the British style—opposing forces lined up facing each other and fired their weapons until one side either fell or retreated to safety. The use of ambush techniques took the British by surprise and contributed to the eventual American victory. Before Stuxnet, the use of computers as weapons was not a mainstream military tactic. The attack on Natanz marked a bridging of the world of cyberwarfare and conventional warfare. The world is now on notice that the weapons of cyberwar are sophisticated and can cause damage in the physical world, similar to that caused by conventional weapons.

The past two centuries have seen a gradual evolution of the way nations fight. The two world wars were highly conventional battles between massive forces, but they saw the introduction of the air domain in warfare. The Japanese surprise attack on Pearl Harbor ushered in American participation in World War II and eventually resulted in the unleashing of atomic weapons on Hiroshima and Nagasaki that ended the war. The introduction of nuclear weapons on the international stage changed the course of history, resulting in a 30-year cold war that was punctuated by conflicts that kept the world on edge—wondering if the theory of mutually assured destruction would prevent the United States and Soviet Union from "nuking each other into the stone age."

This type of evolution will continue over the centuries to come. Societies will continue to engage in armed conflict and will seek to incorporate new technologies on the battlefield. Some of these will enhance physical weapons, but there will be a continuing evolution in the world of cyberwarfare. New weapons and tactics will take the stage, and new ways of fighting will change the future face of conflict between nations and nonstate actors.

The Role of Information in Armed Conflict

Throughout the history of armed conflict, militaries and military leaders have understood the importance of protecting sensitive information. They have also gone to great lengths to obtain the sensitive information of others that may be of strategic or tactical value. The development of cyberespionage techniques is a natural extension of this ancient objective. With large amounts of information stored in computer systems, it is only natural that militaries would seek to infiltrate those systems and gain access to enemy secrets.

Ancient Warfare

One of the earliest recorded attempts to preserve military secrets dates back to approximately 50 BC, when Julius Caesar faced a communications dilemma. As a military leader with forces spread throughout the reaches of the Roman Empire, Caesar needed to communicate with his generals on a regular basis to convey orders and status updates. Without access to any electronic means of communication, Caesar had to rely upon written documents, carried by messengers among his troops. Caesar's adversaries knew that these communications would be sent and would surely want to intercept anyone suspected of being a messenger in hopes of gaining access to Caesar's strategy.

Caesar compensated for this vulnerability in his communications system by using a simple, but effective technology known as the Caesar cipher. He simply went through messages character by character and shifted each character three places to the right. For example, every A in the message became a D. Every B became an E, every C became an F, and so on. He then sent this encoded message on its way with a messenger. Those who intercepted the message were not aware of the encoding system and, lacking knowledge of codes and ciphers, were unable to decipher its meaning.

When a general in the field received a message from Caesar, he knew how to reverse the encoding system: Simply shift each character three places to the left. Convert the Ds back to As, the Es to Bs, the Fs to Cs, and so on; eventually the original message from Caesar would appear.

Although the Caesar cipher was rudimentary, it worked effectively, preserving the security of Caesar's communications system and opening an era of communications security in military operations. Modern militaries share the same objective of both protecting their own communications and intercepting the communications of their adversaries. The difference is only that the tools of information security and cyberespionage have become technologically sophisticated.

World Wars

About 2,000 years after Caesar, military forces continued to find themselves focused on finding methods to preserve the secrecy of communications. The technology used to transmit those communications improved dramatically with the invention of the radio. Unfortunately, the same technology that made it easier for friendly troops to communicate also made it possible for the enemy to intercept those communications. Radio waves travel freely through the air. Anyone with an antenna can intercept them.

The wide use of radios made the use of codes more important. The Caesar cipher got the job done for the Roman army, but it was too simple for modern use. Anyone with a basic knowledge of codes could easily decipher this simple cipher. Specialized mathematicians, known as *cryptographers*, worked hard to develop encryption technology that made it hard for the enemy to decipher communications.

During World War II, the German and Japanese governments developed a specialized encryption device known as Enigma. This system, shown in Figure 1-1, resembles a typewriter. The operator first sets the machine to the code of the day. He then would key in the message letter by letter. As he pressed each key, a different letter would light up on the device. This would be the letter transmitted as part of the encrypted message. When the receiver got the message, he would reverse the process by pressing the keys corresponding to the letters in the encrypted message. The letters of the original message would then light up on the Enigma device.

FIGURE 1-1

One of the Enigma machines captured by the Allies during World War II on display at the National Security Agency's National Cryptologic Museum at Fort Meade, Maryland.

Courtesy of National Security Agency.

FIGURE 1-2

A U.S. Navy *bombe* machine designed to break the Enigma code on display at the National Security Agency's National Cryptologic Museum at Fort Meade, Maryland.

The Enigma system confounded Allied intelligence officials for years. The system was very complex, and military officers were simply unable to break it. British mathematicians, led by Alan Turing, undertook an operation code-named Ultra that eventually broke the Enigma code. They used a very large, special-purpose computer, known as a *bombe* to break the code. An example of one of the bombes used by the U.S. Navy appears in Figure 1-2.

After breaking the Enigma code, Allied war planners gained great insight into German operations. They deciphered bombing targets while enemy planes were in the air. Navy officers read communications intended for German U-boats. The entire German communications system fell into Allied hands. Winston Churchill is famously quoted as telling King George VI, "It was thanks to Ultra that we won the war."

Cold War

The end of World War II marked the beginning of the Cold War. This time of escalated tension between the United States and the Soviet Union lasted almost 50 years. The two superpowers postured with large stockpiles of nuclear weapons but rarely engaged in direct combat. Instead, one of the main characteristics of the Cold War was the battles fought between intelligence officers. Many agencies on both sides developed significant intelligence capabilities. These included the use of spies, eavesdroppers, satellites, and spy planes.

The success of the Enigma program and the wide use of electronic communications after World War II led to the development of sophisticated electronic intelligence capabilities. The NSA led the fight for the Americans, while the Committee on State Security (known by its Russian acronym, KGB) performed a similar function in the Soviet Union. Over the course of the Cold War, both sides developed massive signals intelligence capabilities and became able to spy on each other's communications.

Intelligence played a role in almost every aspect of the Cold War. During the Cuban Missile Crisis, President John F. Kennedy needed evidence that the Soviet Union was placing missiles in Cuba, very close to the Florida coast. He ordered Air Force spy planes to fly over the island, collecting valuable photos. These photos provided unmistakable evidence of missile activity.

Iraq War and Weapons of Mass Destruction

On the other hand, bad intelligence can have serious consequences. This became apparent during the war in Iraq that took place from 2003 to 2011. Before the war, U.S. and British officials claimed to have evidence that Iraq was developing weapons of mass destruction (WMD). Analysis after the war revealed that the programs had ended in the early 1990s, and that there was no credible evidence that Iraq was pursuing the WMD program that the Allies claimed.

In July 2002, several months before the March 2003 invasion of Iraq, British and American officials gathered in London to discuss war plans. In 2005, a copy of the minutes of that meeting was published in *The Sunday Times*. One section of those minutes quotes Richard Dearlove, the head of the British MI6 intelligence agency, as saying:

> Military action was now seen as inevitable. Bush wanted to remove Saddam, through military action, justified by the conjunction of terrorism and WMD. But the intelligence and facts were being fixed around the policy.

This was a damning accusation, as it asserted that military intelligence was being manipulated to tell the story that the government wanted people to hear. The intelligence community suffered reputational damage from this incident, and it took years to recover.

Domains of Warfare

Military planners have traditionally divided war-fighting capabilities into four domains. These domains are used to develop strategies and tactics as well as to organize forces. In fact, most modern militaries are organized according to these four domains of warfare:

- **Land**—The oldest domain of warfare, consisting of any fighting force that remains on the ground. Land forces include infantry, cavalry, armored vehicles, antiaircraft batteries, and artillery. In the U.S. military, the Army primarily controls the land domain.

- **Sea**—The domain of warfare fought on oceans, rivers, and seas. The sea domain includes all of a nation's naval forces. In the U.S. military, the Navy controls the sea domain.

- **Air**—The domain of warfare fought in the sky. The air domain includes fighters, bombers, reconnaissance aircraft, cargo planes, and fuel tanker aircraft. After World War II, responsibility for the air domain in the U.S. military transferred from the Army to the Air Force.

- **Space**—With the advent of space flight, the military added space as a domain of warfare. The primary operations in this domain include satellite operations and the use of intercontinental ballistic missiles. In the U.S. military, the space domain is a mission of the Air Force.

NOTE

Although most people would consider helicopters aircraft, they are considered part of the land domain because they are primarily used in support of ground troops. For this reason, helicopter aviation is an Army mission.

NOTE

The U.S. Marine Corps is a military service that spans domains. Responsible for amphibious warfare, the Marine Corps fights in both the sea and land domains.

During the early stages of cyberwarfare, planners struggled with placing the cybermission into these domains, and each service claimed responsibility for a portion of the mission. In 2010, a panel conducting the Quadrennial Defense Review for the U.S. Department of Defense (DoD) concluded that:

> Although it is a man-made domain, cyberspace is now as relevant a domain for DoD activities as the naturally occurring domains of land, sea, air, and space.

With this statement, the military recognized the **cyber domain** as the fifth domain of warfare, as shown in Figure 1-3. Defense officials concluded that they must organize, equip, and train forces to operate in the cyber domain just as they do for the four traditional domains. Additionally, they recognized that they must be able to conduct their operations across the other domains in cases where use of the cyber domain is degraded by enemy action.

The relationship between the five domains of warfare.

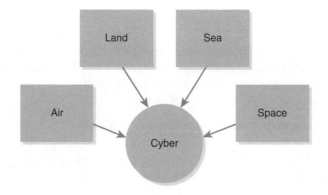

Rather than creating a separate branch of the military to fight in the cyber domain, DoD reacted to this new domain by creating the **U.S. Cyber Command (USCYBERCOM)**. In the USCYBERCOM's commander's vision, *Achieve and Maintain Cyberspace Superiority*, published in 2018, the command outlined five strategic imperatives that they are charged with achieving:

- "**Achieve and maintain overmatch of adversary capabilities**. Anticipate and identify technological changes, and exploit and operationalize emerging technologies and disruptive innovations faster and more effectively than our adversaries. Rapidly transfer technologies with military utility to scalable operational capabilities. Enable our most valuable assets—our people—in order to gain advantages in cyberspace. Ensure the readiness of our forces;
- **Create cyberspace advantages to enhance operations in all domains**. Develop advantages in preparation for and during joint operations in conflict, as well as below the threshold of armed conflict. Integrate cyberspace capabilities and forces into plans and operations across all domains;
- **Create information advantages to support operational outcomes and achieve strategic impact**. Enhance information warfare options for Joint Force commanders. Integrate cyberspace operations with information operations. Unify and drive intelligence to support cyberspace operations and information operations. Integrate all intelligence capabilities and products to improve mission outcomes for the Joint Force and the nation;
- **Operationalize the battlespace for agile and responsive maneuver**. Facilitate speed and agility for cyberspace operations in policy guidance, decision-making processes, investments, and operational concepts. Ensure every process— from target system analysis to battle damage assessment, from requirements identification to fielded solutions, and from initial force development concepts to fully institutionalized force-management activities—aligns to the cyberspace operational environment;

- **Expand, deepen, and operationalize partnerships.** Leverage the talents, expertise, and products in the private sector, other agencies, Services, allies, and academia. Rapidly identify and understand cyberspace advances wherever they originate and reside. Increase the scope and speed of private sector and interagency threat information sharing, operational planning, capability development, and joint exercises. Enable and bolster our partners."

USCYBERCOM is responsible for conducting operations across all of the military services in these areas of responsibility. The director of the NSA, a four-star military general officer, commands USCYBERCOM.

Exploring the Cyber Domain

Cyber is a domain of warfare as significant as the other domains. As the newest domain of warfare, it is the least understood. Military planners specializing in land and sea operations have millennia of military history to draw upon when developing plans and strategies. Air and space have shorter histories as war-fighting domains but still have existed for more than half a century. The cyber domain is much newer and military plans simply have not adapted fully to this new way of fighting.

The discussion of the cyber domain is organized around two major categories of information operations:

- **Offensive information operations**—Actions taken to deny, exploit, corrupt, or destroy an adversary's information or information functions
- **Defensive information operations**—Actions taken to protect your own information and information systems from an adversary's attempt to deny, exploit, corrupt, or destroy them

The combination of these two domains represents a full set of information operations capabilities that may be used by a military force to achieve its own objectives and prevent adversaries from achieving their own.

Offensive Information Operations

Offensive information operations are intentional military actions that are designed to adversely affect an enemy's information or information systems. Offensive information operations objectives fall into four categories:

- *Deny* an adversary access to his or her own information or information systems.
- *Exploit* the sensitive information belonging to an adversary for your own military advantage.
- *Corrupt* information in an adversary's possession.
- *Destroy* the information or information systems an adversary relies on.

Offensive information operations may have one or more of the preceding objectives as goals. In all cases, the operations are designed to achieve specific military, political, or economic objectives that benefit the attacking force.

Defensive Information Operations

As with all domains of military operation, the cyber domain is two-sided. While militaries certainly seek to exploit the cyber domain to their advantage, they must also recognize that their adversaries are doing the same thing. In his 2017 *National Security Strategy*, President Donald Trump recognized this when he stated:

> America's response to the challenges and opportunities of the cyber era will determine our future prosperity and security. For most of our history, the United States has been able to protect the homeland by controlling its land, air, space, and maritime domains. Today, cyberspace offers state and nonstate actors the ability to wage campaigns against American political, economic, and security interests without ever physically crossing our borders. Cyberattacks offer adversaries low cost and deniable opportunities to seriously damage or disrupt critical infrastructure, cripple American businesses, weaken our federal networks , and attack the tools and devices that Americans use every day to communicate and conduct business.

If cyberspace is a national security issue, then the military must defend it as they would any other domain. This requires investing in military and civilian personnel with the skills required to operate in the cyber domain and equipping them with the tools necessary to meet their mission.

One distinguishing characteristic of cyberspace is the fact that military and civilian lines are blurred. The 2017 National Security Strategy cites power systems, banking and financial operations, and the food supply chain as critical assets. None of those assets is under military control. Therefore, a successful defense of the cyber domain requires partnerships between government and the private sector. Planning for defensive information warfare requires this coordination as well as international cooperation between allied countries.

Information Operations Techniques

Information operations are more than cyberwarfare. They include any activity undertaken to attack or protect information and information systems. This chapter considers seven categories of information operations techniques:

- Computer network attack
- Computer network defense
- Intelligence gathering

- Electronic warfare
- Psychological operations
- Military deception
- Operations security

These seven categories were outlined in the *Information Operations Roadmap* developed by the DoD. The *Roadmap* organized information operations into these categories and made specific recommendations about how the military might better organize, train, and equip to wage information operations in the future.

Figure 1-4 illustrates the relationship among these domains, cyberwarfare, and information operations. Notice that computer network defense and computer network attack fall squarely within the realm of cyberwarfare. They correspond to the cyberattack function discussed earlier in this chapter. Some intelligence-gathering activities fall within the cyber domain: specifically those that use cyberespionage techniques. However, while all intelligence gathering fits within the domain of information operations, not all intelligence operations are cyberwarfare.

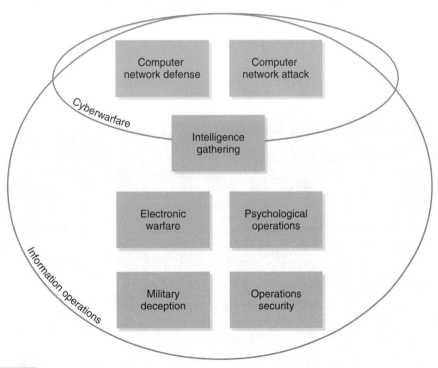

FIGURE 1-4

Cyberwarfare and the seven techniques of information operations.

Computer Network Attack

Computer network attack (CNA) is one of the core capabilities of offensive information operations and cyberwarfare. It consists of actions taken through the use of computer networks to disrupt, deny, degrade, or destroy an adversary's information and/or information systems. As knowledge of a military's CNA capabilities provides helpful information to an adversary's defense efforts, governments are extremely reluctant to openly describe CNA activities, even in a theoretical sense.

The weapons used by military forces engaging in CNA activities are similar (and sometimes identical) to those hackers use in seeking to undermine information systems security. The significant financial and human resources of military forces provide them the ability to create CNA weapons that exploit vulnerabilities in computer systems and networks—vulnerabilities unknown to the outside world. These are known as **zero-day vulnerabilities** and are extremely difficult to defend against.

Computer Network Defense

Computer network defense (CND) activities are designed to protect, monitor, analyze, detect, and respond to unauthorized activity in friendly information systems and networks. This domain closely maps to the civilian field of information security, and talent and tools are frequently exchanged between the military and the private sector in this area.

Edward Snowden and CNA Capabilities

After the Snowden disclosures, reports in the *New York Times*, *Washington Post*, *The Guardian*, and *Der Spiegel* provided unprecedented insight into the alleged then-current CNA capabilities of the United States and its allies. The reports alleged that governments have the following capabilities:

- The ability to intercept communications between commercial data centers operated by Google and Yahoo! by tapping undersea communications cables

- Access to encrypted communications through weaknesses in commercial encryption algorithms

- Direct access to Microsoft, Google, Facebook, Skype, YouTube, and Apple servers

- The ability to reveal sender identity information about some communications sent anonymously through The Onion Router (TOR) network

Although governments do not normally disclose CNA capabilities, the examples described by the media appear to show that CNA capabilities are extremely sophisticated. They reflect the investment of millions of dollars and countless hours of time and expertise in CNA techniques.

Intelligence Gathering

Intelligence gathering is one of the core competencies of information operations. It includes efforts to gather information about an adversary's capabilities, plans, and actions. Military effectiveness is enhanced when leaders and planners have access to information about their adversary. Intelligence operations seek to obtain as much of this information as possible.

The domain of intelligence collection includes a wide variety of activities that collect intelligence using diverse sources and methods. When those methods include the exploitation of computer systems and networks, the activities fall under the category of cyberespionage and are part of both information operations and cyberwarfare.

Electronic Warfare

Electronic warfare includes all military actions designed to use electromagnetic or directed energy to either control the electromagnetic spectrum or attack the enemy. Examples of electronic warfare in practice include the following:

- Jamming enemy radio transmissions
- Disrupting the use of global navigation systems
- Placing false images on enemy radar screens or removing real images from those screens

Computer Network Exploitation

The cyberespionage capabilities of the U.S. military are commonly referred to using the term **computer network exploitation (CNE)**. CNE uses the capabilities of CNA to gain access to information systems and then infects them with malicious software designed to steal sensitive information on an ongoing basis.

As with CNA activities, CNE operations undertaken by governments are not normally publicly discussed. In a 2013 article, *Foreign Policy* magazine alleged that the NSA has a Tailored Access Operations (TAO) group that conducts this type of warfare and had infiltrated more than 50,000 systems.

The magazine reported that the TAO has approximately 600 employees who work at NSA headquarters in 24-hour shifts. Their activities include hacking into systems, cracking passwords, stealing data, and installing malicious software.

In 2020, the Joint Chiefs of Staff released Joint Publication 3-85, *Joint Electromagnetic Spectrum Operation*, where they divided electronic warfare activities into three subdivisions:

- *Electronic attack* involves the use of electromagnetic energy, directed energy, or antiradiation weapons to attack personnel, facilities, or equipment. Electronic attacks intend to degrade, neutralize, or destroy enemy combat capabilities.
- *Electronic protect* includes actions taken to protect personnel, facilities, and equipment from any effects of friendly or enemy use of electronic warfare.
- *Electronic warfare support* includes actions taken to search for, intercept, identify, and locate sources of electromagnetic radiation. This is done for threat recognition, targeting, planning, and engaging in electronic attack operations.

Electronic warfare is waged using a wide variety of weapons systems, including fixed ground stations, specialized aircraft, and ships at sea. The U.S. military has been conducting electronic warfare operations for decades and has entire units dedicated to the electronic warfare mission.

Psychological Operations

Psychological operations (PSYOPs) are defined by the U.S. military as military operations planned to convey selected information and indicators to foreign governments, organizations, groups, and individuals in order to influence their emotions, motives, objective reasoning, and behavior. PSYOPs are the military equivalent of social engineering attacks in the nonmilitary world. In the Army Field Manual *Psychological Operations*, the military outlines five roles for PSYOPs:

- Influence foreign populations by sharing information subjectively. The information shared is designed to influence the population's attitudes and behavior and obtain compliance or other desired changes in behavior.
- Advise military commanders on ways to conduct military actions in a way that attacks the enemy's will to resist and minimizes the adverse impacts on psychological targets.
- Provide public information to foreign populations to support humanitarian activities, restore or reinforce legitimacy, ease suffering, and maintain or restore civil order.
- Serve as the commander's voice to foreign populations to convey intent and establish credibility.
- Counteract enemy propaganda to portray friendly intent and actions in a positive light for foreign audiences.

Propaganda is one of the major tools of psychological operations, and it may take place in oral or written form. Some of the most famous propaganda attacks against the U.S. military include the "Tokyo Rose" radio broadcasts from Japan to U.S. troops during World War II and the "Hanoi Hannah" broadcasts used by the Viet Cong during the Vietnam War.

© Shawshots/Alamy Stock Photo.

FIGURE 1-5

American propaganda leaflet distributed to the population of Germany during World War II.

Figure 1-5 shows an example of written propaganda used by the U.S. government during World War II. This leaflet, intended for the German population, includes the headline question "Do you want total war?" It goes on to tell the German people that they must choose between destruction and total war under the Nazis or normal and peaceful development under the Allies.

Military Deception

Military deception actions are designed to mislead adversary forces about the operational capabilities, plans, and actions of friendly forces. The goal of military deception is to guide the adversary toward taking actions desired by the entity engaging in deception. Military deception may occur at the strategic level, attempting to mislead foreign leaders into making tremendous strategic mistakes. It also may occur at the tactical level, with field commanders trying to mislead each other about their intentions.

Operations Security

Operations security (OPSEC) activities are designed to deny an adversary access to information about friendly forces that would reveal capabilities, plans, or actions. It is designed to prevent the enemy from successfully engaging in intelligence gathering. The OPSEC process has five components, shown in Figure 1-6. They are outlined in Joint Publication 3-13.3: *Operations Security*.

FIGURE 1-6

Steps in the operations
security process.

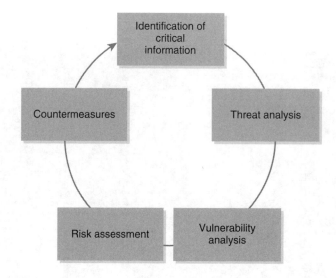

Operation Bodyguard

Both sides made significant use of deception during World War II. One significant military deception campaign was Operation Bodyguard, which took place in 1944. It consisted of a series of coordinated deception efforts led by the Allied Powers designed to mislead the Germans about the time and location of the impending Allied invasion at Normandy.

Several different operations, code-named Fortitude,. Graffham, Ironside, Zeppelin, and Copperhead, engaged in various deception tactics to paint a full, but incorrect, picture for the Germans. They used false radio signals, decoy actors, double agents, and false export restriction requests to mislead the Germans and cause them to structure forces in a manner that facilitated the Normandy landing.

The Operation Ultra intercepts of German Enigma communications mentioned earlier in this chapter also played a role in Bodyguard. The Allies had access to German communications and were able to tell which deception efforts were succeeding in misleading German commanders.

Identification of Critical Information

During the first phase of the OPSEC process, analysts seek to identify the essential information elements that would be valuable to the enemy and cause harm if disclosed. It is crucial that analysts identify this information because it allows the remainder of the process to protect only critical information, rather than trying to safeguard voluminous amounts of nonsensitive information.

The critical information identification process includes a wide variety of military staff representing different military disciplines. The work product of this phase is a document known as the critical information list (CIL).

Threat Analysis

After developing the CIL, operations planners conduct research based upon collected intelligence, knowledge of adversary intelligence capabilities, and publicly available information. The purpose of this analysis is to answer six fundamental questions:

- Who is the adversary?
- What are the adversary's goals?
- What is the adversary's likely course of action?
- What critical information does the adversary already know?
- What are the adversary's intelligence-gathering capabilities?
- Who will share information with the adversary?

The answers to these questions allow OPSEC planners to paint an informed picture of the adversary that may be compared with vulnerabilities in friendly forces during the risk assessment.

Vulnerability Analysis

During the vulnerability analysis, OPSEC planners examine every aspect of a planned operation to identify the ways that an adversary could gather pieces of critical information from the operation. Vulnerabilities exist when friendly forces provide adversaries with the opportunity to collect critical information, analyze it, and take action on it. The vulnerability analysis phase is focused on **indicators**, friendly actions and information that reveal critical information to the enemy.

The vulnerability analysis phase is designed to answer four questions:

- What indicators of critical information will be created by friendly activities?
- Which of those indicators can the adversary actually collect?
- What indicators will the adversary be able to use to the disadvantage of friendly forces?
- Will the use of OPSEC countermeasures actually tip the adversary off to more critical information?

> **Parking Lot Intelligence**
>
> OPSEC indicators come in the strangest places. One tried-and-true vulnerability that an adversary might use is to simply count the number of cars in government parking lots. The presence of an unusual number of people after hours may indicate imminent military activity.

Risk Assessment

During the risk assessment phase, OPSEC planners perform a thorough assessment of the information collected in the first three phases and the countermeasures that may be used to limit the ability of adversaries to collect those indicators. Planners then perform a cost-benefit analysis to identify which, if any, countermeasures should be implemented.

OPSEC planners perform three steps while conducting a risk assessment:

1. Analyze the vulnerabilities identified during the vulnerability assessment and identify possible OPSEC countermeasures for each.
2. Estimate the cost of implementing each OPSEC countermeasure (in terms of time, cost, and impact on operations) and compare it with any harmful effects that would result if an adversary exploits the vulnerability.
3. Select OPSEC countermeasures for execution.

The use of risk assessment allows commanders to make informed decisions about OPSEC countermeasures with knowledge of the associated costs.

Countermeasure Implementation

After selecting appropriate countermeasures during the risk assessment, commanders then execute the OPSEC plan. The overall strategy should meet four criteria:

- Minimize predictability from previous operations.
- Identify indicators that may tip the adversary off to the OPSEC activities.
- Conceal indicators of key capabilities and military objectives.
- Counter vulnerabilities in mission processes and technologies.

Once the OPSEC countermeasures are in place, OPSEC staff should monitor enemy reactions to identify whether the countermeasures are effective and feed this information back into the planning process.

CHAPTER SUMMARY

In this chapter, you learned the fundamentals of cyberwarfare and how information is a military asset. Cyberwarfare includes a wide range of activities that use information systems as weapons against an opposing force. It includes cyberattacks—designed to cause physical or electronic damage to information systems—and cyberespionage—intrusions into computer systems and networks designed to steal sensitive information. Cyberwarfare activities are a subset of the larger field of information warfare and information operations—activities that encompass all of the ways information affects military operations. Cyberwarfare is the latest step in the natural evolution of warfare. As societies have become more dependent upon information technology, so have their militaries, and this technological independence creates an opportunity to militarize cyber as the fifth domain of warfare. In this way, cyber complements the existing warfare domains of land, sea, air, and space.

Information operations techniques are divided into seven major categories. Computer network attack and computer network defense attempt to exploit and defend technology systems. Intelligence gathering has the goal of obtaining information about an adversary. Electronic warfare attempts to use the electromagnetic spectrum as a weapon. Psychological operations and military deception seek to sway and mislead enemy forces and leaders. Operations security seeks to deny adversaries access to critical elements of friendly information.

KEY CONCEPTS AND TERMS

Computer network attack (CNA)

Computer network defense (CND)

Computer network exploitation (CNE)

Cyberattacks

Cyber domain

Cyberespionage

Cyberwarfare

Electronic warfare

Indicators

Information operations

Information warfare

Intelligence gathering

Military deception

Nonstate actors

Operations security (OPSEC)

Psychological operations (PSYOPs)

U.S. Cyber Command (USCYBERCOM)

Zero-day vulnerabilities

CHAPTER 1 ASSESSMENT

1. Information warfare first appeared in the late part of the twentieth century.
   A. True
   B. False

2. Which one of the following is the newest domain of warfare?
   A. Air
   B. Land
   C. Sea
   D. Cyber
   E. Space

3. What are the two major categories of cyberwarfare? (Select two.)
   A. Viruses
   B. Cyberattack
   C. Cyberespionage
   D. Firewalls

4. The U.S. agency most closely associated with waging cyberwarfare is the _____.
   A. FBI
   B. NSA
   C. CIA
   D. DNI

5. All information warfare activities are examples of cyberwarfare.
   A. True
   B. False

6. What was the name of the British and American effort to break German cryptography during World War II?
   A. Ultra
   B. Enigma
   C. Purple
   D. Green

7. The evidence used to prove that the Soviet Union was building secret missile bases in Cuba was gathered by _____.
   A. satellites
   B. human spies
   C. wiretapping
   D. spy planes

8. What unified command of the U.S. military has primary responsibility for operations that take place in the cyber domain?
   A. USCENTCOM
   B. USEURCOM
   C. USSTRATCOM
   D. USCYBERCOM

9. Which of the following techniques include capabilities related to the conduct of cyber-warfare? (Select three.)
   A. Intelligence gathering
   B. Electronic warfare
   C. Computer network attack
   D. Psychological operations
   E. Computer network defense

10. It is extremely difficult to defend against an attacker who is exploiting a zero-day vulnerability.
    A. True
    B. False

11. All intelligence-gathering operations are examples of information operations.
    A. True
    B. False

12. What discipline includes activities designed to deny an adversary access to information about friendly capabilities, plans, and objectives?
    A. OPSEC
    B. Electronic warfare
    C. PSYOPS
    D. CNA

13. During which phase of the OPSEC process does the commander select appropriate OPSEC countermeasures?
    A. Vulnerability assessment
    B. Risk assessment
    C. Threat assessment
    D. Countermeasure implementation

14. What discipline includes actions taken to search for, intercept, identify, and locate sources of electromagnetic radiation?

A. Electronic attack
B. Electronic warfare support
C. Electronic protect
D. Electronic intelligence

15. What name is used to describe the process of gaining access to information systems and then infecting them with malicious software, all designed to steal sensitive information on an ongoing basis?

A. CNA
B. CND
C. CNE
D. CNY

Targets and Combatants

THE HISTORY YOU LEARN IN SCHOOL is filled with stories of military victories and losses. When the word *war* is mentioned, people immediately conjure up images of armies marching off to fight. Over thousands of years of documented human warfare, people have become used to identifying a variety of traditionally accepted military targets. These include military bases, vehicles, groups of soldiers, and infrastructure that supports military operations. Many of the same targets exist in cyberwarfare. An enemy can attack strategically important military units, facilities, infrastructure, vehicles, and, more recently, the control systems for drones.

The variety of targets and combatants who may participate in cyberwarfare is broad. Further, many of the countries considered to have significant cyberwarfare capabilities typically do not make public their doctrines for cyberwarfare. This chapter discusses both operational concepts for the U.S. Department of Defense (DoD) and concepts that expand the ideas of cyberwar beyond those currently accepted as part of U.S. information operations doctrine. Thus, it explores the wider potential boundaries of a constantly changing form of warfare. As you read this chapter, keep in mind that U.S. cyberspace operations doctrine defines three cyberspace missions: offensive cyberspace operations (OCO), defensive cyberspace operations (DCO), and DODIN operations, which are DoD operational actions to secure, configure, operate, and maintain the DoD's systems and capabilities. The United States also defines Information Operation (IO) as "the application of force and the employment of information with the goal of affecting the perception and will of adversaries" and in terms of seven components: computer network attack (CNA), computer network defense (CND), intelligence gathering, electronic warfare, computer network espionage (CNE), psychological operations (PSYOPS), and operations security. Finally, it defines three layers: the physical network, the logical network, and the cyber-persona. The shifting boundaries of cyberwar—and the interaction between all of these parts and layers of information operations—mean that to fully explore targets and combatants, you need to look beyond the current scope of U.S. doctrine to the future as well.

Cyberwarfare expands the boundaries of warfare beyond the traditional targets of conventional warfare. In a connected world where people are surrounded by the Internet of Things (IoT), where network-accessible devices, wearable computing, and embedded computers are part of many individuals' daily lives, attackers now have access to an ever-growing set of new targets. Now, organizations keep all of their business data in network-attached data stores, host those data stores in third-party cloud environments, and run significant portions – or in some cases, almost all of their business via the Internet. Militaries control their weapons, including drones and other significant combat, intelligence, and reconnaissance assets, via networks and often rely on poorly protected electronic systems for communications and other uses. Intelligence agencies and corporate intelligence specialists spy on other countries, corporations, and their own citizens using connected devices, malware, and other cyberwarfare techniques. In fact, advanced persistent threats in the form of cyberwar attacks by nation-state backed attackers are now a fact of life for organizations.

Technological progress has changed the world as you know it. Whereas once people were needed to access data, to control systems, and to gather information, you now live in a world where computers perform each of those tasks. Targets in the information age include smartphones, satellites, laptops, building and home automation systems, embedded systems like smart televisions and appliances, wearable devices, implanted and external medical devices, and many more. These pervasive parts of this computer-based lifestyle can give an attacker access to very targeted intelligence, the ability to shut down or harm critical infrastructure, or the broad media exposure needed to create widespread panic. Previously capabilities that were only within the reach of a national government or large corporation that allow intelligence-gathering and attacks are now within the grasp of small groups or individuals. Those with the resources to develop or acquire arrays of cyberattack tools can pick and choose which data they acquire and how, which populations they influence, and which systems they target and how they impact them.

Unfortunately for civilians, the network and telecommunications infrastructure used for the Internet is shared by the nations and their militaries that are targeted in cyberwar. In traditional warfare, civilian targets are generally expected to receive at least some protection. Cyberwar changes that, not only because of the shared infrastructure of the Internet, but also because cyberwar often ignores the traditional military targets. The greater accessibility of the tools and resources needed to wage cyberwar also means that it is often fought between groups that are not nations. They may not even be organizations that existed in a visible way until they attacked a highly visible target.

This chapter first explores the traditional targets of conventional warfare, their history, and why they are generally accepted as targets. Second, it compares them with their cyber equivalents, emphasizing how cyberwar changes the options available for attacking assets. This includes why those assets are useful targets and how to attack them. Third, it reviews ways in which cyberattacks and information operations expand the potential list of targets for combatants in cyberwar. It compares them with the targets often attacked in asymmetric or guerrilla warfare. The chapter also explores who the combatants are. It explores how other nation-states and civilian or other groups can attack traditional military targets. It also explores how nation-states conduct information operations and wage cyberwar against a wide range of targets through a series of real-life examples.

Chapter 2 Topics

This chapter covers the following topics and concepts:

- What traditional military targets are
- What cyberwarfare targets are
- What targets of information operations are
- Who combatants in cyberwarfare are
- How cyberwar compares with other types of warfare

Chapter 2 Goals

When you complete this chapter, you will be able to:

- Describe typical targets in both traditional warfare and cyberwar
- List common types of warfare
- Explain why cyberwarfare often more directly targets civilian infrastructure
- Describe typical combatants in cyberwar
- Describe methods used by attackers and their similarities and differences from traditional combat

Traditional Military Targets

The basic targets of military strategists and commanders in conventional warfare have not changed significantly for thousands of years. The names and specifics of the units and technologies may have evolved over time from log and stone forts, dirt roads, and combatants on foot or horseback to pillboxes and underground bunkers, railroads and airplanes, and armored vehicles. Throughout these changes the underlying reasons they have been important targets when nations go to war haven't changed. Military action is still conducted against military targets, infrastructure, communications, supply chains, and the research and development systems behind them.

This is because the goal of conventional warfare is to defeat the enemy's military, thus causing the opposing country's surrender or at least to force the opposing sides to negotiate terms. This type of conventional warfare is often more palatable to both sides in the conflict than an all-out, no-holds-barred war would be because it generally avoids directly targeting civilian populations or the total destruction of a country to win. When the combatants are not as well matched in size or capabilities, or there are deep underlying divisions between them, they may resort to other means than just conventional warfare.

Common Types of Warfare

Descriptions of how conflicts are fought frequently include several types of warfare. As you read this chapter, think about how each of these applies to cyberwarfare. Think about how cyberwarfare and information warfare techniques can change how each of these types of combat is waged. The common types of warfare include:

- **Conventional warfare** is fought with traditional military tactics and weapons. It is what most people think of when they envision war. Conventional warfare is fought against an enemy's military, thus forcing surrender or negotiation. Conventional warfare is often conducted in accordance with international treaties, laws, and agreements that set limits on behaviors related to combat and combat-related activities.

- **Unconventional warfare** tries to change how an enemy feels about the conflict by suggesting that peace cannot be achieved without giving in or compromising. Unconventional warfare is typically waged in the minds of an enemy populace rather than with their military and leaders. At times, more traditional military targets are used, but for psychological effect, rather than to win in combat. This type of attack is called psychological operations (PSYOPS) in U.S. information operations doctrine.

- **Asymmetric warfare** is fought between groups whose military power, tactics, or strategies are very different. As world history has shown, small groups that are not militarily stronger can sometimes defeat or at least force a stalemate or withdrawal by

> a larger force. In fact, **guerrilla warfare**, terrorism, and antiterrorism all fit within the description of asymmetric warfare. Many of the conflicts in the modern world can be properly called asymmetric warfare.
>
> - **Total warfare** is fought using all available resources. Traditional rules of war are set aside, and civilian and military targets are considered acceptable.

Conventional warfare between roughly similar **nation-states**—formally recognized countries or nations—is not the only type of warfare. In fact, unconventional warfare and asymmetric warfare have both been increasingly prevalent during the years since World War II. Later in the chapter, you'll examine cyberwarfare in relation to each of these types of conflict. First, though, to better understand cyberwar's place in modern conflict, you need to understand which targets are typically considered acceptable in conventional warfare and why.

Military Targets in Conventional Warfare

The broad variety of legitimate modern military targets in conventional warfare can be broken into a few categories:

- **Military targets**
  - Military forces and other combatants
  - Military leadership
  - Bases and base facilities
  - Fortifications and strong points
  - Supply depots and maintenance facilities
  - Other military installations

- **Infrastructure targets**
  - Communication infrastructure, such as fiber-optic and copper lines, satellites and satellite communications systems, cellular towers, and radio transmitters
  - Transportation infrastructure, including roads, bridges, waterways, railroads, and airports, as well as the vehicles using them
  - Power-generation capabilities primarily supporting military uses

- **Communications targets**
  - Broadcasting and television stations
  - Telephone, network, and telegraph exchanges of fundamental military importance

- **Supply chain targets**
  - Factories that produce military equipment and arms
  - Production, research, and design facilities that are primarily military-related
  - Supporting infrastructure for the supply chain

- **Military research targets**
  - Laboratories
  - Testing grounds
  - Engineering facilities primarily focused on military research and development

Nations have designed strategies around these targets, and battles have been fought over them throughout history for the same reason that modern military tacticians would focus on them: Defeating a nation-state in traditional warfare normally requires defeating its military. That military often cannot be fully defeated without targeting the entire organization behind it. Conventional warfare still relies on a few basic assumptions about the combatants:

1. The combatants are nation-states, or they at least have some form of military that can be separated from the civilian populace. In cases where this isn't possible, such as guerrilla warfare, popular revolutions, and other combat that doesn't fit the norm for traditional warfare, this is much more difficult, if not impossible. Treaties that define the rules of warfare sometimes treat combatants who are not part of a formal military structure differently for this very reason.

2. The supporting infrastructure for the combatants can be determined. In the case of a nation-state, it is simple enough to presume that targeting that nation's infrastructure within its national borders is militarily acceptable or at least may be deemed necessary during warfare. When fighting a guerrilla war, this becomes much more difficult. The supply chain for combatants is often the same as the one supplying noncombatants, and the guerrilla fighters typically blend into the civilian populace.

3. The combatants intend to follow the commonly accepted rules of warfare or those set forth in treaties. The commonly accepted targets presume that there is a desire to fight according to the commonly accepted rules of warfare outlined in international treaties like the **Geneva Conventions.** (These are discussed in more depth later in this chapter.) When that is not the case, the rules and accepted behaviors combatants use and expect can be grossly divergent. During wars fought by the United States and other countries in the latter half of the 20th century and the early part of the 21st century, the rules followed by military units have provided a host of such examples. For example, asymmetric combat, such as that fought against insurgencies, guerrilla fighters, and terrorist groups, may be fought with one side following the laws of armed conflict to a greater or lesser degree, and the other fighting without such limitations.

As mentioned earlier, these assumptions have been written into a number of treaties, laws, and directives that define behaviors and targets between signatory countries. They do not apply, however, when combatants are beyond the bounds of those agreements or are willfully violating them.

Acceptable Targets, Treaties, and International Law

Much like many armed conflicts of the modern era, cyberwarfare is often fought between combatants who haven't formally declared war. Or the combatants cannot formally declare war in an internationally recognized way because they are not nation-states with the ability to do so. Combatants frequently include both nation-state military and intelligence technical assets, nationalist and antinationalist groups, and even small groups and individuals spread around the world. With a variety of combatant types and frequently no formal declaration of war, the choice of acceptable behavior ranges from militarily acceptable combat to electronic warfare that is only restricted by the creativity and the technical capabilities of the combatants.

This wide range of scenarios doesn't mean that limitations aren't in place that could apply to cyberwarfare. International codes of wartime conduct include broad restrictions on targets. These could be interpreted in some cases to apply to digital attacks.

Unlike the rules of traditional warfare, the international community has not agreed to cyberwarfare treaties and accepted practices. Between 2009 and 2012 as cyberwarfare's implications and impacts became increasingly evident, a NATO-sponsored document called the *Tallinn Manual on the International Law Applicable to Cyber Warfare was written,* but no international law is currently in place. For now, the Geneva Conventions and other international treaties on warfare continue to be the best tools available when formal national military assets are being used. Unfortunately for those in the line of fire, most attacks are not being openly conducted as part of a formal, acknowledged, military operation during conventional war.

A Brief History of 100 Years of the Laws of War

Rules for warfare have existed for thousands of years, with examples appearing in the Bible, the Quran, and, in more recent history, the Declaration of Independence. In the United States, a formal code of wartime conduct has existed since President Abraham Lincoln issued the Lieber Code in 1863. The Lieber Code described how to treat captured populations, types of warfare that were unethical or inhumane, such as the use of poison or torture, how wars were started and ended, and what punishments were appropriate for those who violated the rules of war.

On the international front, the concept and custom of declaring military objectives separate from civilian targets has a long history in modern warfare. The First Hague Convention in 1899 and the Second Hague Convention in 1907 established both widely accepted laws of warfare and those activities that would be considered war crimes. The Hague Conventions formed a basis of behavior in warfare that included prohibitions against attacking undefended towns and using specific technologies like chemical warfare and special bullets designed to do greater damage to human bodies. They also set conventions regarding how countries could formally declare war and how neutral powers were to be treated.

In addition to the Lieber Code and the Hague Conventions, the Geneva Conventions are likely the most recognized 20th-century international agreements regarding warfare. Like the Hague Conventions, the term *the Geneva Conventions* actually refers to a series of conventions or treaty agreements, starting with the treatment of sick and wounded servicemen and adding various conventions on handling prisoners, noncombatants, and protection of the victims of conflicts.

Over time, the parties to the Geneva Conventions have added further clarifications regarding when they apply—from their original use during conventional, international-declared warfare to the 1949 addition of coverage for police actions and other nontraditional armed conflict.

FYI

The U.S. DoD requires compliance with the Geneva Conventions as part of DoD 2311.01, the DoD Law of War Program. The policy requires that DoD component organizations and contractors follow the laws of war regardless of the type or characterization of a conflict. It defines responsibilities for compliance, requires reporting of incidents, and establishes when legal review is required. You can read the full text at https://www.esd.whs.mil/Portals/54/Documents/DD/issuances/dodd/231101p.pdf?ver=2020-07-02-143157-007.

Cyber Targets in Unconventional Warfare

Unconventional warfare or psychological operations target the hearts and minds of opponents, attempting to defeat them via psychological means or to bring them to the negotiating table knowing they cannot win. Unconventional warfare techniques have been used for centuries, and while techniques have changed over time, the basic reason for unconventional warfare hasn't. Sun Tzu's *Art of War*, a book that is more than 2,000 years old, includes a reference to unconventional warfare: "One need not destroy one's enemy. One only needs to destroy his willingness to engage."

The Internet has made both mass media and more targeted communications far more accessible to individuals and groups, making unconventional warfare a popular starting place for cyberattacks. Social media creates broad exposure and awareness and provides opportunities to target and manipulate populations in ways that were not possible, or at least were not easily accessible previously. In addition, attacks on Internet services and sites can garner a high level of visibility, which is very attractive to attackers.

The impact of social media as a tool in unconventional warfare has been demonstrated repeatedly, including broad use of social media influence campaigns in the 2016 U.S. election. Modern PSYOPS and information warfare teams use social influence campaigns and carefully craft messages, campaigns, and even false personas for their efforts. In many cases, their ability to conduct these operations are enhanced by the commercial tools and information provided by social media platforms. The impact of this can be seen in the 2018 indictment of Russian operatives as part of the Internet Research Agency, which noted that the individuals involved conspired to defraud the United States by "impairing, obstructing, and defeating the lawful functions of the government...for the purpose of interfering with the U.S. political and electoral processes, including the presidential election of 2016."

The direct impact that social media can have was also demonstrated very visibly in April of 2013. The Syrian Electronic Army (SEA) compromised an Associated Press (AP) Twitter account. The SEA quickly used it to send out a tweet that appeared to be a legitimate AP news item that said, "Breaking: Two Explosions in the White House and Barack Obama is injured." Shortly after the tweet, the Dow Jones Industrial Average (DJIA) dropped by 140 points, although it quickly rebounded. The effect on the DJIA is a great demonstration of the power that a single compromised account can wield as part of unconventional warfare.

The 2020 Twitter compromise resulted in thousands of attempted transfers of cryptocurrency to wallets controlled by the attackers. In fact, almost $1.3 million in cryptocurrency and other transfers worth thousands of dollars were blocked. Fortunately the attackers did not take advantage of the broad range of verified accounts they had compromised or could have compromised to create more chaos and instead limited themselves to a simple scam. Scenarios where this attack could have had a much broader national impact on markets or even international diplomacy are easy to construct.

Common targets in unconventional cyberattacks include both traditional mass media and social media, as well as websites, applications, text messages, and even traditional means of communication between individuals like phone calls. Manipulation conducted via Twitter, Facebook, and other sites using coordinated misinformation campaigns can have a high relative value versus their relatively low cost and complexity to accomplish. In a world with a very short news cycle, few filters that help to ensure that reported events are accurate or true, and broad consumption and trust of social media, attackers have powerful misinformation and PSYOPS tools in their arsenal.

> **NOTE**
> One method used in attacks is *social engineering*, the practice of using human vulnerabilities as part of psychological operations and to help with technical attacks during cyberwar.

Targets in Asymmetric Cyberwarfare

Much like the nearly constant state of guerrilla warfare fought in Afghanistan, Iraq, Iran, and other countries over the past decade, violent nonstate actors are now starting to shape how electronic warfare is fought. Organizations without a nation-state command structure can spring up almost overnight and can wreak havoc on infrastructure, communications, and other connected targets.

As the fabric of everyday life becomes ever more networked, the threats that cyberattacks create make these organizations and individuals far more problematic for defenders. The more advanced your infrastructure is, and the more it relies on networks, telecommunications, and computer control, the more targets are available to potential attackers. Of course, occupying the technological high ground offers advantages too—including greater electronic and cyberwarfare capabilities and more infrastructure to coordinate, host, and manage attacks and defenses.

This can be seen in combat situations like U.S. operations in Afghanistan against the Taliban, which have leveraged cyberoperations to a significant extent as early as 2012 when *USA Today* quoted General Mills as saying, "I was able to get inside [the enemy's] nets, infect his command-and-control, and in fact defend myself against his almost constant incursions to get inside my wire, to affect my operations."

Total Cyberwarfare

Total cyberwarfare is the worst nightmare of those who defend networks and infrastructure. Unfortunately, the reality is that there are no commonly agreed to limits on cyberwar activities, and no real hope of a digital Geneva Convention despite ongoing calls for one to be created. The only restrictions on attacks are the abilities of the attackers, the skills of defenders, and the attackers' willingness to deal with the consequences of discovery. For years, national governments have been reluctant to admit to their capabilities, but as you will find, attacks like Stuxnet, Flame, and Aurora revealed some of the tools major players were able to leverage for attacks. Since then, the ongoing operations of advanced persistent threat groups like APT41, a Chinese-state-sponsored espionage group that targets healthcare, technology, and telecommunications organizations, have shown that nation-state sponsored groups can have broad impact to both national security and the business world.

 NOTE

You can read more in FireEye's detailed report on APT41 at https://content.fireeye.com /apt-41/rpt-apt41/.

Cozy Bear (APT29) lead the news in late 2020 due to their successful compromise of SolarWinds Orion infrastructure monitoring and management tools via SolarWinds' update process. This Russian-sponsored advanced persistent threat (APT) gained access to a variety of major U.S. government networks, as well as those of critical suppliers and businesses. While more than 18,000 SolarWinds customers were estimated to have potentially been impacted, Cozy Bear appears to have focused on specific targets to gain deeper access. At the time of the writing of this book, the U.S. government has responded via sanctions or diplomatic actions, but the scale and reach of the breach make some form of action likely.

In addition to Cozy Bear and APT41, Fancy Bear (APT48) has been active with attacks against German and French elections, multiple Ukrainian targets, and a host of other targets. Cozy Bear is an excellent example of a very active APT that is using both direct cyberattacks and influence campaigns to shift the political climates of their target countries.

These are only three examples of APT groups, and there are many more with varying levels of capability. Unlike nuclear proliferation, there is no worldwide oversight of cyberwarfare capabilities, no international court and law to assign sanctions when attacks are made, and there are no generally agreed on limits to what can be done.

Thus, the military use of **cyberintelligence**, cyberattacks, and even total cyberwarfare to the full extent of the attackers' abilities are all a possibility now. Cyberintelligence refers to intelligence activities related to cyberthreats, including identifying and analyzing their existence and capabilities. Keep in mind that the advanced capabilities of today may become the commonly available tools of tomorrow's guerrilla **cyberwarriors**—combatants in cyberwar, whether formally or informally trained. As society's reliance on comput-erized and network-connected devices grows, the vulnerability to cyberattacks increases.

Cyberwarfare Targets

As mentioned at the beginning of this chapter, cyberwarfare often makes a broader set of targets available to combatants for three major reasons:

- First, cyberwar is often waged outside of the traditional boundaries of armed conflict. Further, combatants who may not be nation-states often declare and covertly fight cyberwar.
- Second, civilian systems and networks are a significant portion of the available route to military and nonmilitary targets and are often easier to gain access to.
- Third, cyberwar does not have an existing body of international law limiting its scope as do the treaties that define targets in traditional combat.

This section looks at examples of cyberattacks and electronic attacks against traditional military targets, targets that are exclusively available to cyberattackers, and targets that combine elements of both.

Cyberwarfare against Traditional Military Targets

Traditional military targets, such as vehicles, bases, and the military supply chain, remain important targets, particularly when nation-states use cyberattacks against opposing forces. In some cases, cyberattacks use completely new modes of attack. In others, traditional warfare techniques are simply updated to use new technologies, leaving attackers and defenders to relearn old lessons in new contexts. The following examples show traditional targets under attack from computer- and signals-based cyberattacks. They also explore new attacks and exploits that are changing how the world thinks about the systems, applications, and devices that are becoming ever-more important parts of people's lives and culture.

Iran versus U.S. Drones

New military technologies are a particularly interesting target for cyberwarfare, as the constant pace of technological development means that they may not be as secure as their creators want them to be. Complex new systems often have the same types of vulnerabilities as civilian systems, despite their creators' best efforts. In addition, nation-states have far more resources to aim at them than do small groups of hackers and individuals.

Multiple examples of U.S. drones being shut down near Iranian airspace, as well as the 2016 capture of U.S. Navy patrol boats point to potential GPS (Global Positioning System) spoofing. Starting in 2011 with the capture of a U.S. drone, these events have continued through 2019 when a U.S. drone was shot down by Iran. Although drones have been shot down or have otherwise ended up in the hands of non-U.S. forces in other circumstances, in these cases the 2011 drone capture and 2019 drone that was shot down may have resulted directly from false GPS signals, as shown in Figure 2-1. When the drone that was captured in 2011 was presented to the world on Iranian state-run TV, the lower portion of the drone was blocked from view. Iranian engineers explained in interviews that the difference in elevation between the drone's base and the place where it actually landed was a matter of meters, resulting in damage to the drone's landing gear as it crashed to the ground while attempting to land.

FIGURE 2-1

A theoretical representation of how a GPS-guided drone could be deceived by fake ground stations using a more powerful or closer GPS signal than that provided by GPS satellites in orbit.

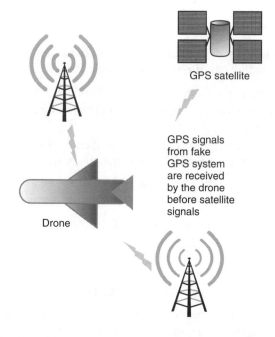

GPS satellite

GPS signals from fake GPS system are received by the drone before satellite signals

Drone

FYI

GPS spoofing isn't an unknown or top-secret military technique. In fact, it has become so commonplace that GPS spoofing software is available to help players with games like Pokémon Go, and GPS spoofers can be purchased in many online storefronts. Nation states continue to use it as well, but the technology is now widespread and easily accessible. You can read more about a variety of attacks at https://www.csoonline.com/article/3393462/what-is -gps-spoofing-and-how-you-can-defend-against-it.html.

Drones aren't the only targets for GPS spoofing. In fact researchers have reported that Russia has interfered with satellite navigation for thousands of ships in areas including Russia, Ukraine, and Syria, and similar interference has been reported in other parts of the world as well.

Of course, physical assets aren't the only target for military cyberattacks. Information remains a major target, and cyberintelligence can be used to replace traditional military intelligence. Organization's like the advanced persistent threat group known at Leviathan, or APT40, target specific information that is valuable for military technologies. Leviathan is believed to be part of China's efforts to modernize its navy, with attacks aimed at research projects and designs for marine equipment and systems.

This behavior isn't new. One of the earliest large scale attacks was seen in 2012, when the Flame malware package was discovered infecting systems in Iran, Syria, Sudan, and other Middle Eastern countries (see Figure 2-2).

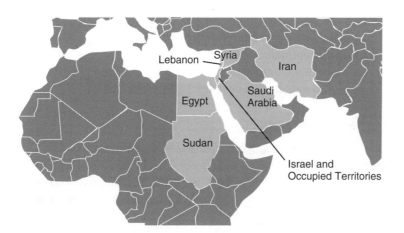

FIGURE 2-2

The Flame malware's infection focused on Middle Eastern countries, with the highest numbers found in Iran.

Flame: Replacing Spies with Software

In 2012, various computer security organizations announced the discovery of a complex malware package on Iranian and other computers. At the time, it was claimed to be the most complex malware that had ever been seen. The Flame malware appeared to be specifically targeted at Middle Eastern countries, and it had been operating since at least 2010.

> **NOTE**
>
> Complex threat actors with strong infrastructure and support are called advanced persistent threats or APTs. APTs are generally nation-state actors or are sponsored by nation-states and have advanced capabilities. They also operate on much longer time horizons than less complex threat actors, with long-term access often seen as a major goal for APTs. This creates significant challenges for many defenders who are not prepared for or capable of addressing such complex and capable threats.

As investigations continued, Flame was found to provide a highly modular design, allowing its controllers to add additional components and capabilities. The complex code included advanced compromise techniques and used a variety of known vulnerabilities to exploit and maintain access to systems. Once on those systems, it could then send information back to controllers, providing a broad intelligence net for the organization that created it.

In 2012, the *Washington Post* claimed that Flame was created by the U.S. National Security Agency, the Central Intelligence Agency, and the Israeli military in an attempt to gather information prior to cyberattacks against the Iranian nuclear program. Later analysis showed similarities between Flame and the Stuxnet malware that attacked Iranian nuclear facilities in 2010. Both the United States and Israel have denied responsibility.

The First U.S. Cyberwar Strike: Serbia and Kosovo

During the 1999 bombing campaign against Serbia as part of Operation Allied Force, the United States reportedly hacked into the Serbian air defense system. It distorted images the system showed in order to deceive Serbian air traffic controllers. This cyberattack was "essential to the high performance of the air campaign," according to John Arquilla, professor of defense analysis at the U.S. Naval Postgraduate School, in a 2003 PBS interview. This was the first publicized use of offensive information operations by the United States and is considered by some to be the first known U.S.

cyberwarfare operation. An Air Force Office of Information report noted that "the secret new arts of disrupting enemy capabilities through cyber-space attacks appeared to have been a big part of the campaign." However, reports at the time also noted that the full capabilities of the Information Offensive team were not used, and the team was not integrated in ways that would have helped it be fully effective.

Nontraditional Cyberwarfare Targets

In addition to traditional military targets, cyberattacks have been aimed at a variety of nontraditional targets. This section explores political activism and hacktivism, industrial espionage, guerrilla cyberwarfare, and military hacking and cyberattacks against nontraditional targets through a series of examples. As you read these examples, keep in mind the differences between their targets and those that the commonly accepted international laws of war allow.

Political Activism and Hacktivism

Political activism isn't typically thought of as part of traditional warfare, but cyberwarfare has created a new place for activists to make their views visible. It has also made it possible for activist groups to target national governments effectively. Activists are no longer limited to sit-ins, marching with signs, giving passionate speeches, and engaging in strikes. Today, a political activist can shut down his or her target from the comfort of home by attacking a website. Or he or she can spread rumors by hacking an official news source and publishing false articles. This online activism is commonly known as **hacktivism**.

> **NOTE**
>
> Omega, a member of the hacking group the Cult of the Dead Cow, first coined the term *hacktivism* in 1996. Later, Dorothy Denning, a noted cybersecurity and information warfare author, helped to popularize the term in her writings.

In addition to compromising servers, online political activist groups have found ways to make individual, Internet-connected PCs part of effective attack engines. Their politically motivated attacks are often characterized by denial of service (DoS) attacks and attacks against social-media accounts. In addition, online political activist groups bring attention to their chosen issues using social-media awareness campaigns.

Anonymous

Perhaps the most famous currently active online activist group that uses cyberwarfare techniques to target its opponents is the group known as **Anonymous** (see Figure 2-3). Anonymous is a nonstructured group without direct leadership, although strong members of the Anonymous community have led it toward many of its major actions. Because of its lack of internal structure, it is difficult to assign specific motives to Anonymous or to be certain which attacks are truly conducted by Anonymous and which are simply claimed to be done by Anonymous by third parties. While motivations vary, its members have tended to be against Internet censorship and have often claimed to act in favor of human rights, particularly around information and Internet-related freedoms. Somewhat uniquely, Anonymous has also had in-person, albeit masked, attendees present as part of its coordinated actions in some cases.

web icon

FIGURE 2-3

One of the logos commonly used by Anonymous.

2

Targets and Combatants

Anonymous and Anonymous-related groups have attacked targets associated with a range of issues, from antipiracy efforts, copyright, and digital rights management issues to human rights abuses, child pornography, and hate crime. Their targets have included corporations, government agencies, law enforcement, and others.

Now, attackers and malicious actors know that claiming to be a member of Anonymous or acting on behalf of a group like Anonymous can help conceal their true intentions or even give them an air of political-activism-related legitimacy.

Syrian Electronic Army

Some activist groups directly align themselves with a country. The Syrian Electronic Army (SEA) is a group of hackers who appeared in 2011 and acted in support of Syrian President Bashar Al-Assad. The SEA targeted websites belonging to human rights groups and news organizations, among others. Like Anonymous, the SEA also engaged in DoS attacks against organizations that it saw as opposing its political views.

In addition to defacement campaigns and highly visible social-media efforts, the SEA has hacked a number of high-profile organizations. SEA's successful attacks included the compromise of the *Forbes* blog site administrative console and a host of defaced websites, compromised social-media accounts, and other attacks. For example, the SEA hacked sites using Gigya's comment system in 2014, resulting in messages being displayed on hundreds of sites, including Forbes, the Independent, the Telegraph, NBC, the National Hockey League, and others. While SEA was highly active in the 2011–2015 time frame, SEA-related hacks have not been prominent since, and suspects were indicted in 2018 in the United States for some of the U.S.-focused SEA hacks.

The Impact of Hacktivism

The 2020 hack of a group of Twitter accounts including Elon Musk, Barack Obama, Bill Gates, Jeff Bezos, Kanye West, and companies like Apple and Uber was used to post messages like "Feeling grateful, doubling all payments sent to my BTC address! You send $1000, I send back $2000! Only doing this for the next 30 minutes," with a Bitcoin address included in the Tweet.

While the attack was relatively unsophisticated, the broad visibility and potential impact if it had been better implemented caught the attention of security practitioners, governments, and individuals around the world. With world leaders and others increasingly using social media to communicate important information, the ability of attackers to take over accounts and have a worldwide impact creates a real threat to national security and to global markets.

In the end, the Twitter hack proved to be the work of teenagers, and their total haul was only $118,000 of Bitcoin. They had used internal Twitter tools and social engineering to gain access, and their unsophisticated use of the hacked accounts limited the damage. The next time this occurs, however, it could be a more significant problem.

Hacktivism and activism remain a highly visible part of cyberwarfare activities and have even been used as a lure in social engineering attacks. 2020's Twitter breach of multiple verified accounts using a Twitter back-end administrative tool used language that note that the actual owners of the hacked accounts were "Giving back" during COVID-19. While the breach was discovered to be the result of successful spear phishing (targeted phishing) of Twitter employees, with 130 accounts targeted and 45 used to tweet, the rudimentary attempts to make the attack appear to be a form of activism did work enough for the attackers to make thousands of dollars in Bitcoin.

Industrial Espionage

Industrial espionage consists of intelligence activities conducted for business purposes rather than for national security reasons. It is a favorite technique of companies and nation-states looking for information outside their borders. Espionage activities against foreign countries targeted at stealing technologies, research, or information are nothing new in warfare. But the ever-increasing role of computers and electronic communications provides a way to attack companies without resorting to traditional human intelligence and spies. Nation-state industrial cyberespionage has an extensive history that dates back to the early 2000s and has continued to increase to the modern day.

Titan Rain: Attacks on U.S. Defense Contractors

The attacks in 2003 on U.S. government defense contractors, including Lockheed Martin, NASA, the Redstone Arsenal, Sandia National Laboratories, were a wake-up call for many. The complex, multi-channel attacks combined technical and social engineering methods to gather sensitive information. In the end, Titan Rain is believed to have allowed China to acquire data from a broad range of industrial sectors, including defense, energy, manufacturing, and aerospace companies as well as others in related fields.

Aurora: China versus Google

In 2010, Google released information via the Google blog about a sophisticated 2009 attack against its systems using what came to be called the Aurora (or Hydraq) Trojan. This was a malware application that allowed remote access to compromised systems. The attack resulted in the theft of source code and other intellectual property from Google, and the attackers attempted to access Google mail accounts of human rights activists. Via its official blog, Google described the attack and noted a few key points:

1. The attack was also aimed at "at least twenty" other major companies in a variety of sectors, including "Internet, finance, technology, media, and chemical" companies.

2. At the time, the attacks appeared to be aimed at "accessing the Gmail accounts of Chinese and Vietnamese human rights activists."

3. Third parties routinely accessed the accounts of dozens of human rights advocates throughout the world. Google suspected this was likely due to phishing scams or malware placed on the individuals' computers.

Although Google never specifically identified where the access occurred, noted security expert Bruce Schneier alleged that the Chinese hackers gained access to a Google-created, backdoor system used to provide wiretap capabilities for Google mail accounts by government agencies.

Over time, further information has come to light suggesting that at least 34 companies were targeted. As of this writing, the security company Symantec had reported that the Aurora attackers are still in operation. Moreover, they appear to have both a large budget and a broad array of unpublished (zero-day) vulnerabilities, which they use to maintain access to compromised systems.

The attack on Google demonstrated the attractiveness of attacks on major cloud communications providers. The presence of a government-required backdoor for legal compliance meant that attackers had a far easier way to gain access to individual emails. It also meant that attackers could use a central tool rather than having to attack dozens or hundreds of individual accounts and passwords.

The sheer number of Google user accounts—and the amount of data that passes through Google and other cloud service providers' data centers on a daily basis—is extensive. This makes Google and other cloud service providers a constant target for attackers seeking the information they possess. Although the civilian communication infrastructure was traditionally a part of intelligence-gathering missions, it is now likely to be directly targeted by military cyberwarfare teams, activists, and other hackers as well.

Saudi Arabian ARAMCO and Shamoon

In August 2012, attackers calling themselves the "Cutting Sword of Justice" targeted Saudi ARAMCO, the most valuable company in the world. They chose Lailat al Qadr (the Night of Power), a major Islamic holy day, for the attack. This is a common tactic attackers use to limit the number of defenders who will be on-site or monitoring systems. On the morning of August 15, attackers used a virus to erase data on three-quarters of

ARAMCO's corporate PCs, replacing the data with images of a burning American flag. The impact on ARAMCO was significant, resulting in a shutdown of its internal network, email, and Internet access.

FYI

Consider the following interesting facts regarding the possible thought process behind how the Shamoon malware was used:

- The malware was set to take action just as ARAMCO staff left for a one-week vacation, meaning that it had a better chance of going unnoticed for a longer period of time.
- Shamoon only attacked business systems, not those used for oil production. In fact, it had no impact on ARAMCO's oil production.

Thus, although this attack was very visible, it did not directly impact the actual core infrastructure of ARAMCO's oil production. The reasons why an attacker might choose to limit targets in this way might offer insight into the identity of the attackers.

Analysis of the virus showed that it had been built to do two things: destroy data and replace it with the burning flag, and to report back. Analysts named the malware Shamoon after a word found in its code and found clues to its origin and capabilities embedded within it. The portion of the code that had been used to destroy data was called Wiper, the same name as that of the Flame virus that had targeted Iranian oil companies earlier in 2012.

Although no direct line was found to link Iran to the attack, many analysts believe that the evidence suggests that Iran was responsible for both Shamoon and a similar attack against RasGas, a natural gas company in Qatar.

Columbian Strategic Business Information Attacks

Strategic level intelligence continues to be a major target, with 2019 attacks on the Columbian government and major companies by an APT believed to be from South America. Intellectual property and business information were both targeted by these attacks.

Industrial espionage attacks can disrupt industry, acquire data to create a competitive advantage, or transfer military or civilian technology. The frequently porous nature of corporate networks presents a relatively low barrier against attack. And such attacks can have an effect out of proportion to the ease with which attackers can implement them. As Shamoon showed, one piece of malware deployed into a corporate network can have a huge impact. If Shamoon had successfully targeted the control systems for ARAMCO's oil drilling, pumping, and other infrastructure facilities, it could have resulted in deaths and destruction of property instead of screens full of nationalist images.

Military Cyberattacks on Nontraditional Targets

As state-sponsored cyberwarfare capabilities grow, their usage against nontraditional targets becomes more likely. Cyberattacks are an increasingly attractive alternative to traditional conflict. Cyberattacks can provide the ability to neutralize an opponent's infrastructure, communications, or other capabilities before, or even instead of, a traditional attack. Because no international law prevents such attacks, they even fall into a gray area if and when they are discovered.

Often, both the attacker and the target keep quiet about nation-states' offensive cyberattacks against nontraditional targets. However, one example was brought to light in 2009, which can help you understand how cyberattacks were seen then and what thought might go into an attack now. This example occurred prior to the invasion of Iraq in 2003.

United States Versus Iraq, a Canceled Attack

According to a 2009 *New York Times* article by John Markoff and Thom Shanker, the United States had prepared a plan prior to the U.S. invasion of Iraq that was intended to "freeze billions of dollars in the bank accounts of Saddam Hussein and to cripple his government's financial system." Worries about collateral damage in the form of financial chaos in the Middle East caused the cancellation of the plan, despite assurances that it would work. Interestingly, the United States did use traditional banking means to freeze Iraqi assets.

Markoff and Shanker point out that the United States targeted Iraq's communications infrastructure using jamming and cyberattacks, and this had unexpected side effects. The attacks "temporarily disrupted telephone service in countries around Iraq that shared its cellphone and satellite telephone systems." This type of attack against communications is far more typical during warfare than disrupting Iraq's financial system via cyberwarfare would have been.

Targets of Information Operations

The targets discussed in this chapter can also be related to the seven major categories of information operations listed in Table 2-1. Remember that the U.S. military's formal definition of information operations include computer network attack, computer network defense, intelligence gathering, electronic warfare, computer network espionage, and psychological operations.

As you can see, both traditional and nontraditional targets are part of information operations. Attacks can themselves be targets of other attacks, as recovering information about the tool used can provide intelligence about the attacker's identity, capabilities, and goals. In some cases, the tools can even be reverse engineered to use against the original attacker.

Combatants in Cyberwarfare

The variety of possible combatants in cyberwar is as broad as the types of cyberwarfare. Military units with network security and cyberwar capabilities, intelligence organizations, businesses, insurrectionist and terrorist organizations, law enforcement, activist groups, and even individuals can all participate in cyberwar-related activities. Their capabilities, motivations, and the rules they operate by vary greatly. Defenders must be prepared to deal with attackers who have significant resources if defenders have valuable data, infrastructure, or access.

TABLE 2-1 Mapping information operations to common cyberwar targets.

INFORMATION OPERATIONS TECHNIQUE	TARGETS
Computer network attack	Networks, computers, and technology systems
Computer network defense	Inbound attacks, malware, and attackers
Intelligence gathering	Stored data, communications, sensor, and other live information
Electronic warfare	Broadcasting capabilities, control channels, GPS signals
Computer network espionage	Networks, systems, data stores, and communications
Psychological operations	Social media, websites, email, and other communications that influence targets
Operations security	Protect information systems and networks as well as the personnel who operate and manage them

In its broadest definition, any aggressor can wage cyberwar against any opponent. To truly be wide-scale conflict, however, cyberwar must typically involve a state-sponsored group and an opponent that it targets, whether it is another nation-state, an alliance, a guerrilla resistance, an insurrection, or a business.

Military Forces

Modern military units focused on cybersecurity and cyberwar have been an acknowledged part of military service since the early 1990s, when the U.S. Air Force used cyberwarfare techniques in Operation Desert Storm. Although Russian, Israeli, French, and Chinese cyberunits are maintained in relative secrecy, at least some U.S. organizations operate in the public eye. Further, the U.S. policy of developing a strong information operations capability that combines electronic warfare with psychological operations, deception, operations security, and electronic warfare is public knowledge.

U.S. Cyber Command

The United States operates the U.S. Cyber Command (USCYBERCOM), a unified command combining elements of the Air Force, Army, Marines, and Navy. It contributes components of the joint command, which includes a number of information- and cyberwarfare-specific units:

- Air Forces Cyber (Sixteenth Air Force) includes reconnaissance, intelligence, and surveillance wings, dedicated cyberspace wings, and the Air Force Technical Application Center, which monitors for nuclear treaty compliance.

- The Fleet Cyber Command (10th Fleet) is composed of 40 Cyber Mission Force units, 28 active commands and 27 reserve commands, and more than 14,000 active and reserve military members as well as civilian staff.
- The Marine Corps Cyberspace Command, the Marine cyberwarfare unit, has approximately 800 dedicated personnel, including the Marine Corps Cyber Operations Group, and the Marine Corps Cyber Warfare Group.
- The Army Cyber Command (2nd Army) is composed of 6,500 soldiers and 10,000 civilian contractors and employees, with five worldwide Cyber Centers, 41 active cyber teams, and 21 reserve cyber teams.

This sampling of the USCYBERCOM's elements shows the breadth and diversity of units that the U.S. Cyber Command includes and demonstrates the strategic importance of cyberwar capabilities for the United States. The U.S. military sees cyberwar as both a threat and a significant part of the United States' war-fighting capabilities and has increasingly used electronic means of attack as a lower-impact capability than traditional warfare activities.

Guerrilla Cyberwarriors and Insurrectionists

One of the biggest challenges in cyberwar can be identifying your opponent. The systems that an attack comes from may simply be compromised and used to hide the attacker's trail. Even if you recover the tools used, clever opponents make sure the clues they leave in their attack take investigators down a false trail. This means that it can be next to impossible to prove an attack is nation-state sponsored as part of its cyberwarfare program rather than conducted by a guerrilla group, insurrectionists, or patriots attacking from their own country due to a real or perceived slight.

> **China's Hacking School**
>
> Although no official confirmation exists, China is believed to have one of the largest and most capable state-sponsored hacking training groups in the world. The Aurora attacks conducted by APT17 and discussed earlier in this chapter are generally believed (albeit, without direct proof) to be state-sponsored due to the targeting of Chinese political dissidents as well as information that points to the network origins of the attack. This type of attack doesn't serve typical traditional military purposes, but it does reflect the cyberintelligence side of cyberwarfare.
>
> FireEye provides an extensive list and description of APT groups at https://www.fireeye.com/current-threats/apt-groups.html. Significant investments by many countries have resulted in highly capable and complex APT groups being deployed around the world.

In some cases, the best way to identify attackers is by following their social-media presence and their interactions with news organizations. Claiming responsibility for attacks isn't a guarantee that the claims are true, but it can help you to understand where and when guerrilla groups have been active.

The options for asymmetric electronic conventional warfare and unconventional warfare that network attacks provide have been repeatedly exposed during the past decade. But unconventional warfare remains the more accessible option for most guerrilla groups. A few examples include the following:

- The Taliban has consistently advocated using Western technology to overcome the West. In 2013, reports came out that the Taliban may have successfully hacked the navigation systems of a German Heron reconnaissance drone, causing it to crash. It isn't a one-sided fight, however, as both pro-and anti-Taliban attacks have occurred—including repeated compromises of the Taliban website. In 2011, for example, one successful attack resulted in the Taliban website announcing the death of the Taliban's leader. In an additional example, attackers in 2013 sent a false email in the name of the Taliban spokesman, declaring a temporary end to hostilities.

- In many ways, 2013 was the year that Iranian hackers revealed themselves to be a believable cyberwar threat, affecting systems across the world. Iranian hackers successfully attacked U.S. energy companies, universities, and financial institutions as well as foreign energy firms and other groups. It isn't clear if the attackers were state sponsored or not, but because the attacks came from Iranian universities and companies, the Iranian government was blamed. The attacks may have been in retribution for the Stuxnet worm, which disrupted the Iranian nuclear program.

- An ongoing cyberconflict has been fought in Ukraine for well over a decade and has served as a testbed for numerous cyberwarfare techniques and tools. Russian-affiliated APT groups like Cozy Bear, Fancy Bear, and Sandworm have all been observed operating in Ukraine. Targets have included 2014 attacks against the Ukrainian Central Election Commission; 2016 attacks against Ukraine's power grid, which resulted in a blackout for 230,000 Ukrainians; and the 2017 NotPetya attack, which caused an estimated $10 billion in damage.

One of the more interesting developments the Internet has enabled is the ability for civilians to attack nation-states without being present in that country. This means that expatriates, activists, and insurrectionists are no longer limited to fundraising and material support from a distance, and they no longer have to travel to the area where the conflict is occurring to contribute to it.

Individuals and Small Groups

One of the biggest differences between cyberwarfare and more traditional types of warfare is the power a single, focused individual—or a small group of individuals—can exert. Although great leaders and tacticians have changed the direction of battles and wars, individual combatants cannot typically defeat an entire nation's military. Cyberwarfare changes that balance of power, with individuals potentially able to significantly disrupt systems or to change the power balance through their own actions. Now an individual can choose to attack a nation-state, conduct his or her own guerrilla warfare, or use targeted attacks to selectively damage organizations, companies, or other groups.

When these solo hackers, cyberwarriors, and activists succeed, it is typically because they are persistent, highly motivated, and highly skilled. They exploit weaknesses that can be leveraged on a large scale. Sometimes they are motivated by simple curiosity or the desire to prove themselves. But most who can be considered in this context have a purpose beyond that. The following examples look at the effects of individual cyberwarriors, which far outweigh the normal reach of a single person.

The Jester

Individuals aren't always working against a national government. In some cases, they might even be described as patriotic, such as in the case of Th3j35ter (The Jester), an American hacker who targeted the hacktivist group Anonymous mentioned earlier in this chapter.

"A small team of A players can run circles round a giant team of B and C players" (Th3j35ter, 2010).

Over time, the Jester targeted a series of more than 200 groups ranging from Anonymous to Wikileaks to the Westboro Baptist Church (an American congregation known for extreme ideologies). He claimed that the Internet provided terrorists with a place to operate safely, and his goal was to prevent that.

The Jester also engaged in psychological operations. He aimed an attack at one of the most common weapons Anonymous uses, a tool called the Low Orbit Ion Cannon (LOIC). This is a **distributed denial of service (DDoS)** application that provided politically motivated Internet users with an anonymous way to use their computers to support Anonymous. In essence, the LOIC was an Internet cannon that fired traffic from hundreds or thousands of systems at its chosen targets. The Jester claimed that he would remove any chance of anonymity by providing a version that intentionally sent out the actual address of the attacker, thus exposing members of Anonymous (see Figure 2-4).

The Jester has continued to be quite active over time, with other attacks including fake news articles, dos attacks against websites, and successful campaigns to expose the leaders of hacktivist and hacker groups. He remains active on social media but has not been publicly identified, despite threats to do so in the past, which led to him create new social media accounts.

> **NOTE**
>
> A DDoS or uses many systems or devices to attack targets. This type of massed attack can overwhelm systems and services simply due to their scale, which can include thousands, tens of thousands, or even hundreds of thousands of systems. Very large DDoS attacks can even disrupt the function of major service providers, making such attacks a real concern for both civilian and military security practitioners.

Comparing Traditional Warfare, Guerrilla Warfare, and Cyberwarfare

This chapter has explored the targets that are accepted both in international treaties and traditional wartime behaviors. It has also delved into how cyberwar targets differ from traditional targets and the ways that non-nation-state actors can participate in

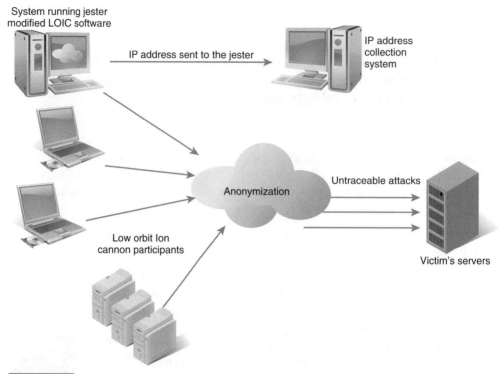

System running jester modified LOIC software

IP address sent to the jester

IP address collection system

Anonymization

Untraceable attacks

Low orbit Ion cannon participants

Victim's servers

FIGURE 2-4

The Jester's method of defeating the LOIC attack software's anonymization abilities replaced the original software with a version that sent its network address, exposing the system.

cyberwar. With a firm understanding of what traditional warfare and cyberwarfare can entail, it helps to add a third type of combat into the mix: police actions, guerrilla, and asymmetric warfare (which this text calls *guerrilla warfare* for convenience). Note that this oversimplifies the concepts each of those descriptions can include, but for the purposes of this text, it is a useful shorthand.

Modern combat often involves groups that are not nation-states, and the ways in which guerrilla warfare is waged are often much closer to how cyberwarfare is conducted. Table 2-2 looks at a few of the most common targets and participants in traditional warfare, guerrilla warfare, and cyberwar. Pay particular attention to the similarities between cyberwarfare and asymmetric warfare.

How Cyberattack Differs from Traditional War

This chapter examined how cyberattacks can be compared with traditional warfare and how combatants in cyberwar line up with those found in conventional war. A few critical differences are important to keep in mind, however. First, consider the different effects current cyberattacks can cause. Second, consider the blurred lines between cyberwar and personally or financially motivated hacking, where the tools can be used for non-cyberwarfare purposes. Third, consider how surprise and deniability factor into

TABLE 2-2 Comparison of traditional warfare, asymmetric warfare, and cyberwarfare.			
	TRADITIONAL WARFARE	**POLICE ACTIONS, GUERRILLA, AND ASYMMETRIC WARFARE**	**CYBERWARFARE**
Combatants	Military personnel Intelligence operatives	Military personnel Intelligence operatives Nonstate actors	Military personnel Intelligence operatives Nonstate actors Small groups or individuals
Leaders	National leaders, governments, or military officers	Frequently cell-based, distributed leadership, often with a formal command structure	Sometimes centrally coordinated, such as organized crime or nation-state groups, but also often led by a group, or composed of loosely affiliated individuals
Military targets	Bases, vehicles, weapons, infrastructure Typically avoid nonmilitary-focused civilian infrastructure	Vary from military to civilian	Network-connected or using electronic signals like GPS or Wi-Fi.
Civilian targets	Civilian infrastructure typically considered a limited target	Civilian infrastructure often considered a legitimate target	Target-rich environment, with many different organizations with differing levels of cybersecurity and capabilities
Civilians	Typically considered a limited target	Often considered a legitimate target	Often a primary target, including systems, networks, and infrastructure
Information warfare	Propaganda targeted at opposing forces and civilian populace	Thematic information campaigns against nation-state or other organizations	Disinformation campaigns, website, Twitter, and other media platform hacks and account compromises

cyberwar. As you consider the difference between traditional warfare and cyberwarfare, keep these differences in mind:

- In traditional warfare, conventional attacks can destroy or significantly damage military units, fortifications, and bases. Cyberattacks are generally not currently capable of that level of physical destruction, although you can see how attacks on infrastructure are possible. It is also easy to predict that attacks that seize control of drones and other remote-or computer-controlled conventional weapons could occur in the future. Thus, for now, the capability for physical destruction by cyberwarfare is lower than that of conventional warfare. But it is likely to increase over time.

- Cyberwarfare activities, particularly low-level cyberwarfare activities, are difficult to distinguish from simple electronic attacks. In fact, the sole difference may be the reason the attack occurs, with cyberwarfare being different simply because is it not conducted for personal or monetary gain.

- Internal political activism doesn't fit with traditional warfare or even typical asymmetric warfare techniques. However, the tools used by political activists to take down or deface political party websites and attempt to change the results of elections are the same tools some actors use in cyberwar.

In addition to these differences, cyberwarfare offers both a new way to use the element of surprise and the chance to disguise attacks, providing deniability.

The Element of Surprise

One of the most effective strategies employed in modern cyberwarfare is technological surprise. Nation-states, commercial organizations, and many other groups build profiles of their expected attackers to assess the risks they face and to plan their defensive strategies. Without detailed information about their opponents' capabilities, however, their guesses are sometimes wrong. That can lead to surprises. The relatively significant impact of Iranian attacks against United States and international businesses in 2013 is an example of surprise in cyberwarfare. Traditionally, the United States, Russia, China, Israel, and France have been seen as the only countries with truly viable cyberwarfare capabilities. Iran's entry into the world of cyberwar was a surprise to many and was more effective because its true capabilities were not known. It is fair to expect that nations with existing cyberwarfare capabilities will continue to be surprised as new opponents bring their capabilities online.

Deniability

Perhaps one of the most interesting differences between cyberwarfare activities and traditional warfare is the ease of deniability for most attacks. It is reasonably easy to cause attacks to come from systems in a country or location that is either a believable attacker or that makes the location of the true attacker very difficult to identify.

Significant attacks like Aurora, Flame, and Stuxnet, allegedly sponsored by nation-states, have been put through in-depth analysis by experts in programming techniques, language, and even analysis of how different cultures tend to craft computer programs.

Although that level of analysis can help identify which countries might have created the tool, it is exceedingly difficult to prove that a country deployed it rather than a rogue element or non-nation-state actor, or that it did so purposely. Because of this, well-crafted cyberattacks provide a strong potential for deniability, which can be much harder to ensure in traditional warfare.

CHAPTER SUMMARY

This chapter compared cyberwarfare activities with those typically seen in conventional, unconventional, asymmetric, and total warfare. You learned that conventional warfare is typically fought between nation-states and has long-standing rules and restrictions found in international treaties like the Geneva Conventions. Unlike conventional warfare, unconventional and asymmetric warfare are frequently fought outside of those rules by non-nation-state groups.

You learned that in modern cyberwar, targets are as varied as the participants. Without broad international treaties or ethical and legal restrictions, those targets are essentially unrestricted. Nonmilitary targets, industry, civilian infrastructure, and even individual noncombatants are all potential targets in a world where there are no accepted cyberwar rules. Nations, guerrillas, insurrectionists, and even small groups and individuals use cyberwarfare techniques to attack each other and to defend their own systems, networks, and infrastructure.

KEY CONCEPTS AND TERMS

Anonymous

Asymmetric warfare

Conventional warfare

Cyberintelligence

Cyberwarriors

Distributed denial of service (DDoS)

Geneva Conventions

Guerrilla warfare

Hacktivism

Industrial espionage

Nation-states

Total warfare

Unconventional warfare

CHAPTER 2 ASSESSMENT

1. Cyberwarfare is addressed by the Geneva Conventions.

   A. True
   B. False

2. Which U.S. joint command coordinates cyberwarfare and security operations?

   A. The Joint Cyber Task Force (JCTF)
   B. The 24th Air Force
   C. USCYBERCOM
   D. Stargate Command

3. Which two types of conflict are typically considered asymmetric warfare? (Select two.)

   A. Cyberwarfare
   B. Guerrilla warfare
   C. Total warfare
   D. Insurrection

4. Which two agreements serve as the basis for international law regarding warfare? (Select two.)

   A. The Geneva Conventions
   B. The Treaty of Versailles
   C. The Hague Conventions
   D. The Lieber Accords
   E. The International Warfare Treaty of 1945

5. Cyberwar can be fought by which combatants?

   A. Nation-states
   B. Insurrectionists
   C. Guerrillas
   D. Corporations and businesses
   E. All of the above

6. Political activism via cyberattacks is often called _____.

   A. Anonymous
   B. Cyberterror
   C. Hacktivism
   D. Politihacks

7. The first major admitted use of cyberwarfare by the United States was during _____.

   A. Operation Enduring Freedom
   B. Operation Red Twilight
   C. Operation Moonlight Thunder
   D. Operation Desert Storm

8. Which attack targeted Saudi Arabia's national oil company?

   A. Shamoon
   B. The Low Orbit Ion Cannon
   C. Aurora
   D. Stuxnet

9. _____'s tweet using a hacked AP account resulted in a 140-point drop in the DJIA.

   A. Anonymous
   B. The SEA
   C. USCYBERCOM
   D. The Chaos Computer Club

10. Which type of warfare do insurrectionists use to change public opinion?

   A. Unconventional
   B. Conventional
   C. Total
   D. Civil

Cyberwarfare, Law, and Ethics

MANY EXPECTATIONS, ACCEPTED BEHAVIORS AND PRACTICES, and commonly agreed-on laws have arisen over thousands of years of recorded warfare. These laws, treaties, and agreements limited the scope, destruction, and casualties of armed conflict. They allowed nations to lose without being destroyed or to pursue war without making their civilian populations and infrastructure an automatic target. During the 20th century, those commonly agreed-upon rules of warfare were written into international law in the form of the Hague Conventions and the Geneva Conventions. These international laws, the formation of the United Nations (UN) and its related judicial bodies, and international treaties and agreements provide a useful foundation of laws that can be taken into account in the context of cyberwarfare.

The Geneva Conventions and the additional protocols that have been added to them since their original creation in 1949 have shaped how modern warfare is fought—describing what is acceptable and what is not. They set forth how armed combat is fought, how prisoners of war are treated, and how civilian populations should be treated and protected from indiscriminate attacks. They also set forth how medical aid and other humanitarian efforts should be treated and preserved. The additional protocols added to the Geneva Conventions have recognized changes in warfare, such as new weapons and the increasing role of guerrilla forces. These protocols also address the potential effects of destroying strategically important facilities, such as nuclear power stations, dams, and the environment itself.

Cyberwarfare tests the boundaries of existing international laws for many reasons. By its nature, it typically requires the use of civilian infrastructure to conduct attacks. The systems from which attacks are conducted are often civilian systems—or the attacks pass through civilian systems as part of their path to government and military targets. In addition, carefully aiming cyberattacks to target only traditionally acceptable targets can be difficult. Those systems can be hard to distinguish from civilian systems or may actually coexist with them on the same hardware in cloud computing data centers.

Cyberwarfare also creates the potential for unrestricted attacks by nontraditional combatants. It places what may be a possibly far more powerful weapon in their hands in the form of malware. Modern advanced malware is designed to disrupt or destroy infrastructure and computer-based systems and networks. It provides a powerful asymmetric weapon in the hands of insurrectionists and guerrillas as well as the militaries and intelligence operatives of nation-states. In traditional warfare, attacks required significant resources and capabilities to strike at physical structures and personnel. Today, however, attacks can be conducted at the push of a button with an army of silent malware-based zombie systems.

This complex environment makes the interpretation of the existing framework of international law challenging. Fortunately, experts in cyberwarfare attack and defense as well as legal experts have provided an analysis for the UN in the form of the Tallinn Manual. The manual examines the existing bodies of law and legal precedent. It couches cyberwarfare in those terms, providing a meaningful way of looking at where cyberwarfare and traditional warfare have strong parallels, and where cyberwar creates real questions.

In addition to legal analysis and responsibilities under international law, the ethics of cyberwarfare are also an important part of the equation for combatants and noncombatants alike. Ethical standards exist for information security professionals, but distinct codes of ethics for cyberwarriors have not been created. Thus, this chapter examines the ethical codes of security professionals to better understand which important elements may be part of the ethics of cyberwar.

Chapter 3 Topics

This chapter covers the following topics and concepts:

- What kinetic warfare is
- What the role of law in cyberwarfare is
- What the ethics of cyberwar are

Kinetic Warfare

Traditional military conflict has been fought using weapons like bombs, tanks, and guns. Warfare between nations has involved military forces that invaded enemy territory or that fought the opposing country's army, navy, or other forces. At times, civilians were involved either because they were caught in the combat area, due to their role in a militia or resistance, or because warfare had grown to involve every possible target.

As new forms of warfare have evolved that do not require the same type of direct, visible force, a new terminology has arisen. It describes the traditional type of military conflict as **kinetic warfare**. High-technology cyberwarfare is called **nonkinetic warfare**.

The international body of laws for military conflict between nations provides a useful starting point when examining cyberwarfare in a legal context. Although cyberwar provides new capabilities and raises new questions, many of the definitions used in kinetic warfare match those that exist when cyberattacks cross into the physical world.

International Law and Kinetic Warfare

Laws of warfare have existed in one form or another since ancient times. But treaties and international agreements in the 20th century have created widely accepted standards for the conduct of war. The first major international agreement on warfare was the 1899 Hague Conventions, which defined rules and customs of war on land, set limits on certain types of weapons, and banned poison-gas projectiles. In 1907, the Hague Conventions added rules on how to start hostilities, naval warfare, and updates to the previous customs and laws of land warfare.

 NOTE

Earlier international laws existed, including earlier Geneva Conventions. However, they did not have the same breadth and widespread adoption as the Hague Conventions and the later Geneva Conventions.

FIGURE 3-1

Civilian, military, and shared
use of infrastructure.

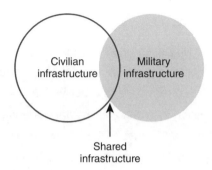

Over time, the need for updates and a newer body of international law on
warfare became obvious, and in 1949 the Geneva Conventions were signed. In
1977, additional protocols were added, clarifying and adding additional rules to the
originals. Since 1977, the Geneva Conventions include rules on the treatment of
prisoners of war; how medical personnel, journalists, children, and civilians should
be treated; and a wide variety of other rules of warfare. They have been updated
to deal with new technologies, including nuclear reactors, poisonous gases, and
other new types of warfare and weapons. They define when **civilian infrastructure** is
protected and when military use of infrastructure results in civilian infrastructure
becoming a valid target (see Figure 3-1).

In addition to the Geneva Conventions, the UN Charter and foundational documents
provide additional rules for member countries, particularly around the use of force and
declaration of war between member countries. The UN also provides international courts,
which have examined conflicts between member nations and have released rulings about
appropriate behaviors and acceptable responses.

The laws of warfare are commonly split into two categories:

- *Jus ad bellum*, Latin for "the right to war." *Jus ad bellum* law determines when
  nations are allowed to enter into war. International law now typically allows
  nations to respond to immediate threats, or active attacks, but not simply to declare
  war. The UN Charter requires nations to receive UN approval before using force
  against another nation except in self-defense. The same rules require that responses
  to attacks be proportionate to the attacks, and that they should not have overly
  broad effects.

- *Jus in bello*, Latin for "the law of war." *Jus in bello* law determines what is allowable
  during wartime. The Geneva Conventions define acceptable conduct during war
  such as that previously mentioned, and thus are *jus in bello*, rather than the UN
  charter's *jus ad bellum* law.

Like most laws and standards, changes in how warfare is fought can lead to changes in
the laws and standards. Cyberwarfare changes how attacks are conducted and can blur
the lines between civilian and military systems and infrastructure. Further, the existing
laws of armed conflict don't fully cover cyberwar. Even when cyberwarfare and kinetic
warfare intersect, the way existing standards apply can appear uncertain.

FYI

The laws of warfare are based on consensus. In other words, they exist only because countries agree to them and agree that they are appropriate. The way that each country interprets the laws of war can vary and often changes over time as circumstances, leadership, and public feelings change. In addition to the varying interpretations of the rules of war, some countries have agreed only to portions of existing treaties or choose to not sign them at all. Thus, although the laws of war are well known, they are not consistently followed, and they are constantly being challenged.

Legal Review and Legality of Actions

Legal authority to determine what is and is not legal according to international law primarily resides with international courts like the International Court of Justice (ICJ), the primary judicial branch of the UN. The ICJ has typically followed the philosophy that "that which is not prohibited is permitted" in its findings on international law and warfare.

Of course, the findings of the court are only binding on those countries that agree to them. This means that at times countries may not agree to be bound by the court's rulings, and states that are not members of the UN are sometimes not subject to the UN's influence.

Cyberwarfare Law

Countries throughout the world have written laws regarding cyberattacks that address their own internal needs. However, the international laws of warfare in the form of the Geneva Conventions and UN decisions have not been updated to address cyberwar techniques and technologies. Major cyberattacks, such as Stuxnet, Aurora, and Flame, as well as the increasing development of cyberwarfare capabilities by nation-state militaries, demonstrate that this is an area that nation-states intend to use more heavily in the future.

Due to the lack of precedent from international courts, and the quickly evolving capabilities of cyberattackers, effective law covering cyberwarfare may be difficult to create. Until the Geneva Conventions are updated, and the UN creates rules around cyberwarfare activities, the existing commonly accepted rules of warfare will be applied where they fit. Nation-states will exploit the gaps between what was possible in kinetic warfare and the new capabilities that cyberwarfare makes available to them.

Cyberwarfare in a Kinetic Warfare Context

The Stuxnet attack offers a look at the growing role of cyberwarfare as an alternative to kinetic warfare. The strike, which was aimed at Iran's uranium enrichment facilities, used a variety of network attack techniques to penetrate non-network-accessible infrastructure. Prior to the advent of techniques like those used in the Stuxnet attack, the only ways that a nation could have disabled an enrichment facility would have been through a kinetic attack—through military action or via intelligence operatives or moles planted in the facility taking action and thus risking their lives. In our post-Stuxnet world, this type of cyberattack has become more frequent, with intelligence, espionage, and industrial secrets as constant targets as well.

Cyberattack methods also have the potential to disable, disrupt, or even redirect military computers and computer-controlled systems. The increasing use of drone-based weapons systems and computer-controlled platforms, including ships, missile systems, and other major weapons platforms, means that a computer network attack could result in the ability to take over kinetic warfare systems. Reliance on tools like GPS as well as Internet connectivity mean that cyberattacks can directly influence or even change the course of kinetic conflict.

Cyberattacks that result in kinetic attacks might also result in escalation between combatants that did not actually intend to be at war. A hijacked drone or missile system that strikes the cyberattacker's target rather than that of the country that owns the weapons system would trigger the target country's right to self-defense against the apparent aggressor. Setting a nation-state up to wrongly bear the blame for an attack is not a new concept, but doing it remotely using the innocent state's own weapons systems and networks has the potential to change international relations quickly.

Kinetic Warfare Law in a Cyber Context

Kinetic warfare can also cross over into cyberwar when the combatants respond to or preempt cyberattacks using traditional means of warfare. A defender may be able to stop an attack in the following circumstances:

- If the location that the attacks are coming from can be determined
- If the network carrying the attacks can be targeted
- If the individuals who are conducting the attack can be captured or killed

The ability to respond kinetically to stop a cyberattack can be challenging. Many of the large-scale cyberattacks up to this point have been hard to track. And some, such as the Stuxnet, Aurora, and Flame attacks, have been very difficult to attribute to specific attackers or physical facilities. In those cases, physical attacks would likely have been overly broad because the individuals and computing resources used to create and launch the attack were not known.

This doesn't mean that kinetic attacks aren't possible in cyberwar. As computer network attack and defense capabilities continue to receive significant investment by military forces in many countries, the likelihood of combining kinetic strikes and cyberwarfare capabilities will increase. Over time, strikes that combine a cyberattack intended to disable systems or to soften defenses prior to a kinetic strike or invasion are far more likely. Countries that lack strong computer network defense and attack capabilities are likely to respond with traditional military capabilities.

With the intersection of the commonly accepted international rules of warfare and cyberwarfare remaining undefined in many areas, the need for a better understanding of where existing laws and traditional rules are useful and where they aren't is increasingly obvious. Various legal experts have worked to develop mappings between the large number of existing laws and treaties as well as common wartime customs and what is possible during cyberwar. One of the best examples of this type of work is the Tallinn Manual.

The Tallinn Manual

The **Tallinn Manual** is an in-depth review and analysis that provides nonbinding advice based on the existing international body of law around armed conflict and related topics. It was created for the NATO Cooperative Cyber Defence Centre of Excellence, an organization sponsored by the United States, Italy, Spain, Germany, and a number of other nations.

The Tallinn Manual is a response to a growing focus on cyberattacks by both nation-states and nonstate actors and the challenges those attacks provide when nation-states attempt to determine if their activities or those of their attackers are covered by international law. International views on whether the existing laws and treaties regarding warfare apply to cyberattacks and cyberwarfare vary, and the threshold conditions that determine when a state of war exists are often a point of contention.

> **NOTE**
> As this chapter describes important parts of the Tallinn Manual, the text often refers to it as "the manual."

The Tallinn Manual points out that the ICJ has stated that existing law regarding armed conflict applies when "any use of force, regardless of weapons employed" occurs. At the other end of the spectrum from the ICJ's take on the use of force, the court's predecessor, the Permanent Court of International Justice of the League of Nations, stated that "acts not forbidden in international law are generally permitted." Because most of the existing international laws of warfare were written before cyberwarfare techniques and capabilities were even imagined, many of the tactics, technologies, and potential impacts of cyberattacks and defense are not accounted for directly in international law.

> **NOTE**
> The NATO Cooperative Cyber Defence Centre of Excellence (CCD COE) was created to improve cooperative cyber defense capabilities for its member nations. It conducts cyber defense research, runs cyber defense (CND) exercises in a game style, provides courses on defense and technical subjects, and conducts conferences on cyberconflict.

In the years since the Tallinn Manual was created, calls have continued for the creation of international law and treaties covering cyberwar. In 2018, the Paris Call for Trust and Security in Cyberspace was drafted and signed by 370 actors, ranging from nations to corporations and nongovernmental organizations. The Paris Call is built around nine core principles:

1. Protect individuals and infrastructure. This principle targets malicious cyber activities that could or will cause significant, indiscriminate, or systemic harm to individuals or critical infrastructure.

2. Protect the Internet. Ensuring that actions that might intentionally and substantially damage the availability or integrity of the Internet is critical for this principle.

3. Defend electoral processes. As the U.S. 2016 election showed, cyberattacks and influence campaigns are now part of the toolkit for those wanting to influence elections. This principle seeks to strengthen the ability to prevent malicious interference aimed at undermining electoral processes using cyber activities.

4. Defend intellectual property. Protecting intellectual property like trade secrets and confidential business information prevents attackers from providing organizations with competitive advantages through cyberattacks or other malicious activities.

5. Nonproliferation. This principle seeks to limit the creation of malicious software and practices intended to cause harm.

6. Lifecycle security. A focus on digital processes, products, and services throughout their lifecycle and supply chain is the core of this principle, with the hope of making the tools and software that make up cyber ecosystems more secure throughout their lifecycle.

7. Cyber hygiene. This principle covers elements like education and planning as well as other cybersecurity best practices.

8. No private hack back. Preventing nonstate actors from conducting attacks or hacking in response to attacks is the core of this principle.

9. International norms. Establishing what responsible behavior is expected to help to build a stronger community worldwide.

While the Paris Call has many supporters including 79 countries, at the time of the publication of this book it has not been adopted as international law or as an international standard.

Much like the differences in views on international warfare laws that exist for traditional kinetic warfare, individual nations have differing views on what is acceptable during cyberwar. The U.S. State Department, for example, states in its pillars of the International Strategy for Cyberspace, intended to build international security: "Build global consensus regarding responsible state behavior in cyberspace including the application of existing international law to enhance stability, ground national security policies, strengthen partnerships, and prevent misinterpretations that can lead to conflict.." The International Strategy for Cyberspace also notes that the right to self-defense is a critical part of existing law and the UN Charter.

Despite varying opinions on how existing laws and treaties should be applied to cyberwarfare, the Tallinn Manual provides a useful framework to understand, interpret, and analyze international law in a cyberwarfare context. The following sections review major parts of the Tallinn Manual and its analysis of existing laws and norms when applied to cyberwarfare, combatants, civilians, nation-states, and the use of force in cyberwar.

NOTE

Germany created a law in 2007 (Section 203(c)) as part of an effort to comply with the COE Cybercrime Convention that banned many hacking and security tools by making it an offense to create, obtain, or distribute computer programs that violated German cybercrime laws. The law as written had a number of unintended consequences due to its breadth, including creating potential challenges for legitimate security practitioners. The nonproliferation principle of the Paris Call for Trust and Security in Cyberspace has a similar intent and, if it was implemented, would need to be drafted in a way to avoid these problems to be successful, such as focusing on intent rather than purpose.

FYI

Critics of the Tallinn Manual have pointed out that the manual doesn't clarify a number of topics, including what a cyberweapon is and how to ensure that targets for counterattacks are truly allowed targets. In addition, Russian critics have been vocal about what they claim are U.S.- and NATO-centric views taken by the experts who wrote the manual.

Sovereignty, Jurisdiction, and Control

One of the first things that must be settled during warfare is the question of where nations have authority, where they can extend that authority, and when they are expected to be in control of actions that they, or others, take on their behalf. These questions are determined by the nations' sovereignty, their jurisdiction, and their control. The Tallinn Manual examines each of these concepts in the context of cyberwar, where each question is made more complex due to the Internet's international reach. In fact, national borders are far less concrete when applied to computer networks and electronic communication systems.

Sovereignty

Sovereignty, or the right to exercise the functions of a state independently, is a key part of law when applied to cyberoperations and infrastructure. Cyberinfrastructure in a nation's territory is thus subject to that state's control, and the infrastructure is protected by the state's sovereignty, regardless of whether it is state or privately owned. That also means that nation-states can control their cyberinfrastructure, including access to the Internet, traffic sent over telecommunications networks, and other computer networks.

National sovereignty is also the foundation of international law that prohibits other nation-states from taking action against the territory or citizens of another sovereign nation. One important part of cyberwarfare law that has not been explored in international law is the effect of nonstate actors in cyberwarfare as a violation of sovereignty.

Jurisdiction

A nation-state's **jurisdiction** is the authority to enforce its will in criminal, civil, and administrative procedures within its territory and outside of its territory where allowed by international law. Cyberwarfare jurisdiction can be especially difficult due to the Internet's international nature and the fact that systems and networks belonging to groups who fall under a nation-state's jurisdiction may not be physically located in that country. Thus, jurisdiction can be difficult to determine during a cyberattack, and the location of a country's sovereign territory in the electronic world can be hard to distinguish.

Mobile and wireless technologies can make jurisdiction difficult to determine as well. The Tallinn Manual's examination of jurisdiction notes that nation-states can claim jurisdiction if individuals operated inside of their territory. Unfortunately, proving that an individual was within a nation's territory when he or she conducted an attack may be difficult to prove.

The manual describes two forms of commonly recognized jurisdiction:

- Subjective territorial jurisdiction, which allows a nation-state to apply its laws if an activity is started within its territory but completed elsewhere—even if it has no effect within the state
- Objective territorial jurisdiction, which provides nation-states jurisdiction over individuals who act against or upon the state, even if their activities were conducted outside of the state's territory

These two forms of jurisdiction provide the opportunity for states to take action against cyberattackers who violate national law from outside of the nation's borders as well as those who attack other nation-states from within their borders. This means that multiple nations may have jurisdiction in some cases, such as when an attack is conducted within one nation's borders against another nation via the Internet.

The Cloud's Problem with Jurisdiction and Control

Cloud computing, which uses shared resources at data centers often spread around the world, creates a host of potential problems for questions related to jurisdiction and control. Although the physical servers used in cloud computing are easy to locate, the individuals in control of a given virtual system that exists in a cloud computing environment can be very difficult to identify if they want to remain anonymous. Thus, an attack from thousands of virtual systems located in data centers in a dozen countries around the world is entirely possible. If that occurs, determining which country is responsible for purposes of an appropriate kinetic or electronic response becomes nearly impossible. Defenders must also take the potential impact of responses to their own country's businesses and infrastructure. They rely on the same cloud providers that may be part of the attack!

The same massive infrastructure that can be quickly repurposed makes cloud computing useful to both attackers and legitimate businesses and government organizations. The massive number of systems available from cloud providers makes them ideal for attackers who can move their attack systems after each use. They can also use the enormous pools of cloud systems to hide next to civilian systems. In some cases, the same data center and cloud servers running outsourced government services could be used by the attackers who are targeting them.

Control

Under existing international law, nation-states are expected to prevent attacks against other countries from resources and individuals that are under their **control** or located within their borders. In cyberwarfare, this expectation can be extended to systems and networks that exist within their boundaries or which they operate themselves. In fact, a nation-state is expected to exercise control over behaviors like this whether the acts are intentionally committed on its behalf or they are committed by organizations or individuals whose actions can't be directly proven to be on that nation-state's behalf.

The concept of control is necessary to support the sovereignty of other states. States are required to have control to ensure that they respect the sovereign rights of other nation-states. In cyberwar, this is difficult because cyberattacks can be difficult to detect and can occur very quickly. The Tallinn Manual points out that stopping such attacks can cause harm to the originating state because of the actions required to stop an unknown, large-scale cyberattack. Preventing large-scale cyberattacks might require a country to entirely disable its Internet connection to the rest of the world or to take major systems offline due to the size of the attack or lack of knowledge of how it works.

> **NOTE**
>
> Control is one of the most difficult concepts to transfer from existing international law into cyberwarfare. Although preventing physical attacks isn't necessarily an easy task, it is typically simpler than preventing cyberattacks, which can have massive impact from the actions of just a few systems and which can be very difficult to detect.

Under international law, nation-states that fail to exercise control can be met with proportional force in return. This means that a country that is experiencing an attack, or is about to experience an attack, can strike back in kind. In cyberwarfare, determining that an attack is about to happen and responding in kind is very difficult. Furthermore, determining what type of response is appropriate can be very challenging.

Responsibility

Under international law, nation-states are responsible for actions when they meet two requirements:

- The action can be attributed to the nation-state under international law.
- The action is a breach of international law.

International law governs both action and inaction, meaning that states that neglect their responsibilities under international law can still bear responsibility for that lack of action. The same definitions of international responsibility apply to wrongful acts that do not reach the level of cyberwarfare but are against other international laws.

> **NOTE**
>
> Responsibility is largely defined by the International Law Commission's Articles on State Responsibility, which can be found at http://legal.un.org/ilc/texts/instruments/english/draft%20articles/9_6_2001.pdf.

Responsibility is automatically attributed to a nation-state if the actions (or lack of action) are performed by a formally recognized part, or **organ**, of that nation-state, regardless of where it occurs. International law doesn't try to distinguish actions that a state intends to conduct or that its organs conduct without instruction. As long as the organ is acting in what appears to be an official capacity, the state bears responsibility for its actions. States may also have responsibility assigned to them in some circumstances if they support nonstate actors after the fact, resulting in actions previously taken by those actors being assigned to the state based on international law.

Proving responsibility can be a challenge. In cyberwar, countries can bring to bear three types of proof:

- Forensic or factual proof with information that they can bring to the ICJ or other international bodies demonstrating that the action was taken
- Technical proof, such as log files, network traffic captures, or other similar information that provides documentation of the action
- Political proof, often in the form of communications or claims by the country that is believed to have taken the action

Proving responsibility in cyberwar will remain a challenge. Determining who was in control of a system when it attacked, or if a system in another country was under the control of that nation, can be difficult if the aggressor wants to cover up his or her actions.

The existing body of laws on international warfare has a useful definition for nonstate actors who are provided with rewards for their participation in warfare. Those individuals and organizations are known as **mercenaries**, and they are carefully defined in existing law.

Nonstate Actors and Responsibility

The actions of nonstate actors can sometimes be attributed to nation-states if those actors are acting on the instructions of the state. The Tallinn Manual notes that Article 8 of the Articles on State Responsibility directly cover this situation, stating, "The conduct of a person or group of persons shall be considered an act of a State under international law if the person or group of persons is in fact acting on the instructions of, or under the direction or control of, that State in carrying out the conduct."

The ICJ previously decided that states bear responsibility for nonstate actors when states have "effective control" over those actors. The manual notes that effective control typically requires that the nation-state provide instruction, direction, or control, and not just financing or technical knowledge or capabilities. This doesn't mean that providing tools may not be a violation of international law. It just means that the state itself doesn't bear formal responsibility for the action taken by the nonstate actors.

Mercenaries

The use of mercenaries, or hired combatants in cyberwarfare, also suffers from the same problems that other nonstate actors do. International law is clear on the definition of mercenaries, as the 1977 addition to Protocol I of the Geneva Conventions lists six key elements in defining if an individual is a mercenary during warfare. By international law, a mercenary is a person who

- is recruited locally or abroad to fight in an armed conflict;
- takes part in the conflict;

- is motivated by his or her desire for private gain, which is promised to him or her by a party to the conflict, and that material gain is substantially greater than what is paid or promised to combatants in the armed forces of the party;

- is not a national of the party and does not reside in territory over or in which the conflict is fought;

- is not a member of the forces of a party to the conflict; and

- was not sent by a state that is not a party to the conflicts as part of his or her duty in that party's armed forces.

Figure 3-2 shows a review process for determining if a combatant is a mercenary based on the Geneva Conventions' definition. Cyberwar's often indistinct boundaries and the potential to be fought across national boundaries mean that identifying mercenaries may be more difficult if individuals or organizations who are not part of a nation-state are employed as part of a nation's computer network attack offensives.

Mercenaries could also be used for computer network defense activities, but this is far less likely to become a matter of international legal importance because computer network defense does not involve the use of force. The use of force and how it can be measured are the next important concepts in cyberwarfare law.

The Use of Force

The use of force is one of the most important points in international law on warfare. Measures of when force is in use, what makes an event a use of force, and what the meaning of a use of force is on an international level are all key elements of existing law.

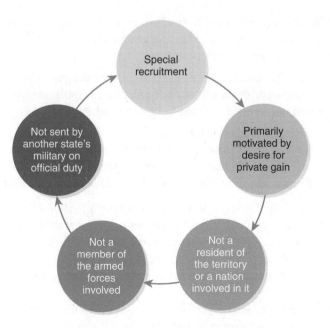

FIGURE 3-2

Identification of mercenary actors in warfare based on international law.

Special recruitment

Primarily motivated by desire for private gain

Not a resident of the territory or a nation involved in it

Not a member of the armed forces involved

Not sent by another state's military on official duty

▶ **NOTE**

The Tallinn Manual points to a ruling by the ICJ, which states that "any use of force, regardless of the weapons used," is prohibited based on Article 2(4) of the UN Charter. The manual therefore presumes that it is reasonable to apply existing international laws and standards on the use of force to cyberwar.

Under existing international law, the concept of an armed attack is the typical measure of when a nation-state can use force in self-defense. Thus, the experts who wrote the manual believe that cyberattacks rising to the same level of impact as an armed attack would qualify as a use of force. Cyberoperations are much more difficult to measure than a direct armed response, so a means of measuring cyberoperations as a use of force is needed.

Measuring Force

The Tallinn Manual notes that determining if force was used is a complex process requiring that the actor be a nation-state. The manual suggests six major criteria for measuring whether force has been used:

- The *severity* of the attack or action. Severity is the most important measure, particularly if the actions taken kill or injure people or destroy or damage objects or facilities. Even if they do not, if the actions have significant scope, duration, or intensity and impact national interests, the manual suggests that the severity may still rise to a level where they are considered a use of force.
- The *immediacy* of the action's results. Thus, if the actions will impact a nation-state quickly, it will be a more significant issue than a long-term consequence.
- The *directness* of the action's impact. If an action does not have a direct impact, and instead the action's results will be felt at a later date—or cannot be immediately and directly linked to the action—the action is less likely to be identified as a use of force.
- The *invasiveness* of the attack or action as measured against the system's security and how deeply the attack penetrates layers of security. The manual distinguishes between successful intrusions into highly secure environments and compromises of common civilian systems with typical commercial security systems in place.
- The presence of *measurable* damage, either in terms of data destroyed or taken, or systems taken offline, or some other measure. If an action's results are not easily measurable, it is much harder to claim that force was used.
- The presence of a *military character*. If the attack has a military character, it is more likely to be seen as a use of force. In fact, existing international law specifically mentions military action, making the involvement of a military group or organization more likely to result in the action being considered a use of force.

Figure 3-3 shows how these individual elements map to a determination of the use of force. Remember that the relationship of all of the elements is important to the determination of the use of force and that the determination is not a simple mathematical calculation. Determination of force by a victim during an attack will be based on what the victim knows. Responses are likely to be based on what the victim thinks an appropriate response might be.

Use of force in cyberwar isn't restricted to government or military organizations. Civilian contractors, mercenaries, and even individuals and groups could reach the

FIGURE 3-3

Determination of the use of force by a nation-state in warfare.

threshold for the use of force. Parts of international law don't
directly apply to those groups if their actions are not attributable to a nation-state. Although their actions may be against
the law, if they are not attributable, they likely fall outside
of the international prohibitions against the use of force by
nation-states themselves.

Threats of Force

In addition to the actual direct use of force, the threat of the use
of force is important to international law. Nation-states can use
the threat of force to influence other countries, endangering their
sovereignty. Due to this, the UN Charter specifically mentions threats
in Article 2(4): "All Members shall refrain in their international
relations from the threat or use of force against the territorial
integrity or political independence of any State, or in any other
manner inconsistent with the Purposes of the United Nations."

Cyberwarfare makes threats of force more complex in some
circumstances. The ability of nation-states with significant control
of the Internet and its underlying infrastructure to damage the commercial and
governmental operations of other countries means that the measures of the use of force
may shift significantly as international standards are created.

> **NOTE**
> Remember that simply
> because an action is not
> immediately attributable to
> a nation-state does not mean
> that it cannot eventually be
> attributed to that state. If a
> nonstate actor is found to
> have been operating at the
> direction of, and with the
> support of, a nation-state at a
> later date, or the nation-state
> takes responsibility for that
> nonstate actor, the previous
> attacks could be considered a
> use of force on behalf of the
> nation-state.

Self-Defense

The right to self-defense is a critical part of international law. The concept of sovereignty and the right of a nation to control and defend its territory lie at the very heart of the laws and accepted customs defining how nations interact. Self-defense in kinetic warfare is often a comparatively simple concept, with aggressors engaging in kinetic attacks against a defender. The laws of *jus in bello* focus on proportionality of responses, requiring defenders to respond in kind rather than with excessive force or against a broader range of targets.

9/11 and the Right of Self-Defense against Nonstate Actors

The terrorist attacks against the United States on September 11, 2001, resulted in a significant shift in how attacks by nonstate actors were perceived in the context of self-defense by nation-states. In traditional international law, an attack by nonstate actors would not have been considered an armed attack that would allow a country to claim the right to self-defense.

The international community largely supported the U.S. claim to the right of self-defense and generally supported U.S. actions taken in response to the attacks, including widespread military action against al-Qaeda, a nonstate actor. The Tallinn Manual's panel of experts believes that this extension of the right of self-defense set a precedent that could be used in cyberwar. If nonstate actors conducted a sufficiently damaging attack, the manual's panel holds that it would be within the rights of the attacked country to respond using force based on the existing rules around the use of force when immediately threatened or under attack.

This becomes more difficult when the use of force in a response violates the sovereignty of another nation that may not be aware of or support the actions taken by the nonstate actors. The use of proportional force may impact the civilian population and could violate the sovereignty of the nation-state or states in which the attackers are operating, resulting in an escalation of force.

In recent decades, guerrilla warfare and combat against insurrectionists have made identifying the attacker in kinetic warfare more complex. Although international law previously recognized only nation-states as potential attackers against whom countries could invoke the right to self-defense, al-Qaeda's 9/11 attack against the United States resulted in much of the international community changing its position on this question.

The manual notes that the right to self-defense against cyberattacks is a reasonable expectation based on both existing law and the international community's reaction to the 9/11 attacks. The more challenging part of exercising the right to self-defense in many scenarios will be identifying the attacker and responding in a proportional way.

International Governmental Organizations

In addition to individual sovereign countries' right to self-defense, the UN bears broad responsibility under its charter when determining if acts by nations are one of three types of aggressive acts:

- A threat to the peace
- A breach of the peace
- An act of aggression

The UN Security Council is tasked with determining what type of act occurred or is occurring, then authorizing nonforceful measures, and escalating those measures to forceful measures if the nonforceful measures fail. The UN also takes into account the behaviors of the aggressors, including whom or what they are targeting, such as civilians, infrastructure, or the natural environment.

Civilians and Infrastructure

The Geneva Conventions specifically call out civilians and civilian infrastructure as protected from direct attack. In fact, the Geneva Conventions declare this in a number of areas, including Article 54 of the Additional Protocol I, which prohibits attacks against the objects "indispensable to the survival of the population." Although this has traditionally meant the food supply, it also includes the infrastructure that supports the food supply. As technology has become more important to food production, the Tallinn Manual points out that although the Internet is not necessary to survival, the cyberinfrastructure that supports survival, such as water supply and electricity generation, likely fits into this category.

Civilians and Military Use of the Internet

Earlier, this chapter discussed the crossover between military and civilian use of the same infrastructure. In wartime, this has often meant the use of the same communication systems, roads, bridges, and other physical infrastructure. Now, shared use of network and Internet resources is also a concern. Further, the line between what is a military use of infrastructure and what is use of the infrastructure by the military for nonmilitary purposes becomes more important. In Figure 3-4, note the use of shared network infrastructure for military purposes, such as communication and cyberattacks, and the line between that use and the use of other systems by the military, but not for military purposes.

An excellent example of this would be use by military personnel of Facebook or Twitter. In fact, many uses of Facebook and Twitter by military units themselves would likely fall outside the accepted rules on the use of civilian infrastructure for military purposes. Although this fine line has not been decided by an international court, attacking Facebook simply because military personnel or units had a Facebook profile or page does not fit the definition, according to the experts who created the Tallinn Manual.

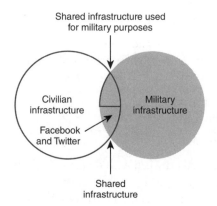

FIGURE 3-4

Facebook and Twitter, civilian, military, and shared use of infrastructure in the cyber world.

Prohibited Targets: Children, Journalists, Medical and Religious Personnel, and Nature

The Geneva Conventions define a number of prohibited targets that are protected as long as they are not taking part in hostilities:

- Children are protected under the Geneva Conventions and are prohibited from involvement in armed conflict. Numerous cases in which children have been involved in cyberattacks demonstrate that this may be a particularly difficult protection for nation-states to apply, much as preventing children from fighting in warfare has presented challenges throughout the twentieth and twenty-first centuries.

- Journalists who are engaged in professional missions in armed conflicts are considered civilians and must be protected as such.

- Medical and religious personnel and material are protected as noncombatants unless they participate in harmful actions. They are to be allowed to perform their duties without interference or harm. The manual interprets this to include medical computer networks and systems, which can be difficult to differentiate in cyberwar. It points out that the rules allowing medical personnel to be armed for self-defense should be interpreted to allow them to have tools that might be useful for cyberwar in their possession without making them a target.

- The natural environment is classified as a civilian object, and the Geneva Conventions specifically prohibit attacks that may cause long-term, widespread, severe damage to the environment.

FYI

The manual notes that "the majority of the International Group of Experts took the position that broadcasts used to incite war crimes, genocide, or crimes against humanity render a journalist a direct participant and make the equipment used [for] military objectives liable to attack, including by cyber means." As with many of the other impacts of international law on cyberwar, the role of journalists and others in cyberwar may be difficult to determine on a consistent basis and will depend on circumstances and actions at the time.

The manual notes that the rules protecting some noncombatants do not mean that their computer systems and networks must be protected beyond the rules that cover civilian infrastructure and systems. Thus, a journalist's activities and equipment are not provided special protection. In contrast, medical personnel are likely to be protected from such attacks if the attacks would prevent them from performing their duties.

The Conduct of Attacks and Indiscriminate Means

A key element of international law is the prohibition of indiscriminate means of attack. According to the Tallinn Manual, indiscriminate means are those that are not "(a) directed at a specific military objective, or (b) limited in their effects as required by the law of armed conflict." In addition to the prohibitions of attacks on civilians, weapons and strategies that target broad populations without concern for who will be harmed are banned. Indiscriminate attacks that successfully target legitimate military objectives are still prohibited because they are not aimed at a specific target. In addition, the manual notes that both military and civilian commanders, as well as their subordinates who carry out their orders, bear responsibility for indiscriminate attacks, and ordering indiscriminate attacks would likely be considered a war crime.

The manual also examines the way in which attacks are conducted as well as the means that are used. The experts on the panel specifically note that based on accepted law and precedent, attacks must not do excessive damage to civilians or civilian objects compared with the military advantage they create. In other words, attacks can't do significant damage to civilians in order to gain a minor military advantage. This is particularly important in cyberwar activities because the shared infrastructure of civilian networks and systems may be an inadvertent target in the path of an attack against military systems.

Espionage, Treachery, and Ruses

Espionage and treachery exist outside of most of the commonly accepted laws of international warfare. The manual notes that the exception to that rule occurs when the action will cause harm or death, thereby meeting the criteria. The manual focuses on the need for the action to be directly linked to the cause of death and not through a series of unforeseen actions.

> **NOTE**
> International law does not directly address espionage acts, meaning that international law does not apply in cases of espionage activity that do not directly violate specific international regulations.

Cyberwar creates the potential to engage in **treachery**—acts that lead an adversary to believe that he or she is entitled to, or obliged to give, protection under the laws of war, while intending to betray that confidence. Falsely identifying computer systems as protected systems, such as those used for medical personnel, or pretending to be civilian systems or infrastructure to avoid attack would constitute treachery. Although pretending to be medical personnel or another protected class of noncombatant is forbidden under the customary rules of war, cyberwar conditions may make separating legitimate protected objects from those falsely appearing to be protected very difficult.

Unlike treachery, a **ruse** is considered a legitimate part of war. A ruse is intended to mislead an enemy without violating the laws of war. Many ruses used in computer network attacks and defenses would fit into this category:

- Creating fake networks and systems
- Providing false information via computer systems
- Launching fake cyberattacks
- Conducting psychological operations
- Redirecting traffic through systems to gain access to the traffic

In essence, the same ruses and deceptions that would be considered a legitimate part of kinetic warfare can be extended to cyberwarfare.

Neutrality

Neutrality is an important concept in kinetic warfare and is perhaps even more important in cyberwar, as many countries in the cyberattack's path will be neutral. Under the customary rules of warfare, neutral states are protected and can continue to conduct their normal activities without interference from combatants.

In cyberwar, this means that neutral states must prevent combatants from using infrastructure that is under neutrals' sovereign control. In kinetic warfare, this is easier to do because national borders can be closed to combatants, whereas closing networks to traffic from warring countries can be difficult if not impossible. If combatants discover that neutral states are allowing the combatants' opponents to use the neutrals' resources, combatants can take action against the neutrals.

The Internet's highly interconnected nature makes the concept of neutrality in cyberwarfare more challenging. This remains one of the areas where international law will require further clarification to provide appropriate protections for neutral countries.

Ethics and Cyberwarfare

Individuals engaged in computer network defense and computer network attacks bring their own ethical standards and those of their profession, military branch, or organization. Although broadly accepted ethical standards specific to cyberwar haven't been developed, common ethical standards do exist in international law. The critical elements of the laws of war are as follows:

- Self-defense, which prohibits nation-states from indiscriminately engaging in war with other nation-states. Cyberwar's blurred boundaries mean that the concept of self-defense can be a challenging one to define, and thus ethical boundaries for what is acceptable to a nation or organization are important.
- Proportionality, which prevents an escalation of warfare when combat does occur. Proportionality can help to ensure that attacks do not increase in scope. A proportional

response is ethically important to help make sure that attacks are not used as an excuse to deploy crippling tools that violate the limitations on acceptable targets and other rules of warfare.

- Limitations on targets, which are important and particularly relevant to cyberwarfare activities due to the shared nature of most major network connections and many computing environments. As the world becomes increasingly more networked, everything from food production to environmental controls and medical systems are all potentially in the line of fire when attackers are focused on government or military systems.

In addition to the ethical concepts that existing and historical laws of war provide, individual practitioners are often trained by information security training organizations that provide their own ethical systems. Professional organizations like those that provide major information security certifications provide codes of ethics that share common concepts:

- Appropriate use of skills and capabilities
- Professionalism
- Adherence to existing laws and standards
- Respect for privacy and confidentiality
- Integrity in actions and access to systems and data

These ethical systems provide a useful starting framework, which may eventually grow into a modern equivalent of the Geneva Conventions for cyberwarfare behaviors and practices.

CHAPTER SUMMARY

This chapter has examined the existing treaties and conventions that make up the laws of warfare, including the Geneva Conventions and the UN Charter. It explored the role of cyberwarfare and traditional or kinetic military action in the context of those laws and how cyberwarfare capabilities and methods go beyond their scope.

Those changes and similarities are well explained in the Tallinn Manual, a NATO-sponsored analysis of existing laws and rules of war. This chapter explored the Tallinn Manual to better understand common concepts and their application to cyberwar. Finally, this chapter looked at existing codes of ethics for information security professionals and military officers who may be called on to participate in computer network attacks and defense actions to better understand the choices they may have to make about their participation in cyberwar.

KEY CONCEPTS AND TERMS

Civilian infrastructure   *Jus in bello*   Ruse
Cloud computing   Kinetic warfare   Sovereignty
Control   Mercenaries   Tallinn Manual
Jurisdiction   Nonkinetic warfare   Treachery
Jus ad bellum   Organ

CHAPTER 3 ASSESSMENT

1. The Geneva Conventions provide specific instruction on what makes up acceptable computer network attack activities.

   A. True
   B. False

2. What of the following terms means the right to war?

   A. *Jus in bello*
   B. *Civile officium*
   C. *Jus ad bellum*
   D. *Pax et bellum*

3. When determining if force was used, what is typically the most important element?

   A. Severity
   B. Military character
   C. Invasiveness
   D. Directness

4. Which of the following three traits are used to help determine if a person is acting as a mercenary during war?

   A. Conscription
   B. Special recruitment
   C. Not a member of the armed forces of an involved country
   D. Sent by another state's military on official duty
   E. Motivated by personal gain

5. Conventional warfare is often called _____ warfare.

6. Deception that results in death or injury is known as _____.

7. The _____ provide(s) international laws for the conduct of war.

   A. Lieber Code
   B. Geneva Conventions
   C. UN Charter
   D. Tallinn Manual

8. The concept of _____ requires that responses to attacks are appropriate to the harm they caused.

   A. proportionality
   B. self-defense
   C. jurisdiction
   D. self-determination

9. The _____ stipulates the rules for actions *jus ad bellum*.

   A. ICJ
   B. Geneva Conventions
   C. UN Charter
   D. Tallinn Manual

10. The Geneva Conventions are an example of _____.

    A. a yearly review of international law
    B. *jus in bello*
    C. U.S. law
    D. *jus ad bellum*

Intelligence Operations in a Connected World

NTELLIGENCE IS THE COLLECTION, analysis, and dissemination of information about the capabilities, plans, intentions, and operations of an adversary. It typically includes the collection of information that an adversary wants to remain secret. Intelligence organizations closely guard the methods that they use to gather intelligence and the sources that provide them with secret information.

Nations consider intelligence critical to advancing their military and political agendas. At the same time, companies, organized crime, and other nonstate actors seek to gain intelligence about their competitors. The information gathered by intelligence sources and methods may give a group a significant advantage over an adversary by providing insight into the adversary's capabilities, thoughts, objectives, and plans. The desire to achieve this advantage leads nations and organized nonstate actors to invest tremendous amounts of money and time in developing advanced intelligence-gathering capabilities.

Intelligence collection is as old as warfare itself. The earliest armies sought to identify the strength and location of opposing forces by using reconnaissance scouts sent out in the field to locate enemy encampments. Early intelligence officers relied upon the use of human intelligence sources. These spies and scouts were sent into the field to gather information that the enemy considered secret. Colonel Thomas Knowlton, considered America's first spy, led the country's first intelligence organization during the American Revolution. George Washington sent "Knowlton's Rangers" out on reconnaissance missions against British troops. One famous member of Knowlton's unit was Nathan Hale, who uttered the famous quote, "I only regret that I have but one life to give for my country" before being executed by the British.

Nations still make significant use of human intelligence today, but this traditional spy craft is now supplemented with advanced methods that take advantage of advanced technological capabilities. In this chapter, you will learn the variety of techniques used by modern intelligence-gathering operations, including the use of imagery intelligence, signals intelligence, open source intelligence, and other capabilities.

Naturally, countries seek to apply intelligence to all forms of warfare. Military units focused on the four traditional domains of warfare—land, sea, air, and space—are able to gather intelligence that assists them in both offensive and defensive operations. The addition of the cyber domain to the military landscape means that military units that engage in this domain now actively seek intelligence to assist them in military operations. As with the other four domains, intelligence support to cyberwarfare can extend to both offensive and defensive operations. In this chapter, you will learn how intelligence can assist cyberwarfare operations.

In 2013, an American military contractor named Edward Snowden released a significant cache of what he claimed were secret military documents to several media outlets. These documents provided the public with unprecedented insight into the abilities, tools, and techniques that governments use when engaging in cyberwarfare operations. In the final section of this chapter, you will learn about these abilities and how they have contributed to cyberwarfare efforts.

Chapter 4 Topics

This chapter covers the following topics and concepts:

- What intelligence is and what the processes to collect it and analyze it are
- What disciplines and techniques are used to gather intelligence
- What types of support intelligence can provide to offensive and defensive cyberwarfare
- How countries have gathered intelligence in a connected world

Chapter 4 Goals

When you complete this chapter, you will be able to:

- Describe the five phases of the intelligence cycle
- Explain the different intelligence disciplines and how they are used to collect intelligence
- Explain how intelligence operations can support offensive cyberwarfare
- Explain how intelligence operations can support defensive cyberwarfare
- Describe examples of the intelligence-gathering techniques used today

Intelligence Operations

As mentioned, **intelligence** is the collection, analysis, and dissemination of information about the capabilities, plans, intentions, and operations of an adversary. Intelligence operations may take place at the behest of nations or nonstate actors seeking to gain an advantage over a military, political, criminal, or business adversary. Every nation around the world engages in the discipline of collecting intelligence. Many nations dedicate significant financial, technical, and human resources to increasing their intelligence capabilities.

As a formal discipline, intelligence operations follow well-defined processes and utilize standard abilities that nations have evolved over time. Using a repeatable, organized approach to intelligence collection and analysis ensures that nations are able to prioritize the use of intelligence resources and gather the information needed by decision-makers at the policymaking, strategic, and tactical levels.

The Intelligence Cycle

The **intelligence cycle**, shown in Figure 4-1, is a model used to describe the process of intelligence operations. It begins when a decision-maker states a need for a specific type of information and concludes when finished information is provided to that decision-maker. The five phases of the intelligence cycle include the following:

- Planning and Direction
- Collection
- Processing and Exploitation
- Analysis and Production
- Dissemination

FIGURE 4-1

The U.S. intelligence community uses the intelligence cycle to describe the sequence of events involved in collecting intelligence.

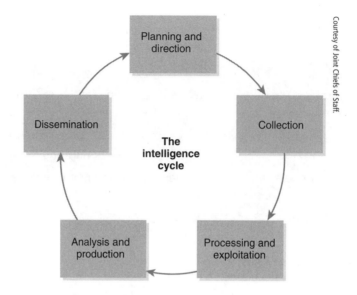

> **NOTE**
>
> Although intelligence professionals describe the process as a cycle and draw that cycle with unidirectional arrows, real-world intelligence collection rarely proceeds in such a clean, organized fashion. Steps in the cycle are often skipped or repeated, as practical demands dictate.

Each of these phases or steps plays a critical role in ensuring that decision-makers receive timely, accurate intelligence that answers the questions they posed. The Joint Publication 2.0: *Joint Intelligence* describes the full intelligence cycle process.

Planning and Direction

The **Planning and Direction** phase of the intelligence cycle consists of several activities designed to ensure that intelligence operations target the appropriate information to assist decision-makers. In this phase, several activities take place:

- **Identify intelligence requirements**—When decision-makers identify a gap in the intelligence available to them, they work with the intelligence community to develop **intelligence requirements**. These requirements are general or specific subjects for which there is a need for information collection or intelligence production. These requirements are then broken down into two components:
  - **Priority intelligence requirements (PIRs)** are the questions identified by decision-makers as needing critical attention from intelligence operations. For example, a PIR might be, "Will the British attack by land or by sea?"
  - **Essential elements of information (EEIs)** are specific pieces of information that may be collected to help answer the priority intelligence requirement. In the case of the PIR above, EEIs might include, "Is the British Navy assembling outside of Boston Harbor?" and "Are British troops leaving New York headed north?"

- **Develop an intelligence architecture**—Intelligence production and dissemination require complex collection, processing, and dissemination systems. During the Planning and Direction phase, intelligence professionals identify the systems and processes that will be used in future phases of the cycle.

- **Prepare a collection plan**—The collection plan identifies the specific intelligence disciplines and methods that will be used to gather the essential elements of information needed by analysts seeking to answer priority intelligence requests.

- **Issue requests to collectors**—After the collection plan is approved, orders are issued to intelligence collectors, directing them to execute the activities described in the collection plan and report results.

NOTE

You may sometimes see a six-phase version of the intelligence cycle. People who use this approach break out Requirements Gathering as a separate phase from Planning and Direction.

The Planning and Direction phase is a critical component of the intelligence cycle. It ensures that intelligence agencies are operating from the same plan of action and are neither conducting redundant operations nor leaving critical gaps that will prevent fulfillment of PIRs.

Collection

During the **Collection** phase, intelligence professionals use the assets at their disposal to gather the essential elements of information. This work is led by collection managers, who are responsible for specific EEIs and have the ability to assign intelligence assets to collect data that may result in the requested information.

NOTE

You will learn more about the types of intelligence assets available to collection managers later in this chapter.

The types of assets used during the Collection phase depend upon the specific information requests made by decision-makers and the availability of appropriate assets. For example, a request to count the number of tanks at a particular military base may be easily fulfilled by satellite imagery under normal conditions. However, if the information is needed on a cloudy day when satellite imagery is unavailable, the collection manager may seek to use an on-the-ground human spy to visit the facility and perform the count.

Processing and Exploitation

During the **Processing and Exploitation** phase, the potentially massive amounts of data collected by intelligence assets must be transformed into information that analysts may use. This phase may include a variety of activities, depending upon the types of intelligence collection performed. Raw data may need to be

NOTE

In cases where information is collected electronically, many processing and exploitation tasks may be partially or completely automated.

decrypted, stored in databases, converted to viewable images, or undergo a variety of other activities so it is useful to intelligence analysts. Translating collected data from one language into another is also an important part of the Processing and Exploitation phase.

If the collection plan was well designed and assets were able to collect the requested information, the output of the Processing and Exploitation phase should include the EEIs developed during the Planning and Direction phase.

Analysis and Production

At the conclusion of the Planning and Direction phase, intelligence agencies now have answers to the EEIs. This information is turned over to professional intelligence analysts. These analysts have access to information from a variety of intelligence sources and correlate the answers they receive to satisfy priority intelligence requests. For example, an analyst seeking to answer a question about whether a missile launch is imminent may have several EEIs available:

- Satellite imagery showing missiles on the launch pad
- Electronic intelligence showing that the missile command systems are activated and in launch mode
- Intercepted communications from the missile group commander ordering that all leaves be canceled
- Report from a human spy that the preparations are part of an exercise

In the **Analysis and Production** phase, the analyst must then interpret this information and develop an assessment that satisfies the PIR. In this case, as in many real-world cases, the analyst has conflicting information. Three information sources indicate that a missile launch is imminent, but one conflicting report from a spy says that the preparations are all part of an exercise. At this point, the analyst must make a judgment about the reliability of the information from the spy. Is it possible that the spy is a double agent feeding false information as part of a deception campaign? Has the spy provided accurate information in the past? Does the spy have access to credible sources?

Dissemination

The final phase in the intelligence cycle is the delivery of finished intelligence products to the decision-makers who made the requests. The **Dissemination** phase can take a wide variety of forms, ranging from intelligence reports provided to the president of the United States to field briefings made to a flight crew preparing to fly into combat. The recipients of intelligence products can then integrate that information into their own processes.

Data, Information, and Intelligence

Intelligence professionals draw important distinctions among the words *data*, *information*, and *intelligence*. These terms are used to describe the products of different phases of the intelligence cycle, as shown in Figure 4-2.

Data are collected from the operational environment by using various techniques available in different intelligence disciplines. Large amounts of raw data are not very useful by itself. For example, imagery data collected by a satellite would include thousands of photos.

Information is created by intelligence operations staffers when they process and exploit data. For example, an image processor might pore over satellite photos and identify all of the military vehicles in photographs by nation and type. This may be useful information.

Intelligence is the final product of the cycle. Intelligence analysts take the information collected through intelligence methods and assess it based upon the operational environment, the context of history, and their professional experience. They correlate information from a variety of sources, draw conclusions, and create forecasts. The resulting intelligence may be quite valuable to decision-makers.

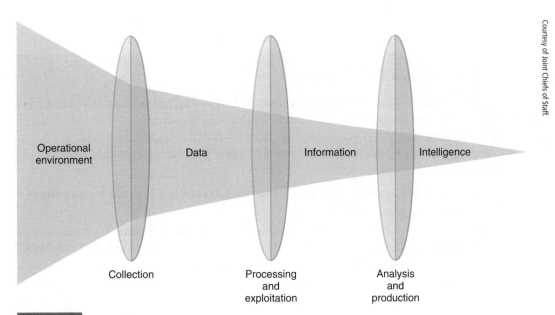

Courtesy of Joint Chiefs of Staff.

FIGURE 4-2

During the intelligence cycle, data are processed into information, which is then analyzed and turned into intelligence for dissemination.

Intelligence Disciplines

When intelligence professionals receive requirements from decision-makers, they choose from a variety of tools and techniques to uncover the information needed to satisfy these requests. These techniques vary widely, from using clandestine spies on the ground to assigning remote electronic capabilities to intercept communications from afar.

Intelligence professionals developing a collection plan must identify the intelligence disciplines most likely to successfully satisfy intelligence requests. They must then work with collection managers to assess the availability of assets in those disciplines and the relative priority of their requests. After that, they assign intelligence assets to work on those assignments, beginning with the highest priority and proceeding down the list. You will almost always have more intelligence requests outstanding than assets available to fulfill those requests, so it is critical to clearly define collection priorities.

Human Intelligence (HUMINT)

The discipline of **human intelligence (HUMINT)** involves the collection of information using techniques that involve interaction among individuals. In the United States, primary responsibility for HUMINT operations belongs to three agencies:

- The Central Intelligence Agency (CIA) has broad responsibility for HUMINT operations outside of the United States.

- The Defense Intelligence Agency (DIA), a combat support agency of the Department of Defense, maintains its own HUMINT clandestine service dedicated to identifying the military intentions and capabilities of adversaries.

- The Federal Bureau of Investigation (FBI) bears responsibility for HUMINT operations within the United States.

When thinking of human intelligence operations, you might immediately imagine the cloak-and-dagger techniques of clandestine spy craft. Although conducting clandestine espionage operations is certainly a component of HUMINT, collectors may exploit many other possible human sources as well. Examples of HUMINT sources include the following:

- Agents are sent clandestinely into an adversary's operating environment to gather information and report back.

- Traitors are recruited from within an enemy's organization to provide information about the enemy. Some traitors (known as *moles*) are recruited long before they have access to desired information in the hopes they will be promoted to positions of power.

- Friendly military forces encounter the enemy and are debriefed afterward to glean as much information as possible about enemy capabilities and tactics.

- Hostile military forces are captured and taken as prisoners of war and may be interrogated for information of intelligence value.

- Businesspeople who travel to a foreign country may be tasked with specific intelligence objectives before they leave or debriefed about their travels after they return.

- Embassy personnel in foreign countries who have diplomatic immunity may be tasked with intelligence-collection objectives.

This is just a small sampling of the types of human intelligence sources that HUMINT collectors may exploit. Creative collectors will develop unusual sources that are unlikely to raise the suspicion of adversary counterintelligence operations.

In some cases, individuals with access to information volunteer to cooperate with HUMINT operations. However, many HUMINT sources must be recruited to participate in these activities. When recruiting HUMINT assets, intelligence professionals rely upon the four primary motivations of spies:

> **TIP**
>
> You can easily remember the four motivations for spies by using the acronym MICE.

- **Money**—Many spies simply want cash. If a spy has access to important information, intelligence services are often willing to provide the spy with cash payments to continue the flow of intelligence.

- **Ideology**—Some spies already possess or later develop strong beliefs that cause them to think that their espionage activities are contributing to the greater good.

- **Coercion**—Spies may be coerced into participating in HUMINT activities. The most common example of this is blackmail, where an intelligence service has embarrassing or incriminating information about an individual. This information often relates to undiscovered criminal acts or embarrassing sexual activity. The intelligence service may even have placed the individual in the situation where the activity took place in the hopes of capturing evidence. It then uses this evidence to force an individual to engage in espionage.

- **Ego**—Some individuals, particularly those with relatively unimportant positions but with access to sensitive information, may see themselves as much more important when they share that information with a foreign power. This desire for self-importance motivates some spies.

Robert Hanssen

Former FBI agent Robert Hanssen is an excellent example of a spy who was motivated by both ego and money. Hanssen spied for both the Soviet and later Russian intelligence organizations over the course of 22 years until his capture and arrest in 2001.

Hanssen's activities were extremely damaging because of the sensitive positions that he held with the FBI. During one period, when the FBI suspected a mole, he actually led the effort to discover a suspected mole within the bureau. Little did his supervisors know that they had actually charged the mole with discovering himself!

After his arrest, a Department of Justice report classified the "vast quantities of documents and computer diskettes filled with national security information of incalculable value" disclosed by Hanssen as "possibly the worst intelligence disaster in U.S. history."

Hanssen's story was immortalized in the 2007 film *Breach*. He is now serving a lifetime sentence in the Florence, Colorado, supermaximum high-security federal prison.

As with any intelligence source, HUMINT collectors are faced with a wide variety of potential sources. The identification and cultivation of a source is a time-intensive, risky process. HUMINT operators must use an approach that prioritizes potential sources based on the likelihood that they will cooperate and the value of the information they possess. The Army assigns each source one of the following three cooperation codes:

- **1**—Responds to direct questions
- **2**—Responds hesitatingly to questions
- **3**—Does not respond to questioning

In addition, the Army assigns each potential subject a knowledgeability level using this three-tiered system:

- **A**—Very likely to possess information that would help satisfy a PIR
- **B**—Might have information meeting lesser intelligence requirements
- **C**—Does not appear to have pertinent information

Notice that the knowledgeability level codes are directly related to intelligence requirements. HUMINT operators place targets that are likely to have information about priority intelligence requirements in the highest category. This approach allows a direct linkage between the source-development process and the intelligence requirements process.

Once potential sources are rated on both cooperation and knowledgeability, they are placed on the graph shown in Figure 4-3. This graph assigns priority levels to each possible code combination, allowing HUMINT collectors to spend their time developing the sources that are most likely to both have important information and cooperate with HUMINT efforts.

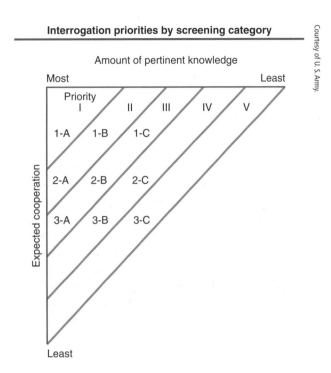

Interrogation priorities by screening category

Amount of pertinent knowledge

FIGURE 4-3

A prioritization chart used to rank potential HUMINT sources by their expected cooperation and level of relevant knowledge.

Signals Intelligence (SIGINT)

Signals intelligence (SIGINT) is the collection of intelligence information through intercepting communications and other electronic signals. Given the tremendous amounts of information exchanged electronically in modern society, SIGINT is often considered one of the most productive intelligence disciplines. In the United States, the National Security Agency (NSA), a supporting agency of the Department of Defense, has primary responsibility for SIGINT operations.

Traditional SIGINT operations are divided into two subdisciplines:

- **Communications intelligence (COMINT)** involves the collection of communications between individuals. COMINT collection sources may include, among other sources, the following:
  - Telephone calls
  - Radio transmissions
  - Text messages
  - Email
  - Web traffic
- **Electronic intelligence (ELINT)** involves the collection of electronic signals generated by nonhuman communications. It includes the electronic emissions generated by radar systems, aircraft, ships, missiles, and other communications systems. Examples of ELINT uses are geolocation of targets and identifying weapons systems' operating characteristics.

4

Intelligence Operations in a Connected World

▶ **NOTE**

Many descriptions of SIGINT also include the capture of signals from foreign devices, known as foreign instrumentation signals intelligence (FISINT). You will find more information about FISINT in the "Measurement and Signature Intelligence (MASINT)" section of this chapter.

In many cases, the communications intercepted by SIGINT operations are encrypted to prevent successful eavesdropping. Even encrypted communications may provide useful information about the location of transmitters and the amount and frequency of transmissions. For example, a sudden burst in encrypted communications between a military command center and outlying command posts may indicate imminent military activity.

However, SIGINT is most useful when intelligence analysts gain access to the contents of communications. For this reason, SIGINT forces retain significant cryptanalysis capabilities. **Cryptanalysis** is the discipline of studying and defeating encryption technology to gain access to the plaintext of encrypted messages.

SIGINT information may be collected from a variety of sources. These include large ground stations with antennas dedicated to receiving communications from specific sources, clandestine facilities located close to targets, and even military ships and aircraft with SIGINT capabilities. As an example, the RC-135 Rivet Joint aircraft is a dedicated SIGINT aircraft operated by the U.S. Air Force. As you can see in Figure 4-4, the Rivet Joint plane is a standard airplane outfitted with a variety of antennas for use in SIGINT collection.

Open Source Intelligence (OSINT)

In some cases, you can glean the information decision-makers require from publicly available information. The practice of obtaining and processing this information for intelligence purposes is known as **open source intelligence (OSINT)**.

FIGURE 4-4

RC-135 Rivet Joint aircraft used by the U.S. Air Force for SIGINT collection missions.

Courtesy of the U.S. Air Force.

A wide variety of information sources are available to OSINT collectors, in English and other languages:

- Newspaper reports
- Radio and television broadcasts
- Internet content
- Presentations made by military and political officials at conferences
- Documents issued by government agencies

The National Open Source Enterprise (OSE) is responsible for the acquisition and analysis of open source materials by the U.S. government. Originally started as the Open Source Center within the CIA, OSE absorbed the agency's Foreign Broadcast Information Service that had been collecting and translating information from foreign media broadcasts since World War II. OSE now operates under the purview of the director of national intelligence.

Geospatial Intelligence (GEOINT)

Geospatial intelligence (GEOINT) includes information gathered through photography, maps, and other information about terrain. The most common subdiscipline of GEOINT is **imagery intelligence (IMINT)**, which consists of photographic information that is collected by either aircraft or satellites overflying an area of interest. Images collected by IMINT sources may be used to assess changes in an adversary's force structure, troop movements, industrial production, and many other areas of intelligence interest.

The earliest IMINT collection programs leveraged aircraft to overfly enemy troops and capture photographs for exploitation. The first widespread use of IMINT occurred during World War II, when Allied forces equipped aircraft with cameras that used photographic film. After an IMINT mission, the film was rushed to the Allied Central Interpretation Unit at Oakington, England, for development and interpretation.

At the conclusion of World War II, IMINT programs expanded through the use of dedicated intelligence collection aircraft in the Cold Car. An example of this is the U-2 spy plane, developed to fly at extremely high altitudes, out of the reach of antiaircraft guns and missiles. The U-2, shown in Figure 4-5, first flew missions in 1957 and remains in use today as part of U.S. intelligence-collection programs.

In addition to aircraft-gathered imagery, the intelligence community relies heavily on images collected by satellites equipped with high-resolution cameras, known as *overhead imagery*. Early satellites, first launched in the 1960s, periodically ejected film canisters that were retrieved by friendly forces and developed for interpretation.

Details of modern satellite imaging programs remain classified, but civilian unclassified satellite imagery programs certainly provide a minimum standard that military technology is likely to achieve. Today, anyone in the world with a computer and an Internet connection can pull up Google Earth and look at recent satellite imagery of almost any area on the globe.

FIGURE 4-5

A U-2 Dragon Lady aircraft used by the U.S. Air Force for IMINT collection missions.

IMINT has played a crucial role in modern military and political operations. It was used extensively during the 1962 Cuban Missile Crisis to gather intelligence about Soviet movements in Cuba. In 2003, then-Secretary of State Colin Powell gave a presentation at the UN Security Council justifying the move to war in Iraq, using satellite imagery to bolster his case. Figure 4-6 shows one of the slides Powell used in his presentation; it contains imagery of a recently bulldozed chemical weapons facility.

IMINT capabilities continue to advance. In September 2019, President Donald Trump sent a tweet stating: "The United States of America was not involved in the catastrophic accident during final launch preparations for the Safir SLV Launch at Semnan Launch Site One in Iran. I wish Iran best wishes and good luck in determining what happened at Site

FIGURE 4-6

A slide from Colin Powell's presentation to the UNs Security Council, February 5, 2003.

Scorching and damage present only on northern side of launch pad

Damaged gantry service tower

Damaged support vehicle

Damaged support vehicle

Damaged propellant burner trailer

Damaged Safir mobile-erector-launcher

FIGURE 4-7

Overhead imagery released in a tweet by President Donald Trump in September 2019.

One." Along with the tweet, he included the high-resolution image shown in Figure 4-7. The technique used to collect this image remains unclear, but it reveals some of the current capabilities of the U.S. intelligence community.

In the United States, GEOINT activities are coordinated by the National Geospatial-Intelligence Agency (NGA).

Measurement and Signature Intelligence (MASINT)

Measurement and signature intelligence (MASINT) makes use of information gathered from unintended electromagnetic emissions generated by a target. MASINT may be used to detect, locate, track, identify, or describe the details of intelligence targets. MASINT may be collected independently or may be gained by applying advanced interpretation techniques to SIGINT and IMINT information.

All of the subdisciplines of MASINT are highly technical fields that require sophisticated technical expertise:

- Radar intelligence (RADINT) collects information from adversary radar systems.
- Frequency intelligence gathers electromagnetic signals generated by adversary military systems.
- Nuclear intelligence monitors radiation and other byproducts of nuclear events for intelligence purposes.
- Geophysical intelligence observes seismic waves, acoustic waves, magnetic field disturbances, and other natural-world alterations resulting from human intervention.

- Electro-optical intelligence captures light emissions, which may be exploited for intelligence purposes.
- Materials intelligence uses the analysis of physical materials to gain intelligence, such as collecting samples of chemical or biological weapons.

MASINT information is often combined with SIGINT information to study the emissions made by enemy weapons systems, such as intercontinental ballistic missiles in flight. This field is known as foreign instrumentation signals intelligence (FISINT).

U.S. MASINT activities are coordinated by the Directorate of MASINT and Technical Collections within the DIA.

Intelligence Support to Cyberwarfare

Over the course of history, intelligence operations have provided support to military operations across all domains of battle. For example, when the air domain became dominant during World War II, intelligence agencies quickly geared up to collect information about enemy aircraft and antiaircraft capabilities. This information proved invaluable to those seeking to conduct offensive and defensive war in the air. Similarly, effective control of the space domain requires information about enemy capabilities in that domain. Intelligence agencies are tasked with collecting, processing, and disseminating information in support of space warfare.

Although little public information is available about how intelligence operations may support cyberwar, it is hard to imagine that operations in the cyber domain would not also draw upon the use of intelligence sources to achieve a battlefield advantage. This support to cyberwarfare likely extends to both offensive and defensive cyberwar.

Supporting Offensive Cyberwarfare

One of the primary ways that intelligence operations support warfare is providing information about potential targets. In the air domain, for example, intelligence operations can identify the military facilities and industrial complexes crucial to the war effort. This information may then be translated into prioritized target lists for bombing operations. Offensive cyberwar also requires target lists. For example, which enemy information systems should be targeted for attack? Intelligence operations can help answer these questions and direct the targeting of cyberwar assets against priority targets.

Another interesting characteristic of cyberwar is the fact that attackers' identities can easily be concealed. The nature of cyberwar sometimes makes it difficult to determine the source of attacks. Clever cyberwar operators may use this to their advantage, collecting information about the cyberwar capabilities of third parties and then using that information to "frame" a third party as the source of an attack.

Supporting Defensive Cyberwarfare

Information gleaned through intelligence operations may also help with defensive cyberwar operations. Building strong defenses requires knowledge of your adversary's capabilities, tactics, and intent. Gathering this information is a traditional strength of intelligence operations, and there is no reason that it cannot be extended to the cyber domain. Intelligence professionals may leverage HUMINT, SIGINT, and other traditional methods to gather information about enemy cyberwar capabilities. They then use this information to build better defenses against those capabilities.

Intelligence operations supporting cyberwarfare may attempt to answer a variety of questions:

- Who are the potential adversaries in the cyberwarfare environment?
- What are those adversaries' technical capabilities?
- What are cyberwarfare opponents' plans and intentions?
- What signs may warn of an impending cyberwarfare attack?
- How will the adversary make decisions about using cyberwarfare techniques?
- How will the adversary communicate orders to engage in cyberwarfare?

These critical pieces of information can provide defenders with advance warning of an attack. They can also provide information critical to preparing cyberwarfare defenses equal to the risk real-world adversaries pose.

Case Studies: Media Reporting on Intelligence Activities

One of the best ways to understand intelligence capabilities is to study media reports about intelligence operations and the tools and methods used to gather intelligence. Until recently, this type of analysis was limited to historic information, such as the Ultra program designed to counter the German Enigma code system in World War II.

When U.S. defense contractor Edward Snowden released a large cache of what he claimed were classified documents to the media in 2013, however, he provided a unique, if unconfirmed, glimpse of U.S. intelligence-collection capabilities. Studying this information shows how intelligence agencies may operate and how SIGINT capabilities, in particular, seem to have evolved to match the communications of a connected world.

Echelon

Although not part of the Snowden disclosures, the public scrutiny of the Echelon program in the 1990s provided much of the world its first glimpse at the intelligence capabilities of the United States and its partners. The system was described in the British *Guardian* newspaper on May 29, 2001 as:

> A global network of electronic spy stations that can eavesdrop on telephones, faxes, and computers. It can even track bank accounts. This information is stored in Echelon computers, which can keep millions of records on individuals.

Purportedly, the Echelon program included SIGINT capabilities targeted against satellites, undersea cables, radio communications, and other signal sources. These alleged capabilities were also mentioned in the Snowden documents. According to both sources, these facilities are operated by the United States and its intelligence partners: the United

Kingdom, Canada, Australia, and New Zealand. These five nations often cooperate on SIGINT activities.

Telephone Metadata

SIGINT operations do not necessarily have to capture the content of communications to be successful. Capturing technical details about the communications themselves can also provide intelligence analysts with important and useful information. Information about communications, known as **metadata**, can provide analysts with details about who is communicating, where they are located, and the amount of information exchanged.

In June 2013, *The Guardian* reported that the NSA was operating under secret court orders to collect telephone metadata from major U.S. cellular carriers and storing that information in a database. Intelligence analysts could then access the metadata by issuing database queries, looking for patterns of communication between suspects and unknown individuals. This provided new targets for investigation.

Data Center Eavesdropping

The efforts of the intelligence community changed during the early part of the 21st century because communications patterns changed dramatically. Although governments and individuals spent most of the 20th century communicating by telephone, radio, and fax, those technologies quickly dried up as communications shifted to the Internet. Use of email, text messaging, and websites quickly eclipsed traditional telecommunications circuits in both quantity and importance. The intelligence community was forced to evolve new capabilities designed to target communications sent via those channels.

The Washington Post reported in October 2013 on a government program that allegedly sought to exploit those electronic communications. This program allegedly bypassed the end user and went straight to the source, targeting the links connecting the data centers of major providers, including Yahoo! and Google. This supposedly avoided the sticky problem of encrypted communications on the Internet by collecting the cleartext communications on private links within the company's internal cloud.

As a result of this disclosure, many Internet companies have since announced that they will be encrypting communications between data centers in an effort to avoid third-party interception and exploitation of customer communications.

Follow the Money

In September 2013, *Der Spiegel* reported an alleged program known as "Follow the Money" that is responsible for **financial intelligence (FININT)**—operations that collect information about financial transactions for use in intelligence operations. The report alleged that the NSA targeted transactions conducted with Visa cards in Europe, the Middle East, and Africa as well as transaction records from the international SWIFT transaction network.

The information collected by the Follow the Money division of NSA was allegedly stored in a database known as TRACFIN. As of 2011, this database purportedly contained information on 180 million transactions.

Quantum

Computers containing desired information are not always connected to the Internet. Another *Der Spiegel* report alleged that the government could access data stored on offline computers. One of these programs, reportedly called Quantum, inserted devices into computer equipment that allowed remote access via covert radio waves. The signals from these devices could allegedly be received by briefcase-sized radio relay sites located up to 8 miles away from the target computer. That relay device would, in turn, retransmit the information to an NSA facility for further analysis.

The *Der Spiegel* report alleged that Quantum was only one of many tools available to NSA agents seeking to gain access to hardware for intelligence purposes. The agency reportedly published a 50-page attack catalog known as ANT, which provided agents with details on the capabilities available for use against many different types of technology.

CHAPTER SUMMARY

In this chapter, you learned how intelligence operations take place in an increasingly connected world. The intelligence cycle allows operators to identify the intelligence requirements of decision-makers and then progress logically through planning, collection, processing, analysis, and dissemination of intelligence to customers. When gathering this intelligence, they are able to draw upon a wide variety of assets in various subdisciplines. These capabilities draw upon human intelligence assets, capture electronic signals, take photographs of areas of intelligence interest, and assess information available in public sources.

Although intelligence information has long been used to assist with operations in the traditional domains of warfare, there is every reason to believe that it is similarly engaged to assist with cyberwar. Intelligence may provide support to both offensive and defensive cyberwarfare, including the development of target lists and the analysis of adversary cyberwar abilities. The release of the Snowden files provided the world with unprecedented reporting about the alleged intelligence-gathering capabilities of the United States and its allies.

KEY CONCEPTS AND TERMS

Analysis and Production
Collection
Communications intelligence
 (COMINT)
Cryptanalysis
Dissemination
Electronic intelligence (ELINT)
Essential elements of
 information (EEIs)
Financial intelligence (FININT)

Geospatial intelligence
 (GEOINT)
Human intelligence (HUMINT)
Imagery intelligence (IMINT)
Intelligence
Intelligence cycle
Intelligence requirements
Measurement and signature
 intelligence (MASINT)
Metadata

Open source intelligence
 (OSINT)
Planning and Direction
Priority intelligence
 requirements (PIRs)
Processing and Exploitation
Signals intelligence (SIGINT)

CHAPTER 4 ASSESSMENT

1. Which phase of the intelligence cycle includes the identification of intelligence requirements?

   A. Planning and Direction
   B. Collection
   C. Processing and Exploitation
   D. Analysis and Production
   E. Dissemination

2. Intelligence operations follow well-defined processes and utilize standard capabilities that have evolved over time.

   A. True
   B. False

3. What is the output from the Processing and Exploitation phase of the intelligence cycle?

   A. Data
   B. Information
   C. Requirements
   D. Intelligence

4. During which phase of the intelligence cycle does an analyst interpret information and develop an assessment that satisfies an intelligence request?

   A. Planning and Direction
   B. Collection
   C. Processing and Exploitation
   D. Analysis and Production
   E. Dissemination

5. Which of the following are the primary motivations of spies? (Select four.)

   A. Ideology
   B. Grief
   C. Money
   D. Coercion
   E. Ego

6. The _____ aircraft is used by the U.S. Air Force for SIGINT collection missions.

   A. RC-135
   B. B-2
   C. F-117
   D. C-5
   E. B-52

7. Collecting and translating foreign media broadcasts for intelligence purposes is an example of _____.

   A. SIGINT
   B. MASINT
   C. HUMINT
   D. IMINT
   E. OSINT

8. Which alleged operation involved covertly planting eavesdropping hardware inside computer systems used by adversaries?

   A. Follow the Money
   B. Quantum
   C. Foxacid
   D. Echelon
   E. Fornsat

9. Which of the following are examples of MASINT? (Select three.)

   A. Monitoring radiation from nuclear explosions
   B. Intercepting communications between submarine commanders
   C. Developing intelligence from information carried by seismic waves
   D. Taking photographs of adversary facilities via satellite
   E. Analyzing samples of chemical weapons to determine their composition

10. Intelligence support may be critical to offensive cyberwarfare operations. However, it is not likely useful for defensive cyberwar.

    A. True
    B. False

11. Which intelligence agency bears primary responsibility for imagery intelligence (IMINT) operations?

    A. NSA
    B. CIA
    C. NGA
    D. DIA
    E. NIA

12. What type of information was allegedly targeted by the internal cloud programs?

    A. Telephone metadata
    B. Communications between end users and email servers
    C. Communications between commercial data centers
    D. Email metadata
    E. Satellite imagery

13. The NSA allegedly developed a database of credit card transactions, which is an example of which intelligence discipline?

    A. MASINT
    B. IMINT
    C. OSINT
    D. ELINT
    E. FININT

14. The collection of information about disturbances in the earth's magnetic field for intelligence purposes is an example of

    _____.

    A. MASINT
    B. IMINT
    C. OSINT
    D. ELINT
    E. FININT

15. Which of the following countries participated in the U.S.-led intelligence-sharing partnership? (Select three.)

    A. Canada
    B. New Zealand
    C. Germany
    D. France
    E. United Kingdom

PART TWO

Offensive and Defensive Cyberwarfare

The Evolving Threat: From Script Kiddies to Advanced Attackers

HACKERS HAVE BEEN AROUND since the early days of computing. In fact, the term *hacker* has not always carried connotations of malicious intent. It was once used to describe anyone proficient enough with computers to find new and innovative ways to use them. After a series of high-profile computer crimes in the 1970s and 1980s, the media began using the term to describe individuals who exploited computer security weaknesses for fame, financial gain, or other purposes. The name stuck, and the world now views hackers as tech-savvy people who are either common criminals or, at best, individuals living on the margins of law and morality.

The history of hacking is filled with many different types of individual actors who occasionally formed loose confederations in pursuit of a common goal. These hackers probed the weaknesses of operating systems, hardware, and applications, seeking out flaws that might be exploited as gateways for future attacks. They sometimes shared information about these flaws and gained credibility within the hacker community when they disclosed previously unknown vulnerabilities. The open disclosure of vulnerabilities led to the rise of a subset of hackers known as **script kiddies**. These individuals did not discover vulnerabilities on their own. Instead, they downloaded the exploit scripts written by others and simply ran them against target systems without a real understanding of the technical details behind the attack.

Gradually, the world began to realize that it was possible to engage in a wide range of malicious activities through hacking. Computers evolved to the point where they controlled the vast majority of financial transactions, physical security systems, business processes, and many other lucrative targets. The darker corners of society began to prize and value the ability to hack into computers. Organized crime began to dabble in hacking, hiring hackers to use their tradecraft to engage in criminal activity that could be monetized. Governments also realized the damage that could be waged through hacking and began to see it as both a military threat and an opportunity. This was the birth of cyberwarfare.

As cyberwarfare capabilities grew, organized hacking began to take on a different character. Instead of being dominated by script kiddies who sought out any system that might be vulnerable to attack, the threat landscape began to include highly organized and capable groups who had specific targets in mind. These groups wielded very sophisticated technical weapons and became known as advanced persistent threats. They embraced a new model of cyberwarfare known as the Cyber Kill Chain. The Cyber Kill Chain takes the traditional model of warfare and applies it to computer-based attacks.

In this chapter, you will learn about the evolution of the cyberwarfare landscape, from the early days of individual actors to the presence of advanced persistent threats. You will learn about the tools and techniques used by these new cyberwarriors, as well as the Cyber Kill Chain process they use to seek out and exploit high-value targets.

Chapter 5 Topics

This chapter covers the following topics and concepts:

- How hacking evolved from a hobby into a tool of warfare
- What the characteristics, motivations, and tools of advanced persistent threats are
- How the Cyber Kill Chain describes the attack techniques of advanced persistent threats

Chapter 5 Goals

When you complete this chapter, you will be able to:

- Describe the changing threat model presented by cyberwarfare
- Explain how opportunistic, semi-targeted, and focused attacks differ
- Explain the characteristics, motivations, and tools used by advanced persistent threats
- Describe the phases of the Cyber Kill Chain

The Changing Threat Model

Over the past three decades, the world witnessed the evolution of hackers from teenagers hunched over a computer in the middle of the night, *WarGames* style, to sophisticated cyberwarfare threats with financial and geopolitical motivations. As the stakes increased, so did the weapons, tactics, and techniques of cyberwarriors. The increased focus on attack mechanisms led to increased investment in defensive forces and tools. This cat-and-mouse game resulted in dramatic changes in the landscape of cyberwarfare and increasing attention from both government and business leaders.

Historical Hacking

The term *hacker* once described an individual who was extremely proficient at manipulating computers. Early computing industry pioneers, including Steve Wozniak, one of the cofounders of Apple Inc., and Bill Gates, the founder of Microsoft, were proud to be counted among the elite ranks of hackers in Silicon Valley.

The use of the term changed after some high-profile computer security incidents took place in the 1980s, and the media quickly co-opted the term **hacker** to describe individuals who exploit computer security weaknesses for fame, financial gain, or other purposes. The stereotypical image of a hacker was a teenage boy who lacked social skills and filled his time with the esoteric pursuit of probing networks for security holes—gaining access to systems that few people knew existed, let alone understood.

One of the first well-known hackers, Kevin Mitnick, quickly became the poster child for the hacker movement. At the age of 16, he hacked into the private network of computer manufacturer Digital Equipment Corporation (DEC) and stole a copy of their proprietary operating system software. He was convicted of this crime in 1988 and was sentenced to a year in prison and three years' probation. While on probation, he was charged with hacking into telephone company voice mail systems and went on the run, remaining a fugitive from law enforcement until his eventual capture in 1995. His exploits are chronicled in the book *Track Down* and the movie *Takedown*.

FYI

Mitnick is now out of prison and employed as a well-known computer security consultant. He frequently makes media appearances in the aftermath of computer security incidents.

> ### Cracking the Cuckoo's Egg
>
> One of the earliest documented hacking investigations took place in 1986 when Cliff Stoll, a systems administrator at Lawrence Berkeley National Laboratory, noticed an accounting discrepancy on one of his computer systems. As he began to investigate, he realized that someone outside the laboratory had gained access to his system. He quickly set a trap, known as a *honeypot*, for the attacker, and cooperated with law enforcement to identify and track the hacker's activity.
>
> When Stoll and the authorities dug further into the attacker's actions, they quickly became concerned. The hacker had broken into a series of military-related targets, including Army and Air Force bases throughout the United States, Europe, and Asia.
>
> Investigators eventually traced the activity back to a German hacker named Markus Hess. The FBI and the West German government conducted a joint investigation, revealing that Hess had been recruited by the Soviet intelligence service, the KGB, to obtain sensitive military information for the Soviet Union. Hess was convicted of espionage and sentenced to a one- to three-year prison term. Stoll later told the story of the investigation in a book titled *The Cuckoo's Egg: Tracking a Spy Through the Maze of Computer Espionage*. The book tells the story of one of the first recorded acts of cyberwarfare.

Modern Hacking

Modern hackers have advanced far beyond the antics of lone ranger hackers like Mitnick and Hess. Early hackers roamed an Internet that resembled the Wild West. There were few security controls and even fewer trained security professionals. Modern hackers face a sophisticated adversary of well-trained information security practitioners armed with sophisticated defense mechanisms. For this reason, they must step up their game and move beyond the world of script kiddies into an environment of advanced cyberwarfare.

The types of attacks waged by modern hackers may be classified into three categories: opportunistic attacks, semi-targeted attacks, and focused attacks.

Opportunistic Attacks

An **opportunistic attack** begins with a tool in mind. The attacker develops an attack technique and then uses a brute-force approach against thousands or millions of targets in an attempt to find a handful of vulnerable systems. The attacker then exploits those systems, typically either for financial gain or to gain access to the computing power of compromised systems.

Opportunistic attackers use the following common attack techniques:

- Malicious software scans networks for systems with a specific vulnerability and then infects those systems. The SQL Slammer and Code Red worms are famous malware outbreaks and examples of opportunistic malware.

- Phishing attacks send out massive quantities of emails soliciting help with a money transfer. These attacks are designed to lure unsuspecting users into providing financial information, which is then used as part of a fraudulent scheme.

- Phishing attacks seek to scare users into providing usernames and passwords. These attacks often threaten that an email account is about to expire and provide a false password-protected link that the user can click to "rescue" the threatened account.

- Password-guessing attacks target websites and other secured servers. These attacks try common combinations of usernames and passwords in an attempt to gain access to legitimate website accounts.

WARNING

Opportunistic attacks occur on the Internet constantly. For this reason, it is essential to maintain basic security controls, such as updated antivirus software and a robust user education program.

The techniques used by opportunistic attackers are not sophisticated. In fact, they are quite simplistic. The idea behind this type of attack is to play a numbers game. Even if only a miniscule percentage of people or systems are susceptible to the attack, the attacker will likely be successful if he or she tries a sufficient number of times.

Semi-Targeted Attacks

A **semi-targeted attack** goes a step beyond opportunistic attacks and seeks to infiltrate a specific organization or type of target. However, semi-targeted attacks continue to use the same brute-force tools used by opportunistic attacks but sometimes with slight modifications designed to better fit the target environment. They do not target very specific individuals or systems but do seek to compromise computers within a specific organization.

NOTE

Opportunistic attacks are sometimes described as "one in a million attacks" because they use the large number of employees at major targets as an advantage. An attack with a 99.9999 percent failure rate will likely succeed against at least one person if you send it to 1,000,000 people.

For example, the classic "webmail expiration" phishing scam described in the previous section might be modified to fit a particular target organization and then sent only to email addresses within the targeted domain. The original phishing message might read:

> Your mailbox has exceeded the storage limit set by your administrator. You may not be able to send or receive new mail until you revalidate your mailbox. To revalidate your mailbox, please click here: mailbox revalidation.
>
> Sincerely,
> System Administrator

This is clearly a generic message, and it has a very low likelihood of success against all but the most naive email user. However, an intelligent attacker seeking to gain access to an Acme Enterprises email account might use knowledge of Acme's organization to revise this message to something more sophisticated:

Dear Acme Employee,

In compliance with the company's recent GOL initiative, all email accounts are being updated with advanced levels of security. To facilitate these changes, all users must change their account passwords before the close of business on Friday.

To ensure the security of your account, you should change your password only by using the Acme password change utility. Never provide your password to anyone via email, telephone, or other insecure means. To access the Acme password change site, please click here.

Thank you for your cooperation. Remember, it is essential to change your password before Friday to retain continued access to your account.

Sincerely,
George Parker
Vice President for Information Technology
Acme Enterprises

This message is much more sophisticated than the previous draft, although it says basically the same thing. It is a threat that the user's account will be terminated, but it has several advantages over the earlier message:

WARNING

Semi-targeted attacks are often successful because of the "one in a million" approach described earlier. It is believed that this type of phishing attack was responsible for the successful 2011 hacking attack against computer security firm RSA.

- It is branded using the Acme Enterprises name and the company's common jargon, giving it the air of legitimacy.

- Users are warned that they should never provide their password via insecure means, even though the email is asking the user to do just that.

- The message uses the name of an individual who is probably well known to Acme employees and has a high rank within the organization.

- Users are given an imminent deadline (the close of business Friday) and are advised of the consequences of failure to comply (loss of access).

Of course, users who actually click the link are taken to a fake password change site, not the actual Acme page. That site will collect the user's password for use in a later attack. The fake site will likely be designed to visually replicate the official site. If it is truly sophisticated, it will accept both the old and new passwords provided by the user and then actually change the user's password on the official site to avoid raising suspicions when the user is unable to use the new password.

Focused Attacks

Focused attacks, or targeted attacks, seek to compromise a specific system or individual user. They use highly sophisticated attack mechanisms that are often custom-designed to compromise a specific account. They may involve the use of

The Many Hats of Hackers

Hackers have many different motivations and use their skills for different purposes. The community has adopted a system of describing these motivations by talking about the hacker's "hat color," using a system adapted from old Western films. The common hacker hats include the following:

- **White-hat hackers** are benevolent security practitioners who use hacking skills for good purposes, seeking to secure the information systems of their employers or consulting clients. They operate with permission.
- **Black-hat hackers** use their skills with malicious intent. They seek to gain unauthorized access to systems for financial gain, to advance a political agenda, or for other purposes.
- **Gray-hat hackers** occupy a nebulous space between white hats and black hats. They have the seemingly noble purpose of informing companies of system vulnerabilities but operate without permission.

As with all areas of human behavior, it is often difficult to neatly characterize the behaviors of any specific individual.

technically advanced malicious software that is able to gain surreptitious access to a system. Alternatively, focused attacks may leverage a variation of phishing attack, known as spear phishing. Spear phishing uses highly specific information about a particular individual to gain access to an account. For example, a spear phishing attack might send a forged email message from the target's spouse saying something like, "Check out these pictures I took at dinner the other night." Seeing familiar language from a familiar address makes it much more likely that the individual will click the link. Whaling attacks are a variant of spear phishing attacks that target senior executives in an organization.

The pinnacle of the focused attack is the advanced persistent threat. These attackers are extremely well-funded organizations with access to advanced technology and specific operational objectives. Advanced persistent threats are discussed in detail in the next section of this chapter.

Inside the Advanced Persistent Threat

The evolution of hackers from script kiddies into sophisticated attackers with focused goals has led to the emergence of a new kind of cyberwarfare threat— the **advanced persistent threat (APT)**. APTs earned their name because they make use of advanced technologies, select specific targets, and then remain focused on those

> **NOTE**
>
> APTs and cyberwarriors can be one and the same. A cyberwarfare group is an APT from the perspective of the attacked organization.

targets until they achieve victory. An APT might be willing to invest months or years of time toward infiltrating and compromising a valuable target.

APTs are highly organized and have significant resources at their disposal. They move beyond the hacker confederations of earlier times because they have defined leadership and a command structure. APTs might be military units, other government-sponsored entities, or under the control of nongovernmental actors, such as corporations, activist groups, or organized crime.

Characteristics of the APT

APTs are a completely different type of attacker than earlier forms of hacker organizations. They share five common characteristics:

- **Sophisticated technical tools**—APTs have access to advanced attack technologies that are simply unavailable to other attackers. These may include the use of vulnerabilities discovered by the APT or APT sponsor that have not been disclosed to anyone else and, therefore, are difficult or impossible to defend against.

- **Use of social engineering**—APTs often worm their way into an organization by using old-fashioned social engineering techniques that manipulate human behavior. These are discussed in more detail in the "APT Tradecraft" section of this chapter.

- **Clear, defined objectives**—APTs operate in a military or paramilitary manner. They have a very clearly defined mission and conduct all of their cyberwarfare operations in support of that mission.

- **Financial and human resources**—The sponsors of APTs provide very high levels of funding and support for their operation. Depending upon the nature of the sponsor, this may include access to talented staff, tremendous financial resources, and access to intelligence-collection mechanisms and finished intelligence products.

- **Organization and discipline**—APTs are not the loosely organized hacker confederations that grew up in the past. Instead, they are well-organized and disciplined organizations that operate in a command-and-control style.

These characteristics differentiate APTs from all other types of attackers. These well-resourced, highly skilled attackers are formidable opponents for information security professionals in the organizations that APTs target.

APT Motivations

Although APTs share common motivations, they may have very different purposes. Some APTs may have military/political objectives, whereas others may be activist organizations seeking to advance a political agenda. In this section, you will learn four common APT motivations: military/political, cybercrime, corporate espionage, and activism.

Military/Political

The most well-known examples of APTs have military or political agendas. These are state-sponsored groups that operate either within or on the fringes of official government agencies. They are typically used to wage cyberwarfare against a nation's adversaries. The cyberwarfare activities of a nation-sponsored APT may be overt actions that are associated with a declared war/military action. Or they may be covert actions designed to advance the nation's agenda in a stealthy manner, avoiding activities that may lead to attack attribution.

Many governments are known or suspected to sponsor APT activities. The Chinese government is widely believed to be behind a group known as APT1 that has engaged in cyberwarfare activities against targets around the world for more than 10 years. The group is believed to be based in Shanghai and associated with Unit 61398 of the Chinese People's Liberation Army.

Cybercrime and Corporate Espionage

A subset of APTs have old-fashioned motivations: They want to make money. The goal is to simply steal items with monetary value or obtain information that may be used as part of a corporate espionage campaign, allowing a company to gain a competitive advantage over another firm. For example, a company might want to gain access to a competitor's bid on a particular project. This would help the company undercut that bid and win an account.

The recent attack against RSA Security is an example of an APT focused on corporate espionage. Attackers used spear phishing attacks to trick company executives into running a malware-infected attachment. They then leveraged this initial access to gain access to sensitive company information stored on internal systems.

Activism

Although many APTs are motivated by financial or military purposes, others simply seek to advance a political agenda while not having an association with a government. They might want to overthrow an existing regime, gain awareness for a political cause, or correct what they perceive to be an institutional injustice.

One of the best-known activist APTs is a group known as the Syrian Electronic Army (SEA). This group conducts social-media spamming, website defacements, and hacking attacks designed to advance its political agenda. SEA's goal is to support the Syrian government and to attack both Syrian opposition groups and Western organizations that it views as opposed to the Syrian government.

APT Tradecraft

APT groups invest tremendous amounts of time and money in developing sophisticated tools to attack their targets and achieve their ultimate objectives. Although they may use common hacker tools, they typically supplement them with attack mechanisms that are unlikely to trip the alert systems of even very sophisticated opponents. APTs use the following tradecraft types: zero-day attacks, advanced malware, social engineering and phishing, and strategic web compromises.

Zero-Day Attacks

Zero-day attacks are one of the most common techniques employed by APTs. In these cases, the APT group or the sponsor of the APT group identifies a new vulnerability in a software package or operating system. Instead of disclosing that vulnerability to the software manufacturer, the APT keeps it secret for use in a future attack. Because it has no knowledge of the vulnerability, the manufacturer never prepares a security patch designed to correct it.

APTs that exploit zero-day attacks are taking advantage of a concept known as the **window of vulnerability**, shown in Figure 5-1. The risk associated with a zero-day attack begins when the attack is discovered and increases until the time that the software vendor releases a patch. Once a patch is available, the total risk diminishes as administrators around the world apply the patch to their systems. However, some residual risk remains until all administrators have successfully patched their systems.

Malware

Malware is one of the staple mechanisms used by attackers of all shapes and sizes. APTs often use malware toolkits to help standardize their attack process. These toolkits provide an easy way to couple a known control mechanism with a novel zero-day exploit. For example, an attacker might discover a new zero-day exploit and use it to gain access to a system. Once he or she has this access, the attacker might install a **Remote Access Trojan (RAT)** to gain permanent access to that system so that it might be

FIGURE 5-1

The window of vulnerability begins when a new vulnerability is discovered and rises to a peak just before a security patch is released.

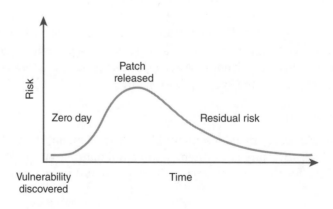

exploited in the future. The APT might use a malware toolkit to develop and deploy the RAT on a compromised system.

Social Engineering and Phishing

Hackers of many different skill levels use social engineering attacks, which are still one of the most successful attack techniques. APT hackers often use highly targeted spear phishing techniques to gain access to a specific individual's system. The hacker chooses a specific individual whom he or she finds of significant interest. For example, the APT might want to gain access to a system administrator's computer because of the unique privileges that the system administrator has on the target network. The attacker might spend weeks conducting careful research about the administrator. He or she might then craft a very clever phishing message designed specifically to fool that administrator into executing attached malware or clicking a malicious web link.

Strategic Web Compromises

Strategic web compromises occur when attackers compromise a website known to be frequented by members of the target organization and install malware on that website. The malware lies dormant until the target visits the website. At that time, it infects the targeted system and gains a foothold in the target environment, allowing future attacks by the APT.

Strategic web compromises are often referred to as *watering hole attacks* because they compromise a location that is commonly visited by the target. Watering hole attacks typically follow a four-step process:

1. The attacker compromises a commonly visited website by conducting an attack against the web servers.
2. The attacker manipulates the code of the compromised website to include instructions to download malicious code, hosted on a second website.
3. The target of the attack visits the website and unknowingly follows the instructions embedded in the watering hole site and downloads the malicious code.
4. The attacker gains access to the target's system through a command-and-control network set up by the malicious code.

Watering hole attacks are dangerous and have become very prevalent in recent years. Several major security vendors have recently documented watering hole attacks that have successfully penetrated both government agencies and large corporations.

The Cyber Kill Chain®

Military targeting specialists use the concept of a **kill chain** to describe how they move from intelligence gathering that identifies appropriate targets to the process of engaging and destroying the adversary. The traditional kill chain is described in the military's *Joint Targeting Manual* and consists of six steps:

1. **Find**—Identify targets and collect intelligence about them.

2. **Fix**—Focus sensors on the target and locate it in time and space.

3. **Track**—Maintain tracking information on the target and conduct additional intelligence, surveillance, and reconnaissance, as necessary.

4. **Target**—Select appropriate weapons, conduct a risk assessment, and decide upon the plan of engagement.

5. **Engage**—Transmit the order to attack and strike the target.

6. **Assess**—Monitor the attack, determine success, and recommend another attack, if necessary.

This kill chain model works very well when describing a kinetic attack against a physical target, such as dropping a bomb on a building. Cyberwarfare planners wanted to adapt this model to reflect the activities of cyberwarfare. Researchers at Lockheed Martin used it as the basis for the Cyber Kill Chain, shown in Figure 5-2.

> **NOTE**
>
> It is important to remember that the steps in the Cyber Kill Chain may occur over dramatically different time intervals. Reconnaissance for an attack may take months in advance of the actual engagement. On the other hand, the entire chain may be compressed into a few hours, depending upon the nature of the attack and external circumstances.

The **Cyber Kill Chain** follows the basic idea of the generic kill chain model. An attack consists of a sequence of events that begins with identifying a target and moves through the stages of selecting and delivering an appropriate weapon. The steps, however, are modified to better describe the actions specific to cyberwarfare attacks.

Individuals participating in cyberwarfare may use the Cyber Kill Chain as an effective way to understand attacks from both offensive and defensive perspectives. Those engaging in offensive cyberwarfare may use the chain to help organize their attacks and ensure that they are progressing in an orderly manner. Defenders may use the chain to gain insight into the minds of their adversaries, recognizing the early signs of an attack in progress and using those indicators to build an effective cyberwarfare early warning system.

Reconnaissance

Attacks begin with the Reconnaissance phase, where cyberwarriors gather information about potential targets. This is extremely similar to the Find phase in the kinetic kill chain model. War planners provide strategic objectives, and targeters search for potential targets that might achieve those objectives. Intelligence professionals gather information about the target's weaknesses that might be exploited in an attack.

In a kinetic attack, this Reconnaissance phase might include using satellite-based imagery intelligence (IMINT) to identify buildings related to the war objectives and determining their precise geographic coordinates to facilitate an attack. The reconnaissance might also include human intelligence (HUMINT) about the construction materials used in a building and signals intelligence (SIGINT) that identifies the hours of occupancy. All of this information may be gathered over an

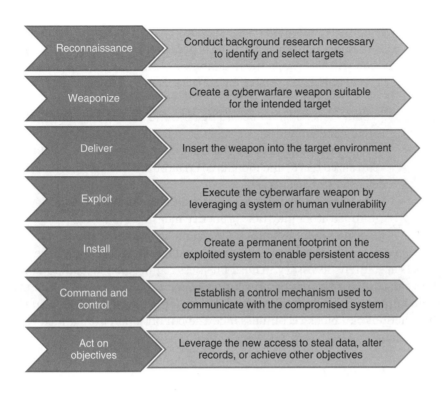

FIGURE 5-2

The Cyber Kill Chain was developed by researchers at Lockheed Martin as an excellent way to understand the process used by APTs to compromise systems.

extended period of time and stored in a targeting database that planners may access quickly in the event of armed conflict.

The cyberwarfare reconnaissance process is quite similar and may leverage some of the same traditional intelligence tools. These tools are normally supplemented, however, with specific tools of cyberwarfare. For example, the Reconnaissance phase might include the following:

- Conducting network sweeps of known IP address ranges belonging to the adversary to identify active servers on those networks

- Running port scans that identify the services running on detected servers

- Using operating system fingerprinting on identified servers to determine both the operating system and version in use on targeted servers

- Accessing web applications running on targeted servers to determine their purpose

- Probing systems for network-based vulnerabilities that may be exploited during an attack

- Conducting web application vulnerability scans to probe running applications for web-specific vulnerabilities

As you look through the preceding list, notice the reconnaissance techniques are listed in order of increasing information utility. You will get more information from the items later in the list. However, the techniques are also increasingly intrusive and are more likely to generate *noise* that alerts the adversary to the fact that the systems are being scanned. Log information will also reveal the IP addresses of the systems conducting the scan, which may reveal important geographic or organizational information about the attackers.

Staff responsible for defending networks should implement robust logging and monitoring capabilities to facilitate the early detection of reconnaissance attacks. This information should be logged to a centralized server and monitored on a routine basis to provide critical defensive intelligence.

In addition to detecting reconnaissance in progress, administrators may take proactive measures to block reconnaissance activities. For example, administrators can configure network and host firewalls to restrict external network access to those servers and ports absolutely necessary to meet business requirements. Additionally, administrators can configure intrusion prevention systems to identify scans that are in progress and automatically block all network access for the systems performing the scan. These actions can block reconnaissance and cut off a potential attack very early in the kill chain.

Weaponize

Once the attacker has identified a target, it is time to develop or customize a weapon that is likely to succeed in compromising the security of that target. The intelligence gathered during the Reconnaissance phase is crucial to this work. Information about the operating system, patch level, and software running on a device allows cyberwarriors to choose a weapon that has a high likelihood of success in compromising that target.

The malicious payload that actually compromises a system is typically embedded in a more innocuous file that is not likely to arouse suspicion. Figure 5-3 shows an example of this, with a malicious payload embedded inside a Microsoft Word document that may be easily delivered via email to the target recipient during the next phase of the Cyber Kill Chain.

Before moving on to the Deliver phase, the attacker will likely test the payload and delivery mechanism to ensure that it will successfully compromise the target system. Highly sophisticated attackers conducting high-stakes attacks will go to great lengths to create a test environment that mimics the actual target environment as closely as possible. The quality of the intelligence gathered during the Reconnaissance phase is directly related to the quality of the test environment. As an example, evidence shows that the Israeli military actually created a physical replica of an Iranian uranium

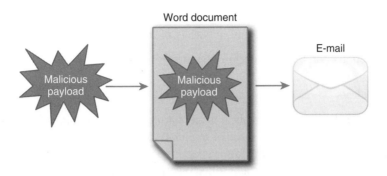

Word document

E-mail

Attackers seeking to create a cyberwarfare weapon first design a malicious payload and then embed it within a delivery mechanism, such as a Word document. In the next phase, attackers deliver the malicious payload to the target, such as via an email message.

enrichment facility during the testing of the Stuxnet worm. This testing effort likely cost millions of dollars, but it made a significant contribution to the eventual success of the attack.

The design of the malicious payload and delivery mechanism typically occurs without any interaction with the target environment. Therefore, it is difficult for defenders to take any measures to either detect or disrupt the Weaponize phase of the Cyber Kill Chain.

Deliver

After developing and testing an effective weapon designed to penetrate the target system, the actual attack takes place. The Deliver phase may occur weeks, months, or even years after the Weaponize phase, depending upon the military or political objectives of the attack. Cyberwarriors may invest significant amounts of time conducting reconnaissance and developing weapons effective against targets that may never be attacked. This is similar to the efforts that kinetic warfare planners put into developing comprehensive war plans that may be used in the event of an actual military engagement.

The delivery occurs by transmitting the embedded payload to the targeted system through any means thought to be effective. A Lockheed Martin analysis of weaponized payloads during the period 2004–2010 revealed three common attack delivery mechanisms:

- Email attachments
- Websites with embedded malware
- USB removable media (flash drives)

Information security professionals seeking to detect and disrupt weapon delivery attempts affecting their users and networks have a variety of tools at their disposal:

- Intrusion detection and prevention systems
- Antivirus software
- Content filters

In addition to these technical controls, it is important to implement and maintain a robust user education program. The first signs of malicious activity are usually visible to end users. If users are trained to identify and report unusual activity on systems, security staff may receive notice of an attack not identified by other detection mechanisms.

Exploit

When the malicious payload reaches the target system, it must then execute and exploit a vulnerability in the target system to gain a foothold. This is the focus of the Exploit phase of the Cyber Kill Chain. The payload may require user intervention, such as opening an infected attachment, as shown in Figure 5-4. Or it may exploit an operating system or application weakness to gain temporary control of the targeted device.

> **TIP**
>
> It is impossible to overstate the importance of quickly installing security patches on systems. This is one of the most important duties of system administrators.

Administrators seeking to defend systems from the Exploit phase of the Cyber Kill Chain must shift their focus to defending the endpoint itself. Once the weapon has been delivered to the target, network-based defenses are no longer effective because the weapon has finished crossing the network. From a detection perspective, host-based intrusion detection software may alert administrators to the presence of an exploit. Antivirus software, either signature-based or behavior-based, may provide a means of both detecting and preventing malware infections.

System administrators may further fortify their devices against attack by ensuring that systems have a full set of current security patches. Applying updates as soon as they are issued reduces the vulnerability of the device to known exploits. Furthermore, the use of

FIGURE 5-4

When the user receives the email with the infected Word document, he or she opens that document, and the payload infects the targeted system.

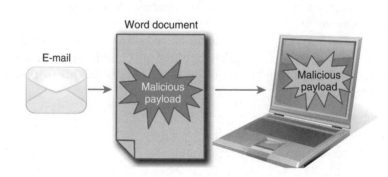

Buffer Overflows and Data Execution Prevention

A **buffer overflow** is a particularly dangerous class of exploit that attempts to force applications to write more data to an area of memory, known as a buffer, than has been set aside for that purpose. If successful, the buffer overflow attack writes into an area of memory designed for program execution. The attacker exploits the attack to push beyond the reserved area of memory and write malicious commands into that executable area. These commands may give the attacker control of the system.

Microsoft's **Data Execution Prevention (DEP)** marks portions of memory set aside for program use as nonexecutable. Even if an attacker is able to exploit a buffer overflow to write commands to memory, he or she cannot force the execution of those commands. Violations of DEP cause the termination of the violating program. DEP is turned on by default in Windows 10.

Many users are frustrated by the pop-up windows associated with DEP. These windows sometimes appear when a user is attempting to install software and prompt the user for administrator credentials. Unfortunately, this inconvenience sometimes leads users to disable DEP, undermining its effectiveness as a security control against buffer overflow attacks.

preventive technology such as Microsoft's Data Execution Prevention may block certain types of attacks from succeeding.

Install

Once the attacker gains access to the target system through the exploitation of a vulnerability, the attacker must move to make that access permanent. This occurs during the Install phase of the Cyber Kill Chain. The attacker leverages, or uses, the temporary access achieved in the Exploit phase and installs a RAT on the system. The RAT allows the attacker to gain permanent access to the infected system. The capabilities of a RAT include the following:

- Obtaining complete control of the infected computer
- Monitoring screen activity
- Logging keyboard and mouse activity
- Stealing data stored on the device or accessible network locations
- Running reconnaissance tools against other systems on the network connected to the RAT-infected system
- Altering system configuration settings

RATs are designed to operate in a stealthy manner, hiding their presence from the end user. They typically communicate back to their controlling user over an encrypted HTTPS channel. This makes it difficult to distinguish the network activity they generate from normal use of secure websites. Attackers using a RAT to control a system must take care to avoid issuing any commands that alert the user to the RAT's presence. For example, a user might be alerted to the presence of a RAT by any of the following indicators:

- Unusual disk activity
- Changes in system configuration settings with user-visible results
- Alterations to data stored on the device
- Slow performance caused by heavy CPU use by the RAT

TIP

The use of behavior-based antivirus software is growing in popularity. This software does not require signatures and is capable of detecting previously unknown malware on a system.

Once an attacker gains access to a system in the Exploit phase, it becomes very difficult to prevent the attacker from continuing through the final phases of the Cyber Kill Chain. Antivirus software may be effective against the use of a RAT, particularly if the RAT is a known piece of malicious software.

Command and Control

In the Command and Control phase of the Cyber Kill Chain, the attacker maintains control of the compromised system through the use of the RAT. This typically takes place through the use of a command-and-control (C2) server, as shown in Figure 5-5. The attacker may either interactively control the infected computer through the C2 server or may store a series of queued commands on the C2 server that will be executed on the infected systems at a later date.

NOTE

A system under the control of a C2 server is commonly referred to as a **bot**. The attacker might have a large collection of bots executing the same commands simultaneously. This is known as a **botnet**. Botnets are commonly used in distributed denial of service attacks.

Most corporate networks block all unsolicited inbound connections from the Internet as a standard defensive measure. They strictly limit inbound connections to a small set of servers designed to handle such traffic, such as public-facing web servers and email servers that accept inbound messages from Internet users. Furthermore, these publicly exposed systems are normally segregated from other systems in a **demilitarized zone (DMZ)** that is unable to access other corporate resources. That way, if a system in the DMZ is compromised, the infection is less likely to spread to other systems on the company's network.

The use of DMZs can spell trouble for C2 connections and is the primary reason that attackers use dedicated C2 servers. Systems operated by end users are almost always on private networks with no external inbound exposure. An attacker may be able to travel through the Install phase by tricking the user into opening an infected email message,

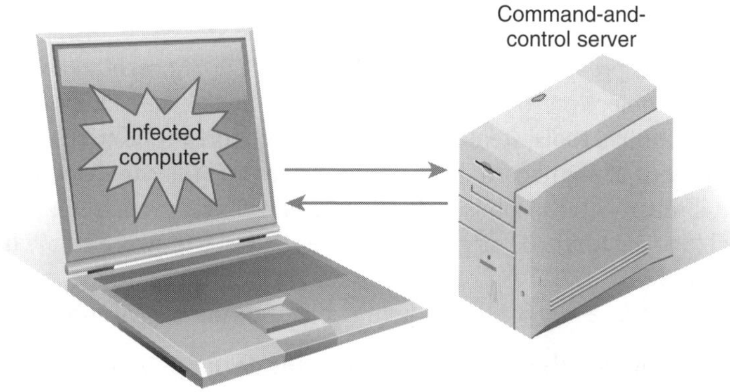

Command-and-
control server

Infected
computer

FIGURE 5-5

After malware is
installed on the target
system, it establishes
a C2 link with a server
maintained by the
attacker.

accessing an infected website, or opening an infected file. However, the presence of a firewall would prevent the attacker from directly connecting to the RAT on the compromised system and issuing commands.

Given this limitation, attackers program RATs to connect to a C2 server once they are installed on a system. The outbound connection takes place over an encrypted HTTPS connection that is allowed by the firewall because it is indistinguishable from routine, secure web-browsing traffic. The C2 server then uses that RAT-initiated connection to issue commands to the compromised system. C2 servers also serve as intermediaries, hiding the true IP address of the attacker from the RAT and frustrating the attempts of investigators to determine the attacker's identity.

Using encrypted connections makes it extremely difficult to detect C2 connections on your network. One technique used to identify these connections is to build a database of known C2 servers and then monitor network activity for attempts to connect to those known malicious addresses. This becomes a cat-and-mouse game, with attackers rotating the IP addresses of C2 servers when they become known to security administrators.

When an administrator detects a connection attempt to a known C2 server, he or she should take two actions. First, block the connection attempt. The administrator typically does this in an automated fashion through a firewall rule or intrusion prevention system. Second, immediately investigate the system that was the source of the infection because it has likely been infected by a RAT. Even if antivirus scans return clean results, it would be prudent to immediately isolate the system from the network and rebuild it to a known secure state.

 TIP

You might want to consider subscribing to a threat intelligence service that offers members access to a C2 server IP blacklist that is updated in real time. This may prevent RATs from connecting to your C2 server and receiving commands.

5

The Evolving Threat

Act on Objectives

After progressing through the previous six phases of the Cyber Kill Chain, attackers may finally act on their ultimate objectives for the attack. Reaching this phase requires careful planning, sophisticated technical tools, stealthy navigation, and fortuitous circumstances. It may require the investment of significant financial resources and take an extended period of time to successfully reach this phase. Many attacks will never reach the point where the attacker is able to act on those original objectives because they are stymied at an early stage of the kill chain.

Attackers may want to achieve any of the following objectives during this phase:

- Stealing sensitive information that is either stored on the device or in a location that the device is able to access
- Monitoring the activity of the device's user for intelligence purposes
- Altering data on the device or a connected database to mislead the enemy
- Using the device as a foothold on the organization's network with the objective of leveraging that access to gain access to yet another system, typically beginning the Cyber Kill Chain anew

Whatever the objectives of the attacker, it has taken a considerable effort to reach this stage in the chain. The advanced technology leveraged to achieve this goal and the sustained effort required to successfully arrive at this point are the reasons that attackers employing the Cyber Kill Chain are known as APTs.

CHAPTER SUMMARY

Hacking techniques and organizations have evolved significantly over the past three decades. Whereas early attacks were waged by lone teenagers exploring the new world of networked computing from their basements, modern attacks may be waged by well-funded and highly organized adversaries with sophisticated technical capabilities. APTs are capable of waging sustained, complex attacks against high-value targets. They use a variety of sophisticated techniques to gain access to systems. They then leverage that access to achieve their ultimate objectives.

Cyberwarfare experts use the concept of the Cyber Kill Chain to describe the phases of a cyberwarfare attack. The Cyber Kill Chain begins with the Reconnaissance phase where the attacker gathers as much information as possible about the target. The attack then moves into the Weaponize phase where the attacker develops a weapon specific to the selected target. This weapon is transmitted to the target during the Deliver phase. The attack then enters the Exploit and Install phases, where it gains access to the target system and installs a permanent footprint on the device. Attackers retain access to the target system during the Command and Control phase. They eventually implement their strategic aims in the Act on Objectives phase.

KEY CONCEPTS AND TERMS

Advanced persistent threat (APT)	Data Execution Prevention (DEP)	Remote Access Trojan (RAT)
Black-hat hackers	Demilitarized zone (DMZ)	Script kiddies
Bot	Gray-hat hackers	Semi-targeted attack
Botnet	Hacker	White-hat hackers
Buffer overflow	Kill chain	Window of vulnerability
Cyber Kill Chain	Opportunistic attack	Zero-day attacks

CHAPTER 5 ASSESSMENT

1. What term is used to describe individuals who simply execute attacks designed by others?

   A. Script kiddies
   B. Cyberwarriors
   C. APTs
   D. Hackers
   E. Information operators

2. During the _____ phase of the Cyber Kill Chain, attackers may use network sweeps to identify active systems.

3. Attackers generally have no interaction with the target environment during the Weaponize phase of the Cyber Kill Chain.

   A. True
   B. False

4. Which of the following mechanisms are commonly used to deliver a malicious payload to a target? (Select three.)

   A. Email attachments
   B. Websites
   C. IRC channels
   D. USB drives
   E. Firewalls

5. During which phase of the Cyber Kill Chain is an attacker likely to make use of a RAT?

   A. Act on Objectives
   B. Deliver
   C. Reconnaissance
   D. Install
   E. Exploit

6. Which technology is an effective way to prevent the successful execution of buffer overflow attacks?

   A. Memory blocking
   B. Firewall
   C. Antivirus
   D. Intrusion detection
   E. Data Execution Prevention

7. Which type of network typically contains all systems that offer public-facing services to Internet users?

   A. Local area network
   B. Internet
   C. Intranet
   D. Demilitarized zone
   E. Buffer network

8. Which protocol is typically used for the command-and-control communications between a RAT and a C2 server?

   A. HTTP
   B. HTTPS
   C. FTP
   D. SFTP
   E. ICMP

9. Cyberwarfare reconnaissance may leverage traditional intelligence methods, such as HUMINT, IMINT, and SIGINT.

   A. True
   B. False

10. Which type of hacker uses his or her skills to find security vulnerabilities in computer systems and advise the owners of the flaws but does so without permission?

    A. White hat
    B. Gray hat
    C. Black hat
    D. Red hat
    E. Blue hat

11. Spear phishing attacks use cleverly written phishing messages that are sent to large groups of people simultaneously.

    A. True
    B. False

12. Which type of attack infects a website known to be visited by a target with malware in the hopes that the target will unwittingly download and install it?

    A. Denial of service
    B. Vishing
    C. Phishing
    D. Distributed denial of service
    E. Watering hole

13. Residual risk continues to exist even after software vendors release a patch for a new vulnerability.

    A. True
    B. False

14. One of the characteristics of APTs is that they typically have access to significant financial and human resources.

    A. True
    B. False

15. Phishing attacks that contact large numbers of people attempting to find one person vulnerable to a scam are an example of _____ attacks.

Social Engineering and Cyberwarfare

GOVERNMENTS, MILITARIES, AND PRIVATE ORGANIZATIONS around the world spend billions of dollars each year on advanced technical controls designed to protect the confidentiality, integrity, and availability of information and information systems. Almost all of these dollars are designed to combat technical threats, such as the zero-day attacks leveraged by advanced persistent threat actors, malware that may seek to infiltrate a network, and probing attacks that search for holes in perimeter defenses. Unfortunately, all of these major security investments may be easily undermined if organizations overlook a single weak link in them all—the human factor.

Unlike computer systems, human beings are complex, nuanced individuals who make decisions based on many different factors, both logical and emotional. Human beings have no firewalls, intrusion detection systems, or antivirus software that may be programmed to prevent undesired behavior. In addition, they often have the ability to override security systems, either by explicitly disabling them or by working around them. For example, a human who knows that a data loss prevention system blocks attempts to email sensitive documents outside the organization may decide to simply place them on a USB drive and hand deliver them to the intended recipient.

Cyberwarfare adversaries understand the complexities of human behavior and use it to their advantage. Manipulating the actions of human beings through psychological influence tactics is a technique known as *social engineering*. Social engineers have a deep understanding of ways to influence the thoughts and actions of individuals and use that understanding to achieve cyberwarfare goals. This may include convincing an insider to divulge sensitive information needed to wage a cyberwarfare attack, perform actions that undermine security defenses prior to an attack, or actually unwittingly become a direct participant in the attack itself.

Security professionals must understand the impact that social engineering may have on cyberwarfare operations and the damage that a victim of social engineering can inflict on the larger organization. This understanding should lead the organization to invest in security controls that protect against social engineering. Most notably, organizations should include comprehensive security education and ongoing awareness programs in their communications and outreach efforts. Everyone in the organization must understand the importance of protecting information assets, the signs that a social engineering attack may be under way, and the proper action to take when he or she suspects that an attacker is engaging in social engineering.

In this chapter, you will learn how social engineers exploit human behavior to achieve their goals. You will learn about the six principles of influence used by social engineers: reciprocity, commitment and consistency, social proof, authority, liking, and scarcity. You will then learn how organizations can prepare themselves for social engineering attacks. The chapter concludes by examining a case study of a social engineer at work.

Chapter 6 Topics

This chapter covers the following topics and concepts:

- How humans are the weak link in the cybersecurity chain
- What social engineering is and how it is used as a weapon of cyberwarfare
- What the key principles of influence are and how they may be exploited by social engineers
- How social engineers use tools to achieve their goals
- How security professionals defend against social engineering attacks
- How a real-world social engineer operates

Chapter 6 Goals

When you complete this chapter, you will be able to:

- Describe how humans are often the greatest weakness in an information security program
- Explain how social engineering is a weapon of cyberwarfare

- Explain the six key principles of influence and how they may be used as social engineering weapons
- Describe the tools of the social engineer
- Explain how to defend against social engineering attacks
- Provide examples of real-world social engineering attacks

Humans: The Weak Link

Humans are the weakest link in the information security chain. Organizations invest significant financial and human resources in building strong layers of technical controls, embracing a defense-in-depth approach to information security. These technical controls are focused on protecting the confidentiality, integrity, and availability of both sensitive information and the information systems that run an organization.

Unfortunately, these controls are not sufficient on their own to maintain the security of information. Organizations must remember that human beings are operating information systems and handling sensitive information, and they often have the ability to bypass or undermine security controls. Consider a few examples:

- Scampi Systems has deployed a cutting-edge network firewall designed to block access to the Scampi network from the Internet. The firewall is configured to allow only electronic mail and web server requests from the outside and to deny all other attempts to access Scampi's systems. Scampi also provides a virtual private network (VPN) to allow employees to bypass the firewall and gain access to the network if they need to work from home or while traveling. If an employee of Scampi provides an attacker with the username and password required to access the VPN, the attacker may then also bypass the firewall.

- Doodle Labs uses a network data loss prevention (DLP) system to monitor outgoing network traffic for the presence of sensitive information. In particular, Doodle Labs is concerned that documents related to a secret product development effort not leave the company's network. An employee with access to the documents would be thwarted by the DLP system if he or she attempted to email the file or upload it to a website. However, if the employee simply copies the file to a removable flash drive and mails the drive to the intended recipient, there is nothing that the network-based DLP can do, and Doodle's secrets may be exposed to the world.

WARNING

It is still important to *have* technical controls. The solution to this problem is to resolve the human issue and complement strong human controls with solid technical defenses. You'll learn more about defending against the human risk later in this chapter.

- Hunter Industries protects employees from visiting malicious websites through the use of a content filtering system that scans the URLs of outbound web requests and compares them with a list of known malicious sites. The system blocks any attempts to access a blacklisted site and creates a log for administrator review. An employee seeking to bypass the system might use the Tor anonymous web browser to hide the details of the destination site through the use of an encrypted connection, rendering the content filtering system useless.

In each of these examples, individuals inside the company took actions that undermine the organization's security. Although they may have done this with malicious intent, it is also possible that they thought they were doing the right thing—cutting through the technical "red tape" to get the job done. Whatever the reason for their actions, the end result is the same: Technical controls are rendered useless by human actions.

Social Engineering

Social engineering is the art of manipulating human behavior through social influence tactics in order to achieve a desired behavior. When employed by a cyberwarrior, social engineering techniques may be used for a variety of purposes:

- Tricking an individual into divulging information about information systems, networks, or other operational details that may contribute to the Reconnaissance phase of a cyberwarfare attack—providing critical information necessary to wage an offensive cyberwar operation
- Influencing an individual to bypass physical security controls, granting an attacker access to a physical facility where he or she might undertake offensive cyberwarfare operations

▶ **NOTE**

Social engineering uses techniques similar to those engaged in during psychological operations (PSYOPS). However, one major difference is that social engineering attacks typically target a single individual, whereas PSYOPS target groups.

- Convincing an individual to disable electronic security controls, such as bypassing a firewall or allowing a VPN connection from an unauthorized source
- Tricking an insider into installing software on a computer located within the organization's protected network that secretly creates a back door, allowing the attacker to gain access to the network

Advanced persistent threats often leverage social engineering as part of a comprehensive attack on an organization. They may use these techniques to perform intelligence gathering, influence user behavior to facilitate an attack, or cover their tracks after an attack takes place.

Kevin Mitnick: Master Social Engineer

Renowned hacker Kevin Mitnick is well known for his mastery of social engineering techniques. He began his social engineering career as a young boy riding the bus system of the San Fernando Valley. While riding the bus, Mitnick realized that the transfer system used by the bus line was ripe for exploitation. The system's security depended upon bus drivers using a special punch to put holes in the transfer slip noting the details of the transfer. Drivers give these slips to passengers who must take multiple buses to reach their destination but are only charged a single fare.

Mitnick realized that if he learned the transfer punch system and obtained both a punch and a book of blank transfers, he could ride the system for free. He learned the punch system by studying the transfer slips that he purchased himself. Poking around the bus depot, he realized that bus drivers often discarded partially used books in the trash at the end of their shifts. This solved the second problem, but he still didn't have access to the actual punch that the drivers used to create the transfers.

He then turned to social engineering and worked his magic with bus drivers to overcome this last hurdle. In his own words, "A friendly driver, answering my carefully planted question, told me where to buy that special type of punch." The masterful use of social engineering techniques is a hallmark of Mitnick's many electronic misadventures.

Influence as a Weapon

The art of influencing people is one of the most studied topics in psychology. Product marketers, salespeople, lobbyists, public relations staff, and many other people use influence as one of the major tools of their trades. Those who possess the power of influence are able to achieve great things and magnify their other abilities. Hackers are no exception, especially when it comes to social engineering. Successful social engineers wield the tools of influence as a weapon.

Robert Cialdini, a marketing professor at Arizona State University, is widely recognized as one of the world's experts in the field of influence. In his book *Influence: Science and Practice*, Cialdini outlines six key principles of influence, illustrated in Figure 6-1:

- Reciprocity
- Commitment and consistency
- Social proof
- Authority
- Liking
- Scarcity

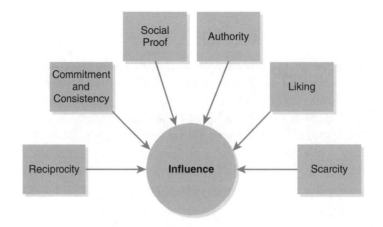

Cialdini's six principles of influence may be used as weapons by cyberwarriors using social engineering attacks.

Each of these tools plays an important role in all types of influence. Social engineers may exploit these techniques to manipulate their targets into performing a desired action or revealing sensitive information that is critical to a successful cyberwarfare attack. In the remainder of this section, you will learn about each of the influence principles and how it might apply in the context of a social engineering attack.

Reciprocity

Life is about give and take. When someone does something for you, you feel a natural obligation to return the favor. In fact, you probably feel like you owe that person a small debt that must be repaid. This is the principle of **reciprocity**—a relationship between two people that involves the exchange of goods or services of approximately equal value.

Certainly, you can think of examples from your life where you've participated in reciprocity and felt the obligation of returning a favor or gift. When you go out to lunch with a friend or colleague and the other person picks up the tab, you naturally feel like you should do the same next time. When someone takes care of your pet while you are traveling, you feel a need to bring a gift from your trip to thank that person for spending his or her time helping you.

Although many examples of reciprocity come from agreements, "I'll do this for you, if you do that for me," not all follow this model. Many are unsolicited and impose an obligation on the recipient that is undesired or even resented. Cialdini describes a striking example of this in his book. He tells of a university professor who engaged in a simple experiment to test the power of reciprocity. The professor purchased a batch of Christmas cards and then mailed them out—to total strangers. Many of those strangers quickly responded with a card of their own without even asking about the professor's identity or his relationship to them.

The power of reciprocity is extremely strong. The social obligation and sense of indebtedness imposed upon you when someone gives you a gift is overwhelmingly strong and may lead you to take actions that you would otherwise never take. Social engineers know this and exploit it to their advantage. In an advanced, well-funded social engineering attack, a social engineer might invite a company executive to an all-expenses-paid conference hosted in a desirable location. After the executive enjoys a day or two of the conference, the social engineer might ask to sit down over a cup of coffee and discuss the company's business. The conversation might begin with very innocent questions and progress into more sensitive territory. For example, the social engineer might ask these questions in order:

1. Can you tell me more about your products?
2. What industries do you typically serve?
3. What product do you think provides you with the greatest competitive advantage?
4. How much of that product do you sell each year?
5. Is it difficult to make that product? Why?
6. That is certainly an impressive product. Who actually designed it?
7. How does the product work?

As you can see, the questions quickly move from information that is probably available on the firm's website into details that are likely confidential. The executive, who has already received and used the gift, likely feels a sense of obligation to the social engineer and may answer many or all of these questions out of a sense of reciprocity.

Commitment and Consistency

Human nature leads people to want to act in a consistent manner. They want to perceive their current and future actions as consistent with what they have done in the past. The principle of **commitment and consistency** states that once someone has made a commitment to a particular course of action, that person becomes confident that his or her action was correct and consistent. This confidence increases the likelihood that his or her future actions will be consistent with that committed decision.

One example, provided by Cialdini, is a study conducted by Canadian psychologists who measured the confidence of individuals placing bets at a racetrack. In the minutes before an individual places a bet, he or she has a measurable sense of uncertainty. The bettor is considering the past performance of the horses competing in the race, the track conditions, any possible weather that may come into play, and a variety of other circumstances. Eventually, the bettor reaches a decision, approaches the betting window, and places a wager with the attendant. When the professors measured the confidence level of the bettor immediately after placing a wager, they found a noticeable increase in confidence from the measurement taken immediately before placing the bet. Nothing had changed about the race. The same horses were still competing on the same track in the same weather conditions. Why had the bettor suddenly become more confident?

This is the principle of commitment and consistency at work. Once the bettor actually made a decision and put money on the line, the bettor desperately wanted his or her thoughts to be consistent with that decision. This led the bettor to develop a sense of confidence in the decision. In fact, if the experimenters had gone on to prompt the bettor to increase his or her wager (which they did not), they may have found that the bettor's newfound confidence would lead to making a substantially larger wager than he or she had originally contemplated.

Social engineers can also use the principle of commitment and consistency to their advantage. Consider the case where an attacker posing as a technical support representative is having a conversation with a target in an attempt to persuade the target to divulge a computer password. The conversation might go something like this:

TARGET: "Hello, Bob Jones here."

ATTACKER: "Hi, Bob. This is Paul. I'm calling from Air Force technical support. I understand that you've been having problems with your computer?"

[*Pause while Bob struggles to recall a computer problem*]

TARGET: "Yes, actually, you're probably calling about the email issue I reported a while back, correct?"

ATTACKER: "Yes, Bob, we've had some new developments, and I think we might be able to fix your problem. Can you describe what you were experiencing?"

TARGET: "My email is pretty slow. Sometimes it takes five minutes to send a message."

ATTACKER: "Oh, perfect. That's exactly the problem that we're fixing. Would you like me to correct your account?"

TARGET: "Yes, please!"

ATTACKER: "OK, I'll just need a little information from you to help identify and fix your account. Is that OK?"

TARGET: "Sure."

ATTACKER: "Great, and what is your email address?"

TARGET: "It's bob.jones@af.mil."

ATTACKER: "Great, thanks Bob. I just need to log on to your account and make a few setting changes. What is your password?"

Now, of course, Bob has been trained to never divulge his password to anyone. In this scenario, however, he might just do that because of the principle of commitment and consistency. The social engineer cleverly led him through a series of questions where he answered, "Yes," committing to the fact that he had a computer problem, that it was with his email, and that he wanted the attacker's help fixing the issue. Then, when the attacker asks for the password, Bob probably knows he should not provide it, but he has already committed to obtaining the help and providing information.

Social Proof

The principle of **social proof** says that an individual in a social situation who is unsure how to act will likely follow the example of the crowd and behave in the same way as the people around him or her. Cialdini points out the example of canned laughter on television as an example of this. When asked, most people consider canned laughter silly and strange and would say that they don't find things funny just because there is a "laugh track" running on a television show. Furthermore, the actors and writers who make television shows also often object to the use of a tactic that seems phony to them. So, if audiences and talent both claim that they don't like it, why do television producers continue to use it? Quite simply, it works. Canned laughter is an example of a social proof cue. It influences the audience to react with laughter because the "crowd" is acting the same way. This is true even though the "crowd" doesn't really exist!

How could a social engineer exploit the principle of social proof? Let's return to the case of Bob and Paul, where Paul is a social engineer seeking to obtain the password to Bob's computer. If Paul were trying to exploit the principle of social proof, the conversation might go like this:

TARGET: "Hello, Bob Jones here."

ATTACKER: "Hi, Bob. This is Paul. I'm calling from Air Force technical support. I'm working through some system updates and need your help."

TARGET: "OK, it's Friday afternoon, though; is this going to take long?"

ATTACKER: "No, it shouldn't be too long. My boss is on me to get 50 of these done before the weekend and I only have two left to go. The others took about 10 minutes each."

TARGET: "Great, what do we need to do?"

ATTACKER: "Let me walk you through the first few steps...."

[*Paul has Bob perform a number of simple steps.*]

ATTACKER: "OK, now I need to apply the security settings through the central server console. I'll add you in with the rest of the users I put in today. What's your password?"

At this point, Bob is in the same situation where he found himself in the earlier scenario. He's being asked to do something that he doesn't want to do, but Paul has introduced some social proof cues that might modify Bob's normal behavior. Paul repeatedly referred to the fact that he's already done 50 similar updates and that the other people all provided their account information.

 NOTE

The principles of social influence are not normally used in isolation. In a typical social engineering attack, the attacker exploits several of these principles at the same time in an effort to persuade the victim to perform the desired action or divulge sensitive information.

Authority

Authority is one of the most powerful weapons of influence. If an individual perceives that someone making a request has power over him or her, he or she is much more likely to accommodate the request, even if it seems unreasonable or counter to policies and beliefs. Social engineers who adopt signs and symbols of authority are sometimes able to achieve significant results.

An individual may possess three main types of authority in any given situation:

- **Legal authority**—The actions of the individual are, or seem to be, backed by the authority of law. Law enforcement officers, judges, and tax collectors all operate under legal authority. A social engineer might attempt to falsely assume the identity of someone with legal authority before making a request. For example, the social engineer might pose as a police officer conducting an investigation that requires access to an organization's network.

- **Organizational authority**—Most organizations operate in a very hierarchical fashion, with lower levels in the organization giving deference to higher levels. A social engineer understands the organizational authority dynamics in place at a targeted organization and uses those dynamics to his or her advantage. For example, the social engineer might conduct reconnaissance to obtain a copy of the organizational chart and learn the name of a senior executive. He may then pose as the executive when making demands of others within the organization.

- **Social authority**—Every group of people has a natural "pecking order" that is determined by the leadership tendencies of the group's members. This natural leadership that emerges generates social authority. Group members are likely to defer to the leader to curry favor and remain in good graces with the group. Social engineers may develop understandings of informal social networks and then exploit them in the same way they would organizational authority.

Individuals seeking to exploit authority as a weapon of influence often attempt to assume the identity of someone with authority. To do so, they often only need to convince the other person that they seem like someone with authority. In the words of Cialdini, "The outward signs of power and authority frequently may be counterfeited with the flimsiest of materials." Social engineers often assume three major symbols of power:

- **Titles**—People are deferential to those with superior titles. Exploiting titles may be as simple as introducing yourself as the "senior vice president of Blahbittyblah" and proceeding to give individuals directions.

- **Clothes**—People are conditioned to interpret someone's clothing as a sign of his or her level of authority. The obvious example of this is that a social engineer posing as a senior executive should dress in the same style as executives from the targeted company. In a less-obvious example, many hackers seeking to gain physical access to a facility dress in the characteristic coveralls of a repairperson and don a tool belt. People see a repairperson working in telecom closets and usually don't even think twice, if they consciously notice it at all. They assume that someone wearing that type of clothing has the authority to work in a maintenance closet.

Using the Title of "Doctor"

Cialdini tells the story of a study into the power of titles that took place in a Midwestern hospital. The researchers placed telephone calls to the nurses stations on 22 different hospital wards, identified themselves as a doctor unknown to the nurse, and gave orders that a patient be administered 20 milligrams of the drug Astrogen.

The nurses were, of course, highly trained medical professionals who should have picked up on several problems with this request:

- Hospital policy forbade the use of telephone prescription orders.
- The drug Astrogen was not approved for use in the hospital.
- The dose the doctor ordered was double the maximum daily dose.

Despite these warning signs, almost all of the nurses hung up the phone and headed straight for the medicine cabinet. Researchers then intercepted them to make sure that the drug was not actually administered to a patient.

This scenario, scary as it may seem, is a perfect illustration of how an individual's title can convey immediate authority. The nurses did not know the doctor but, given the social dynamics of a hospital, chose to unquestioningly obey an order that they should have known was dangerous, merely because a "doctor" told them to do so.

- **Trappings of power**—People also expect individuals with authority to have possessions that match their level of power. For example, if you are impersonating a corporate executive, you'd better pull up to the meeting in a luxury car. No CEO is going to show up driving a clunker!

Social engineers use these signs and symbols of power as a way to project authority that they do not have. Through the use of a combination of titles, clothing, and trappings of power, they may assume the false identity of someone who has legal, organizational, or social power. Once they have successfully adopted the role, they may then use the authority to influence the actions of individuals, achieving their end goals.

Liking

The principle of **liking** says that people are more apt to be influenced by people whom they know and appreciate. It should come as no surprise that people are willing to do things for those whom they like. The principle of liking is closely related to the principle of authority in that they both represent situations where the relationship between two individuals guides the willingness of the individuals to do things for one another. In authority situations, the relationship is based upon power, with one individual having legal, organizational, or social power over the other. In liking situations, the relationship is based upon friendship or affinity, and one individual is willing to help the other because of that relationship.

Social engineers can take advantage of the liking principle by engaging in activities that might make the target like them more. Some examples of this include the following:

- Complimenting the target on his or her personal appearance
- Praising the professional background of the target
- Smiling and projecting a friendly attitude

Many social engineers combine the liking principle with small gifts that also trigger the need for reciprocity. A beautiful woman might approach a male target, smile, and compliment him on his appearance and then offer to buy him a cup of coffee. That's quite a bit of influence all rolled into one small encounter.

Scarcity

The scarcity principle says that people are likely to respond to a situation where they believe they might experience a loss. This might be due to the limited availability of a resource, time limitations on an offer, or an exclusivity of access that would normally preclude them from a resource or experience.

Some of the most glaring examples of scarcity at work are the television commercials that scream, "Time is limited, only 30 seconds remain in this offer. Operators are standing by!" Of course, the same commercial airs over and over again, and there really is no time limit. The advertiser is leveraging the principle of scarcity, trying to create a sense of urgency on behalf of the viewer. They hope that the scarcity will prompt the viewer to purchase now, before the product is never again available at this low, low price.

Imagine again the scenario of Paul and Bob, where Paul is attempting to social engineer Bob into providing access to his system. Here's how the conversation might go if Paul decides to use the principle of scarcity:

TARGET: "Hello, Bob Jones here."

ATTACKER: "Hi, Bob. This is Paul. I'm calling from Air Force technical support. I'm working through a set of urgent system updates and need your help."

TARGET: "How can I help you?"

ATTACKER: "The payroll system was infected by a virus, and we are up against a deadline. If the system isn't corrected by 5:00, the month's payroll won't run, and nobody will be paid!"

TARGET: "Oh my gosh! That would be terrible! You only have an hour left. What do you need me to do?"

ATTACKER: "There is a utility that must be run on your computer to update the accounting software. Once we have your update complete, we can then clear out the master system and remove the virus."

TARGET: "Of course, what do I need to do?"

The attacker now has gained the confidence of the victim and created a sense of scarcity—he must act quickly to save the payroll system or nobody will be paid. Bob certainly does not want to be the person who disrupted the paychecks of his coworkers, so he may be prone to comply. Furthermore, the short deadline reduces the likelihood that Bob will follow up with his supervisor or security officials—he needs to act quickly to save everyone's paycheck.

Tools of the Social Engineer

In the previous section, you learned about the weapons of influence that social engineers use to influence the thoughts and actions of their targets. In this section, you will discover the tools and techniques that social engineers wield to exploit these influence principles. These are the actual tactics they engage in to wage cyberwarfare against unsuspecting human targets.

> **NOTE**
> The tactics used by social engineers aren't that much different from those used by swindlers and grifters throughout history. They've become much more effective, however, with the massive communications power of modern networks.

Pretexting

Pretexting is one of the trickiest weapons of the social engineer. In a pretexting attack, the social engineer creates a false set of circumstances and uses them to convince the target to take some form of action. Essentially, pretexting is weaving a web of mistruth that convinces someone else that he or she should take an action.

Social engineers engaging in pretexting commonly assume the identity of someone with the authority to perform an action and then use that identity to execute the action. Here are some common examples of pretexting at work:

- Calling the customer service line of a bank and requesting that the representative change the home address of an account belonging to the assumed identity. The pretexter might do this in an attempt to delay the discovery of fraudulent transactions.

- Phoning an online retailer and changing the email address associated with an account. The pretexter can then visit the website and use the "forgot password" link to trigger an account reset email to the pretexter's email address, gaining access to a third party's account.

- Emailing the customer service department of a telephone company and requesting that a detailed call history be sent to the assumed identity. This might be an attempt by the pretexter to learn whom the target communicates with on a regular basis.

> **NOTE**
> Notice that the examples all involve the use of the telephone and email. Although not all pretexting attacks take place over the phone or email, the vast majority do. This is because a pretexter is more likely to be caught if he or she shows up in person.

The most important ingredient to a pretexting attack is information, especially when the attack takes place in person or over a live telephone call. The pretexter needs enough information about the victim to be able to answer security questions and intelligently discuss the account. For example, if the pretexter provides the victim's street address and the representative follows up with, "Oh, and what county is that in?" the pretexter had better be able to quickly respond. It would be highly unusual for someone to not know the name of his or her county.

Here are some basic facts that the pretexter will want to have lined up about a victim before conducting an attack:

- Name
- Home and work addresses
- Telephone numbers
- Email addresses
- Mother's maiden name
- Enough personal history to answer most security questions
- Work history
- Information about the individual's relationship with the firm

The specific details will vary depending on the type of account being attacked. Some pretexters conduct "test runs" by attempting to pretext accounts other than the victim's in an attempt to learn the types of questions they will be asked when attempting the true pretexting attack.

Hewlett-Packard Caught in the Act of Pretexting

Hewlett-Packard, one of the largest computer manufacturers in the world, found itself embroiled in a pretexting scandal in 2006. The company experienced a leak of confidential information and suspected that a member of the company's board of directors was responsible for sharing proprietary information with journalists.

Newsweek conducted an investigation, which uncovered that board chairwoman Patricia Dunn ordered an intelligence operation that employed security consultants and private investigators to look into the leaks. They investigated several of the board members as well as journalists. The private investigators engaged in pretexting attacks to obtain the telephone records of the investigation targets and concluded that George Keyworth, a board member, was the source of the leak. Keyworth denied his involvement but resigned from the board.

After *Newsweek* revealed the pretexting operation, both Congress and state and federal prosecutors launched investigations of Hewlett-Packard and the investigative firms. Three investigators pled no contest to state wire fraud charges and received 96 hours of community service. A fourth investigator pled guilty to federal charges of conspiracy and identity theft and was sentenced to serve three months in prison.

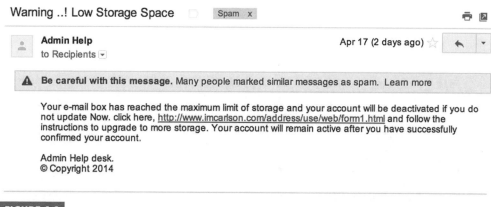

FIGURE 6-2

Example of a phishing email seeking to obtain personal information from unsuspecting individuals.

Phishing

Phishing is one of the most common social engineering tactics in use today. In a phishing attack, the attacker sends the victim an electronic message (via email, text, or other means) attempting to solicit sensitive information from the victim. The message is usually disguised to look as if it originated from an entity that the victim has a relationship with. It is forged, of course.

Figure 6-2 provides an example of a phishing message. Notice that the message uses the social influence tactic of scarcity. The user is warned that if he or she does not follow the instructions at the linked website, the account will be deactivated. After clicking the link, the user is redirected to a website that asks him or her to log on, providing the username and password associated with the account. This information is then sent to the attacker, who can now access the victim's account.

Social engineers commonly use several subtypes of phishing attacks. Some standard phishing attacks may be rudimentary, such as the one shown in Figure 6-2. However, other types of phishing may be quite advanced, including two particular subtypes:

- **Clone phishing** captures an email message being sent from one individual to another, modifies it slightly, and sends a new copy, appearing to be a resend of a message. One clone phishing tactic involves resending a copy of a legitimate message with the header "Sorry, I forgot the attachment" and with a malicious file attached.
- **Spear phishing** attacks are directed at a specific individual. Before engaging in a spear phishing attempt, the attacker gathers personal information about the individual and the organization and uses it to lend credibility to the communication.
- *Whaling attacks* are a subtype of spear phishing attacks that are directed at senior executives and leaders of organizations.

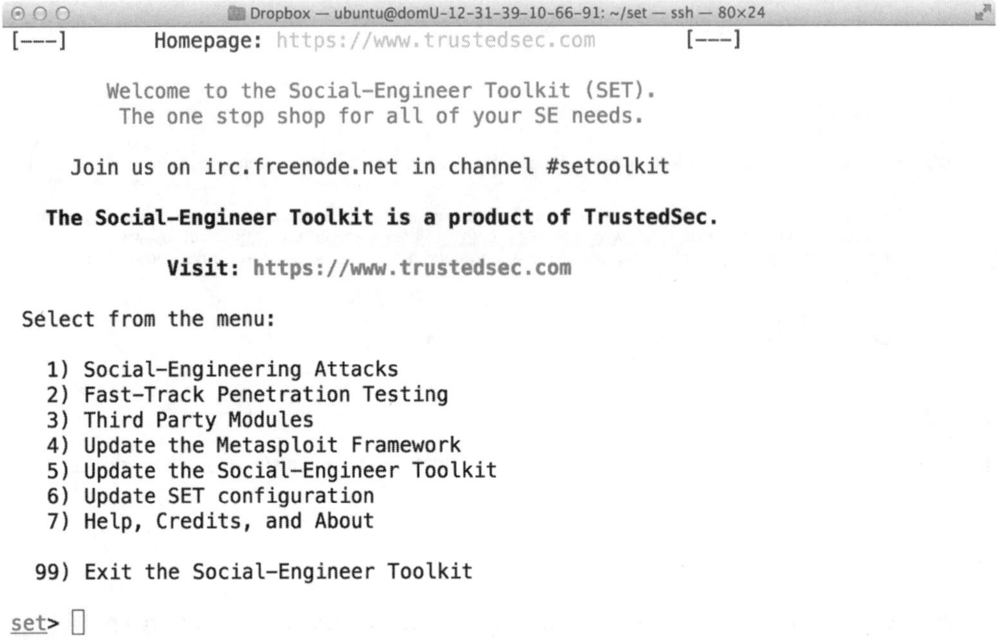

```
 ○ ○ ○             Dropbox — ubuntu@domU-12-31-39-10-66-91: ~/set — ssh — 80×24
[---]          Homepage: https://www.trustedsec.com          [---]

              Welcome to the Social-Engineer Toolkit (SET).
              The one stop shop for all of your SE needs.

         Join us on irc.freenode.net in channel #setoolkit

       The Social-Engineer Toolkit is a product of TrustedSec.

                  Visit: https://www.trustedsec.com

     Select from the menu:

       1) Social-Engineering Attacks
       2) Fast-Track Penetration Testing
       3) Third Party Modules
       4) Update the Metasploit Framework
       5) Update the Social-Engineer Toolkit
       6) Update SET configuration
       7) Help, Credits, and About

      99) Exit the Social-Engineer Toolkit

     set> []
```

FIGURE 6-3

The Social-Engineer Toolkit automates the creation and execution of social engineering attacks.

Social engineers make widespread use of these techniques and, in fact, have a set of tools available to them that facilitate the automation of phishing attacks. These toolkits can automatically scrape the contents of a legitimate website, create a false version of the site that captures password information, and then automatically send out links to that we site in customized phishing messages. Figure 6-3 shows an example of one of these toolkits, the Social-Engineer Toolkit (SET).

Dumpster Diving

Closely related to social engineering is the art of dumpster diving—combing through the trash and recycling bins of an organization seeking out documents that contain sensitive information. This is a time-honored practice of social engineers, who can often use the information that they find to help inform their social engineering attacks by identifying targets, obtaining information about organizational structures, and adding details to their pretexting attacks.

Free Tablets!

In a recent example of baiting uncovered by government investigators, a foreign government sponsored a high-profile keynote speaker at a conference. They provided the speaker with a brand-new tablet as a token of thanks for participating in the conference.

When counterintelligence investigators studied the device, they discovered that the tablet came with more than a standard operating system. It had been infected with several backdoor Trojans that would provide the foreign government with the ability to access content on the device at a later date. This is a very clever attack that could be effective against senior executives.

Baiting

In a **baiting** attack, the attacker creates a flash drive that contains malicious code and leaves it in a parking lot, lobby, or other location where a target is likely to discover it. The target may then plug the flash drive into a USB port on a computer within the organization to view the contents. When this takes place, the flash drive installs the malicious code on the victim's computer, and the attacker gains access to the organization.

It is widely believed that baiting was used in the Stuxnet attack against the Iranian nuclear program. Analysts believe that flash drives seeded with the Stuxnet worm were placed in areas where nuclear facility personnel were likely to find them and that one of those drives made its way into the Natanz nuclear facility. There it triggered a malware outbreak that caused substantial damage to the facility's high-speed uranium-enrichment centrifuges.

Defending against Social Engineering

Defending an organization against social engineering is of paramount importance. Throughout this chapter, you learned about the havoc that social engineering can cause in your enterprise and discovered the dangers of social engineering attacks. You also learned that technical controls are not always effective against this type of attack because the attack relies on the human ability to circumvent security controls. In this section, you will learn some techniques that may be used to defend against social engineering.

Security Awareness and Education

The primary and most effective defense against social engineering attacks is user education. Users must understand the tools of the social engineer and learn to recognize when attacks are under way. This training should involve the use of specific examples and take place in several forms:

TIP

Many organizations already have a wide variety of required training programs. You may want to combine social engineering training with related security training programs to avoid aggravating users with yet another mandatory training session.

- *Initial training* should introduce users to the concepts of social engineering, provide them with examples of social engineering attacks, and help them learn strategies for dealing with suspected social engineers.

- *Refresher training* should take place on a periodic basis and remind users of the lessons they learned during their initial training.

- *Ongoing awareness programs* should keep social engineering at the top of users' minds, reminding them that these attacks take place every day.

Training programs are the most effective way to arm users to live in a world where social engineering is a clear and present danger. Training programs should be on the top of your agenda when it comes to protecting against cyberwarfare threats.

Incident Reporting and Response

One of the messages that your training program should deliver to users is that they need to report any and all suspected incidents of social engineering to appropriate authorities. The reporting mechanism should be easy to use and users should be thanked or rewarded for filing reports.

TIP

When using real-world examples in your social engineering awareness program, it is a good idea to remove the name of the victim as well as identifying details. You don't want to stigmatize the victim because this will result in decreased future reporting.

The information obtained through reporting may be used to immediately bolster the organization's defenses against a specific social engineering threat. If a social engineer is targeting one employee of an organization, it is very likely that other employees are also being targeted. If the original victim cuts off contact with the social engineer, this may prompt contact with other employees as the attacker seeks an avenue of entry into the organization.

When the organization receives a report of social engineering, it should communicate it to security professionals and consider taking actions to block the attacker's continued access to personnel. The example may also be used as part of the organization's ongoing awareness program.

Content Filtering

Many social engineering attacks depend on electronic communication to succeed. Organizations can monitor the content of electronic communications and filter out those believed to be part of phishing attacks before they reach the intended recipient. Organizations can use content filtering in two main ways to combat social engineering.

First, email messages can be screened at the mail gateway to identify those that contain malicious links or match the patterns of previous phishing attacks. This may include verifying the reputation of the sender with a security service and analyzing the content of the message for suspicious content. Messages that resemble phishing attacks may be either quarantined or completely blocked.

Second, web content filtering may be used to block the URLs of websites known to be participating in social engineering attacks. Organizations may subscribe to services that catalog known malicious websites and then use a content filtering tool to block all access to those websites from the organization's network.

The downside of these techniques is that they are only effective against known threats. Although content filtering is quite good at blocking large-scale phishing attacks, it does not work very well against highly targeted attacks in which the social engineer sends an extremely small volume of messages and directs the recipient to a website that has not previously been used in a phishing attack. Therefore, the site would not appear in any malicious website database.

Penetration Testing

Penetration testing is the final weapon in the arsenal of cyberwarriors seeking to block social engineering attacks. In a penetration test, the organization attempts to attack itself using the techniques used by real attackers. This allows the organization to identify potential security weaknesses and improve its security.

Social engineering penetration tests conduct phishing, pretexting, and other social engineering tactics against an organization's own employees. The penetration testing team then educates victims about the risks of social engineering and tells them that, had this been an actual attack, they may have been compromised.

> ⚠ **WARNING**
> Penetration testing carries quite a bit of political risk. You may anger people when you perform these tests, so you should only do so after careful consideration and with appropriate permission.

Robin Sage: A Case Study in Social Engineering

In one of the most stunning examples of social network–based social engineering, Thomas Ryan of Provide Security conducted an operation known as "Robin Sage." In his words, the operation was an experiment that "seeks to exploit the fundamental levels of information leakage—the outflow of information as a result of people's haphazard and unquestioned trust." It provides a telling example of the dangers of social engineering.

Ryan created fake social networking profiles belonging to a fictitious woman named Robin Sage. He used a photo lifted from an adult website as the profile photo and created an entirely false background for her, giving her excellent professional credentials as a Cyber Threat Analyst with the Naval Network Warfare Command. He then sought out social networking connections with members of the military and intelligence community.

In his social engineering attack, Ryan took advantage of several of the principles of social influence discussed in this chapter. First, Ryan made use of the principle of liking. He deliberately created a profile belonging to an attractive female. In Ryan's words:

Choosing a young, attractive and edgy female was a deliberate decision on the part of the creator. Today, the vast majority of the security industry is comprised of males. The heavily male dominated sector allows women to be a commodity in more ways than one.

Second, Ryan took advantage of the principle of social proof. After gaining a few well-respected connections within the security community, Robin Sage quickly attracted the attention of hundreds of other members of that community. When evaluating her friend request, those individuals saw that Robin was already connected to people whom they knew and trusted.

Was Ryan's attack successful? By the end of his 28-day experiment, Robin Sage had gathered hundreds of social networking connections from across the security industry. Her contacts included line staff and senior executives from the National Security Agency, the Department of Defense, the military, Fortune 500 companies, and other sources. Some of the specific contacts made with Robin included the following:

- A lecturer at NASA's Ames Research Center asked Robin to review a paper that he had written and provide her opinion on it.
- A senior executive in the homeland security field requested a telephone conversation to discuss Robin's background in more detail.
- Connections offered her gifts, job opportunities, and invitations to speak at upcoming information security conferences.

Had Ryan been a malicious actor, he could have attempted to take the Robin Sage experiment a step further and actually continued the contact with her new network, seeking to obtain access to sensitive information and systems. It is clear from this short experiment that social engineering attacks pose a clear and definite risk to cybersecurity.

CHAPTER SUMMARY

Social engineering is a substantial weapon in the toolkit of the cyberwarrior. The effective use of influence techniques to obtain sensitive information and modify the behavior of human subjects can yield significant results. Social engineers know that the six influence techniques of reciprocity, commitment and consistency, social proof, authority, liking, and scarcity are powerful weapons for cyberwarfare. They use these techniques while engaging in pretexting, phishing, baiting, and other social engineering activities. Organizations must take measures to defend themselves against social engineering attacks and ensure that they do not fall victim to this threat.

The Robin Sage case study illustrates the effectiveness of combining social influence knowledge with real-world social networks. In the case, a security consultant was able to quickly build a network of respected security professionals by baiting them with the profile of an attractive woman with an impressive security background. The military, government, and corporate officials who fell victim to the Robin Sage project were later embarrassed to discover that they had been the victims of a social engineering experiment.

KEY CONCEPTS AND TERMS

Authority	**Liking**	**Social engineering**
Baiting	**Phishing**	**Social proof**
Clone phishing	**Pretexting**	**Spear phishing**
Commitment and consistency	**Reciprocity**	

CHAPTER 6 ASSESSMENT

1. Which principle of influence states that when someone gives you a gift you are likely to give him or her something in return?

   A. Reciprocity
   B. Social proof
   C. Authority
   D. Liking
   E. Scarcity

2. The use of canned laughter in television programming to prompt an audience reaction is an example of _____.

   A. reciprocity
   B. social proof
   C. authority
   D. liking
   E. scarcity

3. Social engineers typically use multiple influence principles during a single attack, rather than focusing on one principle per attack.

   A. True
   B. False

4. The six principles of social influence were outlined by _____ in his book, *Influence: Science and Practice*.

   A. Sigmund Freud
   B. Kevin Mitnick
   C. Chip Heath
   D. Robert Cialdini
   E. Daniel Pink

5. Which of the following are examples of the types of authority that a social engineer might try to assume in a situation? (Select three.)

   A. Legal authority
   B. Abstract authority
   C. Psychological authority
   D. Organizational authority
   E. Social authority

6. Which of the following are examples of ways that an individual might project the signs of authority? (Select three.)

   A. Title
   B. Trappings of power

   C. Pretense
   D. Clothing
   E. Presence

7. Which of the following activities attempt to influence the actions of individuals or groups? (Select two.)

   A. ELINT
   B. PSYOPS
   C. SIGINT
   D. Social engineering
   E. MASINT

8. When an individual is willing to help another person because of a friendship relationship, this is an example of _____.

   A. commitment and consistency
   B. authority
   C. social proof
   D. reciprocity
   E. liking

9. Placing a telephone call to a customer service department pretending to be someone else is an example of _____.

   A. baiting
   B. pretexting
   C. social proof
   D. elicitation
   E. emotional manipulation

10. What kind of social engineering attack is specifically targeted at the senior executives of an organization?

    A. Spear phishing
    B. Pretexting
    C. Phishing
    D. Clone phishing
    E. Whaling

11. What social engineering attack is believed to have been used in the Stuxnet attack against the Iranian nuclear program?

    A. Baiting
    B. Pretexting
    C. Social proof
    D. Elicitation
    E. Emotional manipulation

12. Security professionals should covertly conduct penetration tests designed to test whether senior executives fall victim to social engineering attacks.

   A. True
   B. False

13. Which of the following are the most likely places to perform content filtering as a defense against social engineering attacks? (Select two.)

   A. Email
   B. Web content
   C. Text messages
   D. Social networks
   E. Database servers

14. What is the most effective defense that an organization can use against social engineering attacks?

   A. Content filtering
   B. Education
   C. Penetration testing
   D. Threat analysis
   E. Incident response

15. What kind of social engineering attack uses modified versions of legitimate messages to fool a recipient into opening an attachment or visiting a website?

   A. Spear phishing
   B. Pretexting
   C. Phishing
   D. Clone phishing
   E. Whaling

Weaponizing Cyberspace: A History

C YBERSPACE MAY BE USED as an arena of warfare. It adds new dimensions to all aspects of warfare, including offensive operations, defending one's own assets, and gathering intelligence. Modern militaries are actively taking note of these developments and building their own cyberwar abilities. In this chapter, you will learn how cyberwar actions and capabilities have grown over the past three decades from a fringe area of interest for military strategists to a core part of modern operations. In fact, a number of recent military actions have taken place only within the limited domain of cyberspace.

The first major cyberwarfare attacks took place in the late 1990s on the heels of a military exercise known as Eligible Receiver. Eligible Receiver revealed significant weaknesses in the U.S. military's ability to detect and defend against cyberwarfare attacks. Shortly after the exercise took place, the U.S. government suffered a series of cyberattacks that confirmed the exercise's findings. The events of Solar Sunrise and Moonlight Maze pointed out that the vulnerabilities in the nation's cyberinfrastructure were both real and exploitable.

The activities of the late 1990s were harbingers of larger things to come. The first decade of the new millennium was a turning point in the world of information security and cyberwarfare. In addition to the continued exploitation of government computer systems by foreign actors, computer systems around the world fell victim to malicious code that rapidly spread from system to system. The worms of the 2000s worked by exploiting previously known, but undefended, weaknesses in common operating systems and applications. This wave of activity brought information security to the forefront of the national consciousness. It resulted in increased attention by governments and military forces as well as corporations and private citizens. Information security jobs and departments began to spring up in places where the subject was previously unaddressed.

In the 2010s, cyberwarfare continued to evolve into a very real component of warfare. During those years, sophisticated, organized cyberattacks were waged both by and against the United States and its allies. The Stuxnet worm, perhaps the best-known example of an advanced military cyberattack, disabled an Iranian uranium enrichment facility that was critical to that nation's nuclear-weapons development program. Also in the past few years, the advanced persistent threat (APT) and the involvement of many nations in organized, sophisticated offensive cyberwarfare activities have continued to evolve.

In recent years, cyberwarfare has continued to grow. The Russian government is alleged to have used cyberwarfare techniques extensively in their military engagement with Ukraine. In 2019, *The New York Times* reported that USCYBERCOM had extensively penetrated components of the Russian electrical grid.

As you peruse the examples in this chapter, you may find that some involve a blurred distinction between cyberwarfare and the actions of nonmilitary groups. This is because it is often difficult to distinguish between the two. In many cases, the perpetrators of cyberattacks are of unknown origin. In others, it is possible to identify the probable country of origin but difficult to attribute the attack to state-sponsored activity. This attribution problem plagues military planners as they attempt to identify and counter threats in the cyber domain.

Chapter 7 Topics

This chapter covers the following topics and concepts:

- How early cyberattacks awakened the world to the risks of cyberwarfare
- What malicious code (worm) outbreaks in the 2000s brought cyberwarfare activities into the public eye
- How the Stuxnet worm and other actions by nation-states in the 21st century contributed to the evolution of cyberwarfare

Chapter 7 Goals

When you complete this chapter, you will be able to:

- Explain how the Solar Sunrise and Moonlight Maze attacks brought to light major weaknesses in U.S. cyberwarfare readiness
- Describe the types of malicious code outbreaks that occurred in the early 2000s
- Explain how nations have become involved in waging offensive cyberwarfare activities
- Describe how the Stuxnet worm marked a turning point in cyberwarfare
- Understand the risks of cyberwarfare in the contemporary military environment

Early Attacks: The 1990s

In the early 1990s, few people were aware of the potential threats posed by cyberwarfare activities. The Internet was new, and the minority of citizens who had even heard of the global network only considered it a novelty designed for a very limited audience of computer geeks and academics. One author, Winn Schwartau, wrote a prescient book in 1994 that described how cyberwarfare might materialize. In it, he wrote of conflicts between enemies in which information was the prize—in which computers were offensive weapons and, along with communications systems, were targets forced to defend themselves against invisible attack.

Schwartau foresaw a type of warfare in which information weapons replaced conventional arms. Anyone—not just governments, intelligence agencies, and militaries—would be able to obtain these weapons from readily available commercial and other sources. And potential attackers could build these weapons in the comfort of their homes.

The end of the 1990s saw the rise of organized cyberwarfare activities around the world that Schwartau predicted. Shrouded in secrecy, governments quietly organized new military units composed of cyberwarriors who were responsible for developing plans, tactics, and weapons for use in this emerging domain of warfare. Cyberwarfare capabilities were not publicly discussed, but a public debate about the future of warfare began to slowly simmer as some details of cyberwarfare capabilities began to surface in national media reports.

Solar Sunrise

According to a report on the *InformIT* website, in February 1998, U.S. military-run intrusion detection systems began to pick up signs of attacks against military computer systems located around the country. The attacks first surfaced on February 3, 1998, with the successful compromise of a government computer system at Andrews Air Force Base (AFB) in Maryland, the home of Air Force One. Over the course of the following days, the Air Force system detected similar successful attacks against systems located at installations around the country:

- Lackland AFB in San Antonio, Texas
- Kirtland AFB in Albuquerque, New Mexico
- Tyndall AFB in Panama City, Florida
- Maxwell AFB's Gunter Annex in Montgomery, Alabama
- Columbus AFB in Columbus, Mississippi

FYI

The Solar Sunrise incident took place at a critical time in military history. Saddam Hussein's regime still ruled Iraq and was rebuffing efforts by United Nations inspectors to investigate Iraq's nuclear program. The nation was not yet at war, but it was becoming clear that military action was imminent. Solar Sunrise investigators worried that the attacks might be related to war preparations.

It quickly became apparent to investigators on the Air Force's Computer Emergency Response Team (AFCERT) that they were experiencing an organized attack directed specifically against military computer systems. After conferring with other federal agencies, they realized that the attack was more widespread and involved systems run by other branches of the military: the National Aeronautics and Space Administration (NASA), federal laboratories associated with the military, and foreign military forces. They quickly organized a government-wide investigation into the attacks and mobilized a force of cyberwarfare investigators that was unprecedented in size and ability. They gave their investigation the code name **Solar Sunrise**.

> **NOTE**
>
> University computer systems have historically been jumping-off points for cyberwarfare attacks because of their significant computing and network resources and relatively low levels of security. Attackers able to compromise university systems could then leverage that computing power to their advantage—both hiding their true location and magnifying the power of their attacks.

The investigators quickly traced many of the attacks back to university computer systems located around the United States and Israel, including the Massachusetts Institute of Technology. Further investigation revealed that these institutions were not the true source of the attacks. They were merely relay points used to obscure the true identity of the attackers. Fortunately for investigators, the hackers had made a crucial mistake—they transferred some data from the Andrews AFB attack directly to accounts that they controlled at a California Internet service provider.

Investigators quickly obtained a series of warrants and court orders that allowed them to trace the activity back to two teenage boys in California with the hacker names Stimpy and Mak. They collected direct evidence of the boys' hacking activities and raided their bedrooms, finding the boys working in filthy conditions. Special Agent Chris Beeson of the San Francisco Federal Bureau of Investigation (FBI) led the raid and described the scene: "Their rooms were a mess.... The scene was typical of teenagers. Pepsi cans. Half-eaten cheeseburgers." The two boys, whose names were not revealed due to their age, were arrested, but the investigation was not yet over.

Federal agents realized that a third individual was involved in the attacks and, after reading communications between the boys involved in the attacks, identified him as a hacker using the name The Analyzer. They discovered that The Analyzer was actually the more skilled hacker and had been mentoring Stimpy and Mak as they broke into systems around the nation. After tracing The Analyzer's activity, they learned he was a 19-year-old Israeli named Ehud Tenenbaum. U.S. and Israeli law enforcement worked together to arrest Tenenbaum and charge him with computer attacks against systems located in both countries. Tenenbaum later pled guilty to charges relating to the Solar Sunrise attacks and served eight months in prison.

The Solar Sunrise incident was a serious threat to the national security of the United States. Senator John Kyl, then chair of the Senate Subcommittee on Terrorism, told the BBC that the Defense Department described it as the most serious intrusion into the United States up to that point.

The Analyzer Strikes Again

The Solar Sunrise attacks weren't Ehud Tenenbaum's last stand. After his release from prison, he founded a computer security firm and stayed off the public stage for about 5 years.

In 2008, Canadian authorities working with the U.S. Secret Service arrested Tenenbaum, charging The Analyzer with masterminding a series of attacks against financial institutions throughout the United States. Tenenbaum's second prosecution dragged on for years, eventually resulting in a conviction on bank fraud charges. Tenenbaum was sentenced to time served and ordered to pay $503,000 in restitution while serving a 3-year probation.

Fortunately, the attacks did not inflict significant damage on government computer systems. Instead, they pointed out to both government leaders and the general public the real risk of cyberattack. Solar Sunrise led to a significant increase in funding for information security activities and a new high-level commitment to safeguarding the nation's cyberinfrastructure.

Moonlight Maze

Solar Sunrise was one of the first major attacks against the U.S. cyberinfrastructure. Another followed the very next month, in March 1998, although it wasn't detected until more than a year later. This stealthy attack, code-named **Moonlight Maze**, involved reconnaissance and infiltration of computer systems owned and operated by government agencies, universities, and research laboratories located around the United States.

Although few details of the attacks have been publicly released, some limited details have made their way into the public domain and were reported in a PBS *Frontline* investigation:

- The attacks began in March 1998 and were not detected by the government until the spring or summer of 1999.
- Some of the attacks were traced back to a computer system located in Russia.
- Sources indicate that thousands of files may have been stolen during the attacks and that some of these files were likely sensitive, but not classified.
- Government investigators suspected that the Russian government was involved in the attacks.
- The Russian government denied responsibility for the Moonlight Maze attacks but did not cooperate with the U.S. investigation.

John Arquilla of the Naval Postgraduate School told PBS's *Frontline* that the Moonlight Maze intrusion into Defense Department computers took place over a long period of time and was an indication that modern computer systems are vulnerable to both disruption and exploitation by adversaries. Those adversaries, if successful, could gain access to very sensitive information.

Arquilla also pointed out that the intrusion could easily have been prevented. In his words, "Had the data in question that was being pilfered been strongly encrypted, it would have been of no use to the intruders." Unfortunately, it was not. The information stolen during the attacks was located in a printer queue where it was neither encrypted nor protected by a firewall. "And so," said Arquilla, "it was simply plucked."

Moonlight Maze also demonstrated the difficulty of attributing attacks to their original source. There was some indication that computer systems in Moscow were involved in the attack, but investigators had no way of knowing whether the Russians were complicit in the attack or whether someone elsewhere in the world was relaying an attack through Russian systems in an attempt to throw investigators off the track.

Confirming the Findings of Eligible Receiver

Solar Sunrise and Moonlight Maze weren't the first time that the U.S. military pondered the impact of a cyberwarfare attack on their computer systems. According to *Global Security.org*, a 1997 exercise code-named **Eligible Receiver** purposely tested military computer systems.

During the Eligible Receiver exercise, computer analysts from the National Security Agency (NSA) conducted actual attacks against government computer systems located at the Pentagon's National Military Command Center, U.S. Space Command, U.S. Transportation Command, U.S. Special Operations Command, and U.S. Pacific Command.

The attackers were quite successful. Some of their many achievements during the exercise included the following:

- Breaking into systems and networks, gaining administrative access
- Tampering with and eavesdropping on email messages
- Disrupting operational systems

Summing up the findings of the exercise, then-Deputy Secretary of Defense John Hamre stated: "What Eligible Receiver really demonstrated was the real lack of consciousness about cyberwarfare. The first three days of Eligible Receiver, nobody believed we were under cyberattack."

Official details about the Moonlight Maze attacks remain sketchy to this day. Arquilla's statement equivocating about Russian involvement may be factually correct, but it is not the opinion expressed by military leaders at the time. A 1999 *Newsweek* story quoted unnamed Pentagon officials classifying Moonlight Maze as "a state-sponsored Russian intelligence effort to get U.S. technology." Then-Deputy Secretary of Defense John Hamre put it even more bluntly: "We're in the middle of a cyberwar."

Moonlight Maze did mark a major milestone in cyberwarfare's evolution. Whether the rumors of Russian government sponsorship are true or not, Moonlight Maze was the first time that the U.S. government suspected that it was being systematically attacked by a serious cyberwarfare adversary with military/political motivations. The Solar Sunrise attackers were children playing with dangerous toys. The Moonlight Maze attackers, on the other hand, had significant resources at their disposal and focused on the pilfering of large amounts of information.

Honker Union

The Honker Union is an organized group of Chinese hackers who have waged cyberwarfare against targets whose views and actions conflict with those of the Chinese government. It is difficult to pinpoint when the group formed, but analysts believe it has early origins, perhaps dating back to the 1999 American bombing of the Chinese embassy in Belgrade, Yugoslavia. In subsequent years, the experts say the group has masterminded a series of malicious code outbreaks and cyberwarfare attacks against U.S. government targets.

The precise details of Honker Union activity are not publicly known. The group's involvement is suspected in three prominent actions:

- Creation of the 2003 SQL Slammer worm, discussed later in this chapter
- Cyberattacks launched against the Tibetan dissident Tsering Woeser that took place in 2008, including impersonations of her on Skype. These forced Woeser to issue a warning to her colleagues to verify that the person they were talking to on Skype was really her.
- Ongoing attacks against Japanese targets in the wake of territorial disputes between China and Japan over the ownership of a set of islands

These attacks continue to this day and, if waged by the same Honker Union group, mean that the group is one of the oldest, if not the oldest, organized hacking groups in existence.

The 2000s: The Worm Turns

Although the attacks of the 1990s raised the profile of cyberwarfare, the attacks were relatively unsophisticated. They led the way, however, for an increasingly sophisticated series of attacks that took place during the next decade. The 2000s saw the advent of sophisticated malware that spread around the world under its own power. The decade also marked an uptick in the suspected use of cyberwarfare tactics by nations advancing military and political agendas.

 NOTE

The Code Red worm earned its name from the drink that security researchers were enjoying when they discovered it—Mountain Dew Code Red!

Code Red

On July 19, 2001, the Internet erupted with activity as a new piece of malicious software quickly spread among vulnerable computer systems. This code, known as a **worm**, was especially dangerous because, unlike viruses, the worm was able to spread on its own power—jumping from system to system without human intervention. In a single day, the **Code Red** worm infected more than 350,000 computer systems around the world.

Code Red impacted only systems running Microsoft's Internet Information Server (IIS) web server software. It used a buffer overflow technique that flooded a small location in the server's memory, seeking to write executable commands into a protected area of memory. Code Red simply filled up a buffer with the character *N* and then wrote commands that allowed the worm to infiltrate the machine. The HTTP log entries from one infected system showed the attack in progress:

```
GET /default.ida?NNNNNNNNNNNNNNNNNNNNNNNNNNNNNNNNNNNNNNNNNNNN
NNNNNNNNNNNNNNNNNNNNNNNNNNNNNNNNNNNNNNNNNNNNNNNNNNNNNNNNNNNNN
NNNNNNNNNNNNNNNNNNNNNNNNNNNNNNNNNNNNNNNNNNNNNNNNNNNNNNNNNNNNN
NNNNNNNNNNNNNNNNNNNNNNNNNNNNNNNNNNNNNNNNNNNNNNNNNNNNNNNNNNNNN
NNNNNNNNN%u9090%u6858%ucbd3%u7801%u9090%u6858%ucbd3%u7801%
u9090%u6858%ucbd3%u7801%u9090%u9090%u8190%u00c3%u0003%u8b0
0%u531b%u53ff%u0078%u0000%u00=a HTTP/1.0
```

When the worm infected a vulnerable web server, Code Red changed the contents of the website to display a simple message:

HELLO! Welcome to http://www.worm.com! Hacked By Chinese!

The worm then directed the system to take several actions, depending upon the day of the month:

- On days 1–19 of every month, systems infected with Code Red spread themselves farther, scanning random IP addresses and trying to attack any system they could find.

- On days 20–27 of the month, Code Red launched denial of service attacks against several other servers, including the web server operating *www.whitehouse.gov*.

- On days 28–31 of the month, Code Red performed no actions.

TIP

The Code Red worm demonstrates the importance of applying patches to systems promptly. Microsoft released a security patch for IIS that resolved the vulnerability exploited by Code Red one month *before* the worm wreaked havoc on the Internet.

One research firm estimated the total damage caused by Code Red at $1.2 billion. Who launched the attack? That remains an open question. There is speculation that the worm began in the Philippines, but no perpetrators were ever identified or brought to trial.

> **Lights Out at NSA**
>
> In 2000, the U.S. NSA suffered an extremely serious computer network outage that resulted in a days-long disruption of intelligence-processing activities. It turned out that the disruption was not the work of hackers or cyberwarriors but was the result of an outdated network infrastructure.
>
> In an address to the Kennedy Political Union at American University, the then-director of the NSA, Lt. Gen. Michael Hayden, said, "You may have heard about the recent network outage at NSA. Due to a software anomaly, our aging communications infrastructure failed, and our ability to forward intelligence data, process that data, and communicate internally was interrupted for 72 hours. Thousands of man-hours and $1.5 million later, we were able to resume normal operations."

SQL Slammer

SQL Slammer was a malicious worm that spread rapidly across the Internet in January 2003, infecting systems running vulnerable versions of Microsoft's SQL Server 2000 database software. In addition to rendering the infected database servers unusable, SQL Slammer had a more insidious effect on the Internet as a whole. When a system became infected, SQL Slammer immediately used it to send as many infection requests to other systems as possible. Essentially, the server became dedicated to sending as much traffic across the Internet as possible.

The effect of this traffic was dramatic. The approximately 75,000 systems infected by SQL Slammer represented a relatively small percentage of the total systems connected to the Internet, but they were able to generate enough traffic to affect the network's infrastructure. Routers responsible for transmitting network traffic across the Internet backbone collapsed under the heavy load generated by SQL Slammer. The result was a massive Internet slowdown that affected virtually all Internet users on January 25, 2003.

SQL Slammer was strikingly similar to Code Red in many ways. Like Code Red, it exploited a buffer overflow vulnerability that allowed the worm to execute its own commands on the targeted server. Also like Code Red, Microsoft had released a patch that corrected the buffer overflow before the worm outbreak took place. The worm was released on January 25, 2003, but the patch had been available since July 24, 2002. Unfortunately, administrators of the 75,000 infected systems simply hadn't gotten around to applying the patch at the time of the outbreak.

Titan Rain

In 2005, U.S. government officials disclosed to reporters from the *Washington Post* and other outlets that they were investigating a series of computer intrusions that had been ongoing for at least two years. This investigation, code-named **Titan Rain**, revealed that attacks launched from Chinese web servers had successfully infiltrated hundreds of networks used by the Department of Defense and other government agencies.

As with previous attacks, officials remained tightlipped about the intrusions. Even the FBI refused to confirm its involvement. One official, speaking anonymously to the *Washington Post*, said, "It's not just the Defense Department but a wide variety of networks that have been hit. This is an ongoing, organized attempt to siphon off information from our unclassified systems."

It remains unclear whether the attacks were sponsored by the Chinese government or merely used Chinese servers as a jumping-off point for the attacks. What is clear is that the U.S. government considered the Chinese People's Liberation Army (PLA) a considerable cyberwarfare threat. In a report to Congress released in the midst of the Titan Rain investigation, Department of Defense officials wrote:

> China's computer network operations (CNO) include computer network attack, computer network defense, and computer network exploitation. The PLA sees CNO as critical to seize the initiative and "electromagnetic dominance" early in a conflict, and as a force multiplier…. The PLA has likely established information warfare units to develop viruses to attack enemy computer systems and networks, and tactics to protect friendly computer systems and networks. The PLA has increased the role of CNO in its military exercises. Although initial training efforts focused on increasing the PLA's proficiency in defensive measures, recent exercises have incorporated offensive operations, primarily as first strikes against enemy networks.

The Titan Rain investigation continued the work begun during Solar Sunrise and Moonlight Maze, revealing significant potential for a cyberwarfare attack against the United States. Titan Rain confirmed that the vulnerability to the type of attack conducted during those events in the late 1990s continued to exist through 2005.

Stakkato

From 2003 until 2005, a mysterious hacker waged cyberattacks against servers belonging to the U.S. military, educational institutions, corporations, and government agencies. The attacker was able to gain credentials to access those systems and then exploit privilege escalation vulnerabilities that allowed him to gain administrative access and full control of the systems.

Targets of the attack were widespread and included the following:

- White Sands Missile Range
- NASA
- Cisco Systems
- Stanford University
- The California Institute of Technology
- The San Diego Supercomputer Center

Investigators traced the attacks back to Sweden's University of Uppsala and learned that the perpetrator was a 16-year-old boy named Philip Gabriel Pettersson. Pettersson used the hacker name Stakkato. A grand jury in San Francisco indicted Stakkato on five felony charges in 2009.

Poison Ivy

Poison Ivy is the name given by security analysts to a Remote Access Trojan (RAT) first released in 2005. The tool, often used by script kiddie attackers, is a point-and-click infection utility that allows the user to retain control of infected systems. Some of the features of Poison Ivy include the following:

- Keystroke logging capability that monitors user keystrokes
- Screen capture capability that monitors user activity on the screen
- File scanning and exfiltration
- Relaying of traffic destined for other systems

Notably, Poison Ivy is a tool with amazing longevity. It was first released in 2005 and has not been updated since 2008. However, it remains a widely used tool in the hacker community today. A high-profile attack against security firm RSA in 2011 made use of the tool. Many attacks associated with Poison Ivy are of Chinese origin, leading to speculation that the tool may have been developed by the Chinese military. However, there is little evidence to substantiate these claims. What is certain is that APT actors continue to make use of the tool.

> **NOTE**
>
> FireEye, an information security firm, released a comprehensive report on Poison Ivy in 2013. The company also released a set of tools to help organizations identify systems infected by Poison Ivy. They amusingly named these tools Calamine.

Senior Suter

The Chinese government is not alone in developing tools used by hackers attempting to gain access to systems. Numerous reports circulate around the Internet and news media about the U.S. government's own offensive cyberweapons development capability. According to a 2007 report in *Aviation Week & Space Technology*, one such program was named Senior Suter.

The *Aviation Week* report claimed that **Senior Suter** is a program managed by an Air Force unit known as Big Safari that worked with defense contractor BAE Systems to develop three different versions of the Suter tools for use in offensive cyberwarfare operations. According to the report, the three versions of Senior Suter allowed U.S. operators to monitor what enemy radars could see, to take control of enemy networks, and to invade links to time-critical targets, such as missile launchers or surface-to-air missile launchers.

The report claimed that the Suter tools are specifically designed to manipulate enemy air defense systems. It also stated that several classes of U.S. Air Force aircraft are involved in the program. It is unclear from reports which techniques the Suter tools use to manipulate those systems.

Stuxnet and the 21st Century

If the first decade of the 21st century was the adolescence of cyberwarfare activity, the second and third decades are shaping up to be one of continued maturity of these techniques. The cyberattacks that took place during the first half of this decade are increasingly sophisticated and have had significant effects on their targets. This century has seen cyberwarfare become one of the dominant domains of conflict between nations.

Stuxnet

In perhaps the most significant example of cyberwarfare to date, a computer worm named **Stuxnet** attacked and destroyed the uranium enrichment centrifuges located at an Iranian nuclear facility in the city of Natanz. The worm was carefully designed by a sophisticated adversary to specifically target the Siemens Step 7 programmable logic controllers used to control the expensive centrifuges at the Natanz facility.

Once Stuxnet gained access to the controllers, the software modified their activity to rapidly accelerate and decelerate the centrifuges, causing them to destroy themselves. Reports by the *New York Times* and other media outlets attribute the attack to cyberwarfare operations jointly sponsored by the United States and Israel. The worm is believed responsible for the destruction of approximately one-fifth of the centrifuges in Iran's nuclear program.

Neither government has publicly claimed responsibility for the program, but the *New York Times* reported that one senior U.S. official smiled when questioned about the origins of Stuxnet. According to the newspaper, Gary Samore, a White House advisor on weapons of mass destruction, stopped short of admitting U.S. involvement, but said, "I'm glad to hear they are having troubles with their centrifuge machines, and the U.S. and its allies are doing everything we can to make it more complicated."

The Stuxnet worm is believed to have entered the Natanz facility through a USB drive carried into the facility unintentionally by an employee. The worm then likely spread around the facility's basic network, infecting many systems and the USB drives used with them. Eventually, an individual unknowingly carried a USB drive used on one of those systems to a system located on the isolated nuclear centrifuge control network. The Stuxnet worm was then able to infect the system running Siemens' Step 7 software and gain control of the centrifuges.

Stuxnet is significant because it marks one of the first times cyberwarfare attacks were used to cause physical damage to a significant military target. The damage caused by Stuxnet was equal to, or perhaps of greater impact, than the damage that would have been caused by a conventional bombing attack on the facility. In the eyes of many military analysts, Stuxnet marked the opening of Pandora's box and the beginning of an era of legitimatization for the cyber domain of warfare.

Operation Aurora

In January 2010, Google made a startling public announcement about a major cyber-attack against its operations. Google described the attack in a blog post that revealed the company was one of more than 20 major organizations targeted by China-based attackers and that the attack was targeted against free speech. The announcement read, in part:

> Like many other well-known organizations, we face cyber attacks of varying degrees on a regular basis. In mid-December, we detected a highly sophisticated and targeted attack on our corporate infrastructure originating from China that resulted in the theft of intellectual property from Google. However, it soon became clear that what at first appeared to be solely a security incident—albeit a significant one—was something quite different.

The announcement went on to describe three ways that this security incident differed from previous cyberattacks suffered by Google. First, the attack targeted 20 other large companies in addition to Google. Adobe Systems, Rackspace, and Juniper Networks later confirmed that they were also targeted by the attackers. Second, the attackers were attempting to access the personal email accounts of Chinese human rights activists. Finally, the investigation into the attack revealed that the accounts belonging to dozens of human rights activists in several countries had been "routinely accessed" by third parties.

The media quickly dubbed these attacks **Operation Aurora**, and they became known worldwide. In a detailed report outlining the Aurora attacks, Alex Stamos of iSEC Partners described the process used by the attackers to gain access to a targeted company:

1. The attacker uses social engineering techniques to trick a victim at the targeted company into visiting a malicious website.
2. The website contains malicious code that attacks the user's browser and installs malware on the victim's computer.
3. The malware contacts a command-and-control server operated by the Chinese attackers.
4. The attacker uses the malware to gain administrative access to the targeted company's entire network through the victim's computer.
5. The attacker locates an active directory server on the target company's network and attempts to obtain a password file.
6. The attacker uses password-cracking techniques to extract passwords from the file.
7. Once the attacker obtains a valid password, he or she uses that password to obtain access to the target company's virtual private network (VPN).
8. The attacker executes his or her original objectives, stealing sensitive information from the targeted company through the VPN connection.

The Operation Aurora attack against Google and other firms was a highly sophisticated attempt to infiltrate American businesses and the personal accounts of Chinese activists. In the wake of the attacks, Google announced a major policy change in which it would no longer censor search results at the behest of the Chinese government.

Duqu

In 2011, Hungarian computer security researchers released a report detailing their discovery of a new computer worm that they dubbed **Duqu**. Follow-up research by Symantec Corporation concluded that Duqu was "the precursor to the next Stuxnet" and that it was "nearly identical to Stuxnet, but with a completely different purpose."

Security researchers analyzing the Duqu worm concluded that the worm was written either by the same individuals who wrote Stuxnet or by individuals who had access to the source code used by Stuxnet. The purpose of the worm is believed to be to gather intelligence. The worm appears to be designed to gather information about industrial control systems that would facilitate a future attack against those systems.

Like Stuxnet, Duqu made use of "bridge" systems located on insecure networks to gain access to targeted systems located on more secure networks. Figure 7-1 shows a model of the activity conducted by Duqu.

Flame

A series of reports in mid-2012 announced the discovery of a new piece of malicious software dubbed **Flame** that targeted Microsoft Windows systems. The software was believed to have been operating since at least 2010, possibly as early as 2007. Analysts studying the source code of Flame discovered similarities to some of the

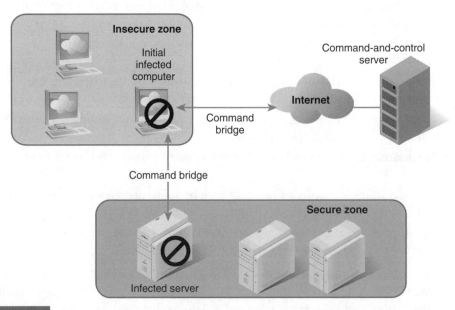

FIGURE 7-1

Duqu uses a bridge system located on an insecure network to relay commands from the attacker's command-and-control server to infected systems located on a secure network.

code used in the Stuxnet attack, leading to speculation that Flame was the product of another joint U.S.–Israeli effort.

Flame was highly targeted and is believed to have infected only about 1,000 systems in the Middle East. It appears to have targeted the Iranian oil ministry. An attack against that ministry in April 2012 caused the ministry to disconnect its computer systems from the Internet.

FOXACID

In 2013, security researcher Bruce Schneier published a blog post offering insight into a program allegedly run by the NSA designed to leverage secret Internet servers for hacking into the systems of carefully selected targets.

Schneier's summary of the **FOXACID** system paints a scary picture of the U.S. government's alleged capability to compromise systems:

> By the time the NSA tricks a target into visiting one of those servers, it already knows exactly who that target is, who wants him eavesdropped on, and the expected value of the data it hopes to receive.

Schneier's description of the FOXACID approach takes cyberwarfare to the next level, allowing the use of automated decision-making to carefully select weapons based upon the value of the target, the likelihood of detection, and the nature of the target's defenses. This risk assessment allows the strategic selection of weapons and reservation of zero-day attacks for times when they are necessary.

Careto

In 2014, researchers at Kaspersky Lab revealed the existence of a previously unreported APT they dubbed **Careto**, a Spanish word meaning "Mask." According to Kaspersky's report, Careto had been surreptitiously operating since 2007, targeting a wide variety of organizations:

* Government agencies and diplomatic missions
* Energy companies
* Activist organizations
* Research organizations
* Financial/private equity groups

The Careto threat compromised more than 1,000 unique IP addresses controlled by 380 victim organizations in 31 countries. Careto's source remains unknown, although evidence in the code suggests that the authors may speak Spanish.

When Careto infects a system, it gains total and complete control using very sophisticated malicious software to monitor the system, control infrastructure elements on the target network, cover its tracks, and steal massive amounts of information. Kaspersky believes that the level of sophistication points to a state-sponsored attack, stating that the use of sophisticated code "puts it above Duqu in terms of sophistication, making the Mask one of the most advanced APTs at the current time."

Russia's Ukraine Campaign

In 2017, ransomware given the name NotPetya began to spread around the world, disabling thousands of systems in power plants, manufacturing facilities, and logistics companies. Later analysis of the attack by malware researchers revealed that approximately 80 percent of the targets were in the Ukraine and that the attack was launched using a special tax software package used by Ukrainian companies. These facts suggested that the attack was specifically targeted to affect Ukrainian systems.

WIRED magazine journalist Andy Greenberg dug into this attack and wrote a book-length investigative report titled, *Sandworm: A New Era of Cyberwar and the Hunt for the Kremlin's Most Dangerous Hackers.* In that book, he provided evidence tracing the source of the attack to a group of hackers sponsored by the Russian government, alleging that the cyberattack was part of an ongoing military conflict between Russia and the Ukraine that began with the Russian invasion of Ukrainian-controlled Crimea in 2014.

This attack provides an excellent example of the use of cyberwarfare techniques as an adjunct to other forms of armed conflict.

USCYBERCOM Action Against Russia

In 2019, the *New York Times* reported that United States forces had engaged in cyber-warfare actions against Russia as a demonstration of power and that those actions had begun as early as 2012. These actions came in response to allegations of Russian interference in U.S. politics and presumably occurred under the authority of a still-classified U.S. cyberspace policy document. This document, National Security Presidential Memorandum 13 (NPSM-13), covers U.S. Cyber Operations Policy, and the details have not been shared with the public. Paul Ney, the general counsel for the U.S. Department of Defense, described the contents briefly in public comments he made in March 2020, saying that NSPM-13 "allows for the delegation of well-defined authorities to the Secretary of Defense to conduct time-sensitive military operations in cyberspace."

A series of reports by the *New York Times* and the Washington Post described two different offensive actions taken against Russian cyber targets:

- Targeting of the Russian Internet Research Agency (IRA), an organization that allegedly led election interference activities against the United States. To protect the integrity of the 2018 election, sources said that the U.S. military took direct action to disable the IRA's Internet connection on the day of those mid-term elections.
- The insertion of malicious code deep within Russia's power grid, designed to provide U.S. forces with the ability to cripple portions of the Russian electrical infrastructure on demand.

When asked about these actions at a conference sponsored by the *Wall Street Journal,* then-National Security Advisor John Bolton said, "We thought the response in cyber-space against electoral meddling was the highest priority last year, and so that's what we focused on. But we're now opening the aperture, broadening the areas we're prepared to act in."

CHAPTER SUMMARY

Cyberwarfare has evolved significantly over the past three decades. Threats that were seen as somewhat remote in the 1990s have become major instruments of modern warfare today. During the 1990s, the world saw the evolution of a new threat environment. A series of attacks code-named Solar Sunrise and Moonlight Maze and waged by relatively unsophisticated adversaries awakened government leaders and the public to a new threat landscape. This resulted in the development of significant offensive and defensive cyberwarfare capabilities.

Over the ensuing years, cyberwarfare capabilities and attacks grew significantly. Organizations of all kinds reported falling victims to cyberattacks that appeared to originate from state-sponsored sources. The first decade of the 21st century saw the advent of attacks by computer worms such as Code Red and SQL Slammer. The Stuxnet attack in 2010 marked a major turning point in the world of cyberwarfare when a news reports alleged that a joint U.S.–Israeli cyberwarfare operation destroyed 20 percent of the nuclear centrifuges in use by Iran's nuclear program. Since 2010, several more sophisticated threats have appeared on the cyberwarfare radar. It would be hard to conceive that even more advanced weapons aren't sitting unused in cyberarsenals, awaiting the appropriate time for their appearance on the world stage.

KEY CONCEPTS AND TERMS

Careto	FOXACID	Solar Sunrise
Code Red	Moonlight Maze	SQL Slammer
Duqu	Operation Aurora	Stuxnet
Eligible Receiver	Poison Ivy	Titan Rain
Flame	Senior Suter	Worm

CHAPTER 7 ASSESSMENT

1. What code name was given to the investigation of a series of computer break-ins that took place in 1998 as the United States was preparing for the second war in Iraq?

   A. Solar Sunrise
   B. Eligible Receiver
   C. Code Red
   D. Stuxnet
   E. Duqu

2. When waging a series of attacks against U.S. government computer systems, the hackers Mak, Stimpy, and The Analyzer used powerful systems located at _____ as jumping-off points for their attacks.

   A. military bases
   B. research labs
   C. universities
   D. corporations
   E. private homes

3. The Moonlight Maze series of attacks was traced back to network servers located in _____, the nation believed to be the sponsor of the attacks.

   A. China
   B. Israel
   C. United States
   D. Iraq
   E. Russia

4. The Honker Union is an organized group of hackers waging cyberwarfare attacks against opponents of their home country, _____.

   A. China
   B. Israel
   C. India
   D. Iraq
   E. Russia

5. What software was affected by the 2001 Code Red malicious code outbreak that impacted over 350,000 systems worldwide?

   A. SQL Server
   B. Internet Information Server
   C. Oracle Database Server
   D. Apache Web Server
   E. System Management Server

6. What type of vulnerability did the SQL Slammer worm exploit to spread around the Internet in 2003?

   A. SQL injection
   B. XML injection
   C. Cross-site scripting
   D. Brute force
   E. Buffer overflow

7. The Titan Rain investigation uncovered a series of computer network intrusions that originated from web servers located in _____.

   A. China
   B. Israel
   C. United States
   D. Iraq
   E. Russia

8. Poison Ivy is an example of what kind of hacking tool?

   A. RAT
   B. Vulnerability scanner
   C. Port scanner
   D. Obfuscator
   E. Rainbow table

9. The Stuxnet worm is widely believed to be an attack developed by the United States and Israel in an attempt to disrupt the nuclear program of what country?

   A. Iran
   B. Iraq
   C. Egypt
   D. Russia
   E. Libya

10. What type of organization was targeted by the Operation Aurora attacks allegedly waged by the Chinese government?

   A. Military
   B. Educational
   C. Corporate
   D. Government
   E. Nonprofit

11. Which of the following malicious code objects are believed to be created by the same source? (Select three.)

    A. Stuxnet
    B. GhostNet
    C. Duqu
    D. Flame
    E. Careto

12. FOXACID's major distinguishing feature was the alleged use of _____.

    A. risk assessment
    B. buffer overflow
    C. worms
    D. compromised antivirus
    E. bridge systems

13. The Flame virus was highly targeted, only infecting a small number of systems in the Middle East.

    A. True
    B. False

14. It is believed that the creators of the Mask originated in a country with what native language?

    A. English
    B. Hebrew
    C. French
    D. Russian
    E. Spanish

Nonstate Actors in Cyberwar

CYBERWAR IS PERHAPS THE MOST COMPLEX DOMAIN of modern warfare. Waging war on land, sea, or air requires a significant commitment of financial, human, and political resources. Building and maintaining a potent military force are generally outside the reach of anyone other than a nation-state or an extremely well-organized and funded group. A very small group, on the other hand, can wage cyberwar on a shoestring budget. Although it is certainly true that a well-funded adversary can wage very complex and organized cyberwar, even an individual can engage in cyberwar actions with the computing equipment already at his or her disposal.

A wide variety of nonstate actors play some role in cyberwarfare. These actors include nongovernmental organizations, organized crime, corporations, terrorists and activists, and even individuals and the media. Their roles may vary widely. Some may wage cyberwarfare activities on their own initiative, while others may simply play supporting or adversarial roles in cyberwarfare. Each adds to the complexity of the cyber environment, and each plays a role that cyberwarriors must keep in mind as they plan cyber domain activities.

Nonstate actors may become involved in cyberwarfare for a variety of reasons, typically related to the reason the nonstate group was organized in the first place. For example, an activist group seeking to gain territorial recognition for a group of displaced persons might use cyberwarfare activities to draw international attention to its cause. Similarly, a group organized in opposition to a nation's belligerent activity might target those cyberwarfare activities as a focus of its protests.

Individuals may also play roles in cyberwarfare, either as participants or agitators. Individual hackers can wage fairly sophisticated cyberattacks with a bare minimum of hardware and software. Although they might not have the resources of a nation-state or a larger organized nonstate actor, they can still engage in significant cyberwarfare activities. Those seeking to defend themselves against cyberattacks must keep such actors under close watch.

Over the past two decades, a number of individuals have risen to particular prominence by acting as whistleblowers, bringing the cyberwarfare activities of governments into the public light. These individuals also can have outsized effects on nations by disclosing confidential activities. These disclosures have two major effects on their victims. First, they may compromise some of the specific cyberwarfare weapons a government uses, allowing other adversaries to tailor their defenses to the threat. Second, they may cause governments great embarrassment as their activities are brought into public view for the first time, prompting a debate on their legality and appropriateness in which the government would rather not engage.

Chapter 8 Topics

This chapter covers the following topics and concepts:

- What types of nonstate actors may participate in cyberwarfare
- What roles a nonstate actor may play in cyberwarfare activities
- What motivates nongovernmental organizations to participate in cyberwarfare
- How organized crime participates in cyberwarfare
- What role corporations play in cyberwarfare
- How terrorists and activists use cyberwarfare activities
- How individuals and the media contribute to the cyber environment

Chapter 8 Goals

When you complete this chapter, you will be able to:

- Explain how nonstate actors participate in cyberwarfare
- Describe the different types of nonstate actors
- Explain the motivations that nonstate actors have for cyberwarfare participation
- Describe the roles that individuals may play in cyberwarfare
- Understand the influence of the media on those engaging in cyberwarfare

Understanding Nonstate Actors

Many different types of nonstate actors play various roles in modern society. Each is organized, either tightly or loosely, to fulfill a specific purpose. Some of these causes are quite noble, such as eradicating human suffering, while others are more questionable and may include the use of terrorist tactics. In this section, you will learn the basic types of nonstate actors that may be involved in cyberwarfare activities. This first section focuses on the characteristics and nature of the actors; the remainder of the chapter discusses the ways that they may engage in cyberwarfare.

Before beginning a discussion of nonstate actors, you must have a clear understanding of the term. In this text, nonstate actors are defined as any entity other than a nation-state that participates in cyberwarfare in any way. This definition is purposefully broad and is intended to encompass the wide range of participants found in the cyber domain, including both large terrorist networks and individual hackers.

Nongovernmental Organizations

Nongovernmental organizations (NGOs) are perhaps the most varied group of nonstate actors in existence today. They consist of essentially any organization that exists outside of the context of a government or a for-profit corporation. They may be organized for a wide variety of purposes, including humanitarian activity, religious purposes, economic development, diplomatic relations, or even armed intervention. They engage in many different types of activities, depending upon their particular mission.

NGOs receive their funding from many different sources. Although they are not part of a government, they often receive funding from one or more governments. NGOs also often solicit private donations to support their activities, calling upon the philanthropic nature of society and the emotional appeal of their missions to raise funds.

There are millions of NGOs in the world, with estimates placing the number in the United States somewhere around 1.5 million. Few of these play any role in cyberwarfare whatsoever, but a small minority does participate.

Organized Crime

Organized crime, perhaps familiar from mobster movies glamorizing gangster activity and famous figures like Al Capone and the Gambino family, continues to exist today. Although the days of Tommy gun-toting and "made men" may be behind us, organized crime still flourishes around the world. The U.S. Department of Justice defines **international organized crime** as "self-perpetuating associations of individuals who operate internationally for the purpose of obtaining power, influence, monetary and/or commercial gains, wholly or in part by illegal means, while protecting their activities through a pattern of corruption and/or violence."

8

Nonstate Actors
in Cyberwar

The organizational structure, business ventures, and crimes committed by international organized crime groups vary, but in a 2008 overview of its strategy to combat organized crime, the Justice Department describes the groups as sharing some or all of these characteristics:

- "In at least part of their activities, they commit violence or other acts which are likely to intimidate, or make actual or implicit threats to do so;
- They exploit differences between countries to further their objectives, enriching their organization, expanding its power and/or avoiding detection and apprehension;
- They attempt to gain influence in government, politics and commerce through corrupt as well as legitimate means;
- They have economic gain as their primary goal, not only from patently illegal activities but also from investment in legitimate business; and
- They attempt to insulate both their leadership and membership from detection, sanction, and/or prosecution through their organizational structure."

The activities of these organized crime groups are increasingly sophisticated. They include activities far outside the scope of the traditional image of organized crime. According to the Federal Bureau of Investigation (FBI), some of the recent actions of international organized crime groups include the following:

- Wielding pervasive influence in the energy field
- Cooperating with terrorists and foreign intelligence services
- Smuggling illegal contraband and people into the United States
- Conducting money laundering in both the U.S. and international financial systems
- Using cyberwarfare activities to target both individual victims and infrastructure
- Manipulating security markets and engaging in sophisticated fraud
- Corrupting both foreign and domestic public officials
- Using violence and/or threats of violence to maintain and increase their power base

The activities of organized crime remain sophisticated enough that federal, state, and local governments dedicate significant resources to tracking and prosecuting such groups.

Corporations

Corporations are for-profit businesses that are organized by individuals and officially registered by a government. Corporations are granted specific legal rights and, in fact, are recognized as "persons" under U.S. law. They are owned by shareholders, and their activities are controlled by boards of directors who are elected by, and act on behalf of, the shareholders. Privately held corporations may have a very small number of owners, whereas shares in public corporations are sold on the open stock market and are available to any buyer.

Terrorists and Activists

Many groups seek to actively effect change in the political arena through activities designed to draw attention to their causes and influence public opinion. They fall into two main categories: **activists** and **terrorists**. Activists seek to use peaceful methods, such as demonstrations, boycotts, political campaigns, and nonviolent protests to achieve this goal. Terrorists, on the other hand, use unlawful, violent means to create fear among those they seek to influence.

 NOTE
Terrorists and activists may also be classified as NGOs. However, due to their unique objectives and similarities, this text treats them as a separate category.

Individuals and the Media

In the realm of cyberwarfare, individuals bear unprecedented power. In all other domains of warfare—land, sea, air, and space—an individual actor can do little to shift the balance of power. In the cyber domain, however, an individual can wield significant power and have a dramatic effect on the actions and capabilities of nation-states and nonstate actors. You need to look no further than the many stories of both hackers and leakers of classified material discussed throughout this text to understand the changing balance of power.

The media plays important roles in magnifying the effect of individuals in cyberwarfare. First, in the case of leakers, journalists may serve as the conduit used to take sensitive information out of government control and place it in the public eye. This was certainly the case with Edward Snowden, who used *The Guardian* and the *Washington Post* to publicly disclose thousands of classified documents from the U.S. National Security Agency (NSA).

The idea of the U.S. press corps serving as a conduit for disclosing information embarrassing to the U.S. government is not new. A famous historical example of this was in 1971, when the *New York Times*, the *Washington Post*, and others disclosed the *Pentagon Papers*. These documents, leaked by government contractor Daniel Ellsburg, disclosed for the first time that the United States had conducted military operations in neighboring countries during the Vietnam War. Figure 8-1 shows a map from the *Pentagon Papers* documenting CIA activities in Laos, Cambodia, and Vietnam.

The publication of these documents led to a firestorm of public debate that was mirrored in the wake of the Snowden disclosures four decades later. The battle led all the way to the U.S. Supreme Court, which ruled in June of 1971, in the case of *New York Times Co. v. United States*, that the papers could continue to be published freely. In his opinion on the matter, Justice Hugo Black stated:

> Only a free and unrestrained press can effectively expose deception in government. And paramount among the responsibilities of a free press is the duty to prevent any part of the government from deceiving the people and sending them off to distant lands to die of foreign fevers and foreign shot and shell.

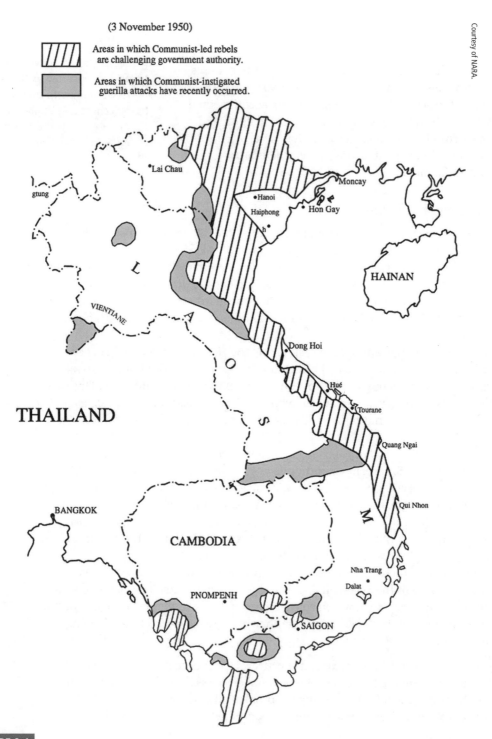

FIGURE 8-1

Map of CIA activities in Southeast Asia first disclosed as part of the *Pentagon Papers* by the *New York Times* in 1971.

This Supreme Court ruling is often cited in cases where the media leaks classified government materials as part of its reporting efforts.

The media also plays a role in the rise of hackers as individual participants in cyberwar. The exploits of hackers are of significant public interest, and the media reports cases of hacking extensively. John Markoff of the *New York Times* extensively profiled the famous hacker Kevin Mitnick. Some people credit the publicity generated by the *New York Times* and investigation by reporters as contributing to Mitnick's capture. It is certainly true that the media drew Mitnick into the public eye, perhaps encouraging other budding hackers to follow in his footsteps.

The Roles of Nonstate Actors in Cyberwar

How do nonstate actors participate in cyberwarfare? In some cases, these groups may be direct participants, either by serving as the targets of cyberwarfare or by engaging in cyberwarfare activities themselves. Nonstate actors also serve as critics of cyberwarfare activity, exposing acts of cyberwarfare to public scrutiny or actively protesting the emergence of cyberwarfare.

Amnesty International Targeted by Cyberattackers

In 2011, American journalist Brian Krebs reported that the NGO Amnesty International was the target of repeated cyberattacks. The human rights group's UK website was compromised and used to distribute malicious software to site visitors. The software exploited a known vulnerability in the Java programming language to install additional malicious software on site visitors' computers.

Krebs pointed out that Amnesty International had been the target of previous attacks as well—with its Hong Kong–based website compromised twice in 2010 and 2011. He speculates that the attackers were probably not trying to gain personal financial information from site visitors. "It appears more likely that the exploit may be part of an ongoing campaign by Chinese hacking groups to extract information from dissident and human rights organizations," Krebs said.

This illustrates the important role that NGOs play in the cyber domain. Amnesty International's role in society makes it a natural gathering point for Chinese dissidents and, if this truly was a Chinese-sponsored attack targeted at those dissidents, marks a major exploitation of that group's role for intelligence purposes.

Targets

The rise of nonstate actors in many different types of warfare leads to them becoming targets for cyberwarfare campaigns. The attacks against nonstate actors may be perpetrated by nation-states or by other nonstate actors. Nonstate actors may be targeted in many ways:

NOTE

The targeting of nonstate actors as the victims of cyberwarfare attack requires that they build and maintain the same types of cyber defenses businesses and nation-states use. The criticality of cyber-infrastructure to the missions of nonstate actors justifies an investment in adequate defensive technologies.

- Website defacements (such as the Amnesty International attack profiled in the nearby brief, "Amnesty International Targeted by Cyberattackers")

- Surveillance operations against nonstate actors (such as the NSA targeting of Red Cross communications discussed later in this chapter)

- Computer network attacks against computing devices used by nonstate actors

The continued influence of nonstate actors on international relations is likely to lead to a continued focus on them as targets for cyberwarfare activities.

Participants

In addition to serving as targets for cyberwarfare activities, nonstate actors may also be the aggressors, initiating acts of cyberwarfare against nation-states or other nonstate actors. They conduct these actions to advance their primary goals:

- Terrorists may engage in cyberwarfare to incite fear among the population of a targeted country. For example, a terrorist group may threaten to attack, or actually attack, the power grid in a city or release dangerous levels of water from a hydroelectric dam.

- Organized crime groups engage in cyberwarfare to generate profits for their organizations or incite fear among those they seek to influence in their quest for profits. For example, an organized crime group may use ransomware to extort payments from unwitting victims.

- Activists may engage in cyberwarfare to deface the websites of their opponents with political messages sympathetic to their causes. This has two desirable outcomes: spreading their message to the intended audience and embarrassing the legitimate operator of the website for having substandard cyber defenses.

Nonstate actors bring an added complexity to the cyberwarfare world. It is often difficult to determine the sources of their funding and whether they are actually state-sponsored. For example, a recent research report by the security consulting firm Mandiant disclosed the existence of an advanced threat group known as APT1. Although there is no concrete evidence, widespread speculation and circumstantial evidence suggests that APT1 is a state-sponsored group funded by the Chinese government. This makes it difficult for an attacked organization to effectively tailor its response. Is it legitimate to target the Chinese government in response to an attack waged by APT1?

Critics

Nonstate actors may also emerge as vocal critics of cyberwarfare activities. The Snowden case illustrated this, prompting a large number of activist organizations to publicly speak out against the U.S. government's surveillance activities.

For example, the American Civil Liberties Union (ACLU), a powerful civil rights lobbying group, strongly opposed the NSA's activities and filed numerous lawsuits objecting to them in court. This activism increased the public scrutiny of NSA activities and presented a legal challenge that the government must respond to in court. In a blog post published several months after the Snowden leaks, the ACLU summarized its position:

> When the government operates in secret, it is hard to know anything with confidence. There is, however, one thing you can say with 100 [percent] confidence: we need to know more.

The organization wants to know what information the government is collecting about Americans, what judicial rulings have authorized this collection, and what the government is doing with the information.

The role of the critic in cyberwarfare is to bring to light those activities that may not survive public scrutiny. This is especially effective in democratic governments, where the voice of the people has substantial influence. Critics draw public attention to issues, forcing politicians to take active positions supporting or opposing controversial activities. They also, as in the case of the ACLU, may file lawsuits directly challenging the constitutionality of government activities.

NGOs in Cyberwar

In the preceding sections of this chapter, you learned about the types of NGOs and how they engage in cyberwarfare. In this section, you will learn about specific examples of the ways that NGOs have found themselves the targets of cyberwarfare activities. From aid groups to diplomatic and religious organizations, many different types of NGOs have found themselves at the receiving end of cyberwarfare attacks.

Aid Groups

Aid groups seem an unlikely target for cyberwarfare activities. After all, their humanitarian missions seem like the type of work that would be inoffensive to all. What type of organization would want to target them with cyberwarfare attacks? Furthermore, would any government or nonstate actor run the risk of public embarrassment if they were seen as the aggressors against organizations with humanitarian missions?

Is it realistic to believe that intelligence agencies would actually target an aid group in cyberwarfare activities? In 2000, the *Cryptome* website published an article alleging

---

Echelon

The *Cryptome* article alleging that the U.S. government targeted the Red Cross claimed that the Echelon system was used to conduct this surveillance. This system was the subject of disclosures about NSA activities in the media around the turn of the century. The system is believed to consist of collection stations located in Australia, the United States, the United Kingdom, Japan, New Zealand, and Canada.

In its 2001 summary of allegations regarding Echelon, *The Guardian* called it "[a] global network of electronic spy stations that can eavesdrop on telephones, faxes and computers. It can even track bank accounts." The newspaper said the program stores information on millions of individuals on its computers. It noted that the U.S. government has denied the program's existence, while the British government has given evasive answers to questions.

---

exactly that. The authors published what they claimed were excerpts from an unclassified PowerPoint presentation created by an officer in the U.S. Air Force's 544th Intelligence Group. On one of those slides, titled "Our Changing World," the officer stated that there were "[a] lot of new fish, in a lot of unfamiliar ponds. They are mobile, diverse, and technology has made them advanced." The slides called out NGOs, and specifically the International Committee of the Red Cross, as being in that category and asked the question, "Friend or foe?"

Diplomatic Organizations

Diplomatic organizations, such as the United Nations, play an important role in international relations. They serve as forums for nations to come together in a neutral environment to discuss pressing issues. They also allow international cooperation on significant issues of human concern, such as the plight of refugees, the welfare of children, and concern for the environment.

Could the United Nations be the target of cyberwarfare activities? It appears quite certain that it already finds itself at the center of actions in the cyber domain. In 2004, the British Broadcasting Company reported that spies in the United Kingdom had eavesdropped on the conversations of then-U.N. Secretary General Kofi Annan. Clare Short, a U.K. cabinet minister, told the BBC that she had seen direct evidence of this, saying, "Well I know—I've seen transcripts of Kofi Annan's conversations."

Perhaps a larger role for the United Nations might be found in the realm of developing guiding principles for cyberwarfare. The International Telecommunication Union (ITU), a branch of the United Nations, has called on governments to do the following:

- Give their people access to communications.
- Protect their people in cyberspace.
- Promise not to harbor terrorists or criminals.
- Promise not to be the first to launch a cyberattack.
- Commit to international cooperation to guarantee peace in cyberspace.

It is likely that diplomatic organizations will continue to play an increased role in cyberwarfare. In addition to being potential targets for cyber domain activities, these organizations will likely mediate disputes and otherwise become entangled in cyberwarfare issues.

Religious Organizations

Religious organizations are some of the wealthiest, most politically active, and widely connected institutions in the world today. The Roman Catholic Church, for example, claims more than 1.2 billion members, represented in every country around the world. It is among the oldest and wealthiest institutions in the world, with a net worth that likely reaches into the billions of dollars. An organization of this size and significance makes for an interesting cyberwarfare target.

In 2012 and 2013, the cardinals of the Catholic Church gathered in Rome for a conclave that eventually elected Pope Francis. The Italian magazine *Panorama* alleged in a report that the U.S. government monitored the Church leaders' conversations in the period before Francis's election, classifying intercepted communications into four categories:

- Leadership intentions
- Threats to the financial system
- Foreign policy objectives
- Human rights

Leaders on both sides of the table were quick to dismiss these allegations. The Vatican issued a statement: "We don't know anything about this, and in any case we don't have any concerns about it." A spokesman for the U.S. NSA also denied the allegations, saying, "The National Security Agency does not target the Vatican. Assertions that NSA has targeted the Vatican, published in Italy's *Panorama* magazine, are not true."

Organized Crime

Organized crime, seeking to develop its tactics to meet the challenges of the information age, also plays an active role in cyber activities. These activities include individual extortion attempts, identity theft rings, and using organized hacking groups to infiltrate computer systems around the world.

One tactic often attributed to organized crime is the use of **ransomware**, malicious computer software that takes over a system, encrypting files with a secret key and rendering

FIGURE 8-2

Screenshot of a ransomware payment demand.

them inaccessible to the legitimate user until he or she pays a ransom. Figure 8-2 shows an example of a message displayed by ransomware after infecting a system.

Along with the intimidation tactics favored by organized crime, the ransomware often makes embarrassing and completely unfounded allegations that the victim was targeted because of involvement in child pornography or other illicit activity. The victim, seeking to avoid publicity and regain access to his or her files, often pays the ransom demand, which usually ranges in the hundreds of dollars. This amount is enough to be significant to the organized crime group, but small enough that the victim is likely to pay it without involving the authorities.

The risks associated with organized crime are not limited to attacks against individuals and are not constrained by international borders. The Department of Justice recently had this to say about organized crime cyberspace activities in Romania:

One example of the intersection between organized crime and cybercrime is found in Romania. There, traditional Romanian organized crime figures, previously arrested for crimes such as extortion, drug trafficking and human smuggling, are collaborating with other criminals to bring segments of the young hacker community under their control. They organize these new recruits into cells based on their cyber-crime specialty, and they routinely target U.S. businesses and citizens in a variety of fraud schemes.

The adaptation of organized crime to changes in economic activity is well documented throughout history. When the government outlawed alcohol during Prohibition, organized crime stepped in and filled the demand for an illicit product. When energy prices skyrocketed, organized crime found ways to generate a corrupt profit. Now that business is moving to cyberspace, organized crime is adopting the principles of cyberwarfare to continue to generate returns.

Corporations

Corporations are completely entangled in the affairs of nations. They are transnational in nature and often pressure governments around the world to make policy decisions favorable to their business interests. In addition, they conduct activities that they want to remain secret from each other, creating an environment where cyberwarfare is arguably inevitable. Corporations become involved in cyberwarfare in two major ways: through industrial espionage and through cooperation with various governments' intelligence agencies.

Industrial Espionage

Corporations engage in espionage activities against each other as they seek to advance their own competitive business interests. These activities are known as *industrial espionage*, and they may take place based upon the independent actions of corporations or with government support. The objectives of these activities include stealing product plans, learning about competitive bidding processes, and gaining confidential information.

In a report to Congress on industrial espionage, the Office of the National Counterintelligence Executive warned, "The willingness of U.S. scientists and scholars to engage in academic exchange makes U.S. travelers particularly vulnerable not only to standard electronic monitoring devices—installed in hotel rooms or conference centers— but also to simple approaches by foreigners trained to ask the right questions."

The FBI cites several examples of known industrial espionage attacks against U.S. business travelers:

- Searching of hotel rooms and belongings while travelers were away from their rooms
- Hotels installing monitoring software to eavesdrop on the Internet activities of guests
- Inspection and theft of laptop computers at airports and security checkpoints
- Hacking cell phones to steal contacts, usernames, passwords, and usage history

In addition to these attacks against travelers, foreign intelligence operatives are known to have engaged in cyberwarfare operations against networked computers operated by corporations. The Operation Aurora attacks disclosed by Google targeted U.S. business interests with cyberattacks originating in China. These attacks were focused on gaining access to confidential information stored on corporate information systems.

Game of Pawns

In a 2014 video, the FBI warned students of the risk posed by foreign intelligence operatives. They specifically described targeting efforts against students studying abroad.

The video tells the story of an American student who spent a year studying in Shanghai and was recruited by the Chinese government agents in a very gradual manner. They began by offering him payments of several thousand dollars to write innocent-sounding white papers on international relations.

The matter escalated out of hand when they paid him $40,000 and asked him to apply for a job with the Central Intelligence Agency (CIA). The story, based upon real-life events, ends with the arrest of the student by federal agents.

Cooperation with Intelligence Agencies

Corporations may also participate in cyberwarfare activities through their cooperation with intelligence agencies as those entities conduct cyberwarfare. One example of this is a program known as PRISM, which *The Guardian* alleges collected information from Internet companies. In PRISM, the government allegedly gained access to information directly from the servers of many major Internet companies:

- Google
- Facebook
- Yahoo!
- Microsoft
- Apple
- YouTube
- Skype
- America Online (AOL)

The Guardian's report states that the NSA is able to use this system to obtain many sources of information about individual users of the companies' products. This data includes email messages, video/voice chats, videos, photos, stored data, Internet usage activity, social network usage, and more.

Although the companies named in the NSA document denied active participation in PRISM, the reporting around this activity prompted a serious national debate on privacy. In response to public pressure, companies began issuing *transparency reports* that disclose the number of government requests for personal information they received.

Terrorists and Activists

Terrorist and activist groups often turn to cyberwarfare tactics because of their disruptive capability. Groups with relatively small memberships and budgets can wage war on a scale that draws international attention to their activities.

Estonia

In April 2007, the Estonian government moved a war memorial erected by the former Soviet Union. Outraged by a move they deemed offensive, many Russians protested the relocation of the memorial and rioting occurred in Estonian streets.

The more interesting outcome of these riots from a cyberwarfare perspective is that they were accompanied by a series of cyberattacks that appeared to originate in Russia. These attacks were distributed denial of service (DDoS) attacks that flooded servers with traffic, attempting to disrupt legitimate use of the servers. The targets included the following:

- The president of Estonia
- Parliament
- Government offices
- Political parties
- News organizations
- Financial firms
- Telecommunications firms

In response to the attacks, the Estonian government was forced to dramatically limit communication with the outside world, hoping that cutting off Internet access from foreign addresses would allow the systems to recover and begin serving legitimate domestic requests.

Syrian Electronic Army

The Syrian Electronic Army (SEA) is an activist organization composed of hackers who support the Syrian government. It is unknown whether it is actually state-sponsored or independent in nature. SEA is well known for a series of attacks against popular media outlets where it defaced websites, replacing content with anti-American and/or pro-Syrian items. For example, in April 2013, the SEA hijacked the Twitter account of the news program *60 Minutes* and issued the following tweet:

Exclusive: Terror is striking the #USA and #Obama is Shamelessly in Bed with Al-Qaeda

In an even more serious Twitter hijacking attack that took place that same month, the SEA used the Associated Press account to tweet:

Breaking: Two Explosions in the White House and Barack Obama is injured

This tweet caused an immediate reaction in the financial markets, with the S&P 500 index falling 1 percent in the course of 3 minutes before quickly recovering. That dip was equivalent to the loss of $136 billion in equity.

Anonymous

The loosely organized group known as Anonymous is a collection of activist hackers who orchestrate DDoS attacks against targets they select based upon ideological concerns. Recent targets of Anonymous have included the following:

- The Church of Scientology
- PayPal

Operation Payback

One of the best-known actions undertaken by Anonymous was Operation Payback. In this attack, Anonymous hackers took on organizations associated with antipiracy efforts on the Internet. The attack included a public relations campaign, which used materials such as the graphic shown in Figure 8-3.

During the campaign, Anonymous launched denial of service attacks against the websites of the Recording Industry Association of America and law firms associated with copyright complaints.

Anonymous later redirected these efforts in support of WikiLeaks founder Julian Assange. When financial organizations stopped processing donation transactions for the WikiLeaks site, Anonymous targeted Visa, MasterCard, PayPal, and other firms they felt were complicit in denying WikiLeaks funds.

Courtesy of The Anonymous hacker group.

FIGURE 8-3

Poster used by Anonymous to protest copyright protection efforts during the Operation Payback campaign.

- Visa
- Sony
- Government agencies
- Child pornography sites
- Copyright protection agencies

The group often releases statements, in text or video form, explaining its actions and attempting to justify its attacks.

Individuals and the Media

Individuals and the media play important and interrelated roles in cyberwarfare. By acting as hackers or leakers/whistleblowers, individuals can wield outsized power against larger forces. The media can trumpet these actions, bringing attention to individuals and their causes. Recent years have seen many cases where both hackers and whistleblowers have caused massive changes in public perception and government policies.

Individual Motivations

Many of the same motivations that encourage them to take part in other illegal and controversial activities cause individuals to take on large institutions and participate in cyberwarfare. Hackers and leakers generally have some common motivations:

- **Greed**—Individuals may be motivated purely by financial gain and attempt to steal information or compromise systems they will be able to profit from.
- **Technical challenge**—Many hackers seek to demonstrate their technical proficiency to other hackers and the world at large. They engage in cyberwarfare attacks to demonstrate the possibility of successfully attacking large organizations and to gain "street credibility" within the hacker community.
- **Ego**—Many hackers are motivated by simple ego. They want to show their power and demonstrate to the world that they can wield influence.
- **Ideology**—Activists and leakers are often motivated by ideological concerns, seeking to undermine governments or other institutions they oppose or to call attention to what they perceive as illegal or immoral activities.

The true motivations of any individual engaged in cyberwarfare are often mixed and hard to define. In reality, most individuals have a mixture of motives, some noble and some less so. It is important to understand that often simple causes drive individuals to undertake risky, sophisticated attacks against establishment targets.

Hackers

Hackers are the most common example of individual actors in cyberwarfare. The actions of hackers can often be categorized into three groups, based upon the motivation and authority of the individuals waging the attack:

- *White-hat hackers* conduct their attacks under the official sanction of the organization being attacked. These security professionals use hacking techniques to test the security defenses of organizations and point out opportunities for improvement.
- *Black-hat hackers* conduct attacks without permission and with malicious intent. They may be trying to achieve any of the hacker goals, including financial gain, notoriety, or activist objectives.
- *Gray-hat hackers* conduct attacks without permission but without malicious intent. Although many consider their actions illegal, they usually inform the attacked organization of the security risks they uncover.

Kevin Mitnick is one of the best-known examples of a black-hat hacker. Mitnick conducted a series of computer intrusions in the 1990s and was arrested, convicted, and sentenced to prison for his actions. He has been heavily profiled by the media and has garnered business as a computer security consultant (white-hat hacker) since his release from prison.

Time to Change Hats?

The use of these colored-hat terms to describe hackers has a long history in the world of cybersecurity. Recently, many people have begun to move away from these terms for reasons of cultural sensitivity.

As part of this shift, some people know refer to these attackers as:

- Authorized attackers (for white-hat hackers)
- Unauthorized attackers (for black-hat hackers)
- Semi-authorized attackers (for gray-hat hackers)

Because there is not yet consistency around the use of these terms, we continue to use the traditional terms in this book because you will hear them in practice.

Leakers and Whistleblowers

The actions of leakers and whistleblowers often cause great embarrassment to government agencies that would prefer to keep their secrets from public view. Throughout history, whistleblowers have come forward to call attention to questionable government actions and have been met with mixed reactions. In recent years, several individuals have come forward with classified documents belonging to the U.S. government and released the documents to media fanfare, only to later face criminal prosecution for their actions:

- Julian Assange is the founder of the WikiLeaks website. On that site he disclosed a huge quantity of information from government sources in several nations. He rose to prominence when he used the site to publish the Manning documents.

- Chelsea Manning is a former Army soldier who leaked hundreds of thousands of documents to the WikiLeaks website in 2010. Those documents included allegations about the Iraq War that deeply embarrassed the U.S. government. They also included more than 250,000 alleged diplomatic cables containing details of relations between the United States and other nations around the world. In 2017, President Obama commuted her sentence and she was released after serving seven years in prison.

- Edward Snowden is a former defense contractor who released a massive amount of information about alleged operations of the NSA in 2013. Snowden worked with reporters from *The Guardian* and the *Washington Post* to release these materials to the public, prompting an international debate about the propriety of government surveillance operations. Snowden sought asylum in Russia in 2013 and remains there as of 2020.

NOTE

The distinction between a leaker and whistleblower is a fine one and depends upon the perspective of the person applying the label. Although many civil rights activists have branded Edward Snowden as a hero whistleblower, progovernment activists have just as vociferously dubbed him a traitor to his country.

The role of the whistleblower has increased significantly since the time of Daniel Ellsberg and the *Pentagon Papers*. The main reason for this transition is the ease with which technology allows modern whistleblowers to steal large quantities of government documents and release them to the media. It is likely that Assange, Manning, and Snowden are not the last names to go down in history for releasing embarrassing information about clandestine government activities.

NOTE

Chelsea Manning previously went by the name Bradley Manning before changing her name and gender identity in 2013.

8

Nonstate Actors in Cyberwar

CHAPTER SUMMARY

The role of the nonstate actor in cyberwarfare is significant. The nature of cyberwar makes it possible for groups of all sizes, and even individuals, to participate in conflict on an unprecedented scale. The tools of the cyberwarrior are simply a computer and an Internet connection. Cyberattacks may be launched from any location against systems located anywhere on the globe. These attacks can target other nonstate actors or larger nation-states and draw international attention when successful or high profile.

Students of cyberwarfare must study the role of the nonstate actor carefully and understand how different types of actors play the roles of participants, targets, and critics of cyberwarfare. NGOs, organized crime, the media, individuals, corporations, and other groups play roles in cyberwarfare to advance their political, ideological, and business agendas.

KEY CONCEPTS AND TERMS

Activists	*New York Times Co. v. United States*	*Pentagon Papers*
Corporations		Ransomware
International organized crime	Nongovernmental organizations (NGOs)	Terrorists

CHAPTER 8 ASSESSMENT

1. The International Committee of the Red Cross is an example of what type of nonstate actor?

   A. NGO
   B. Religious organization
   C. Activist
   D. Terrorist
   E. Media

2. Although aid organizations and religious organizations may express views on issues related to cyberwarfare, they are not normally the targets of cyberwarfare activities.

   A. True
   B. False

3. A corporation is recognized as a person under U.S. law.

   A. True
   B. False

4. Which of the following types of nonstate actor seeks to effect political change through unlawful, violent means?

   A. NGOs
   B. Terrorists
   C. Activists
   D. Hackers
   E. Diplomatic organizations

5. Which one of the following characteristics is *not* commonly used to describe international organized crime?

   A. They attempt to gain influence in government, politics, and commerce via *both* corrupt and legitimate means.
   B. They exploit differences between countries to further their objectives.
   C. They always engage in armed violence against their adversaries.
   D. They have economic gain as their primary goal.
   E. They attempt to insulate their leadership and membership from detection or prosecution.

6. What U.S. Supreme Court decision came in the wake of the *Pentagon Papers* disclosure and affirmed the constitutional basis for publishing classified government information in the public interest?

   A. *Plessy v. Ferguson*
   B. *McCulloch v. Maryland*
   C. *Reynolds v. United States*
   D. *New York Times Co. v. United States*
   E. *United States Department of Justice v. Reporters Committee for Freedom of the Press*

7. What branch of the United Nations created five principles for creating an environment of cyber peace?

   A. UNICOM
   B. UNICEF
   C. UNESCO
   D. IETF
   E. ITU

8. What type of malicious computer software is specifically used by organized crime?

   A. Ransomware
   B. Trojan horse
   C. Virus
   D. Logic bomb
   E. Worm

9. What group is responsible for hijacking the Twitter accounts of the Associated Press, *60 Minutes*, and other media organizations?

   A. Anonymous
   B. SEA
   C. APT1
   D. WikiLeaks
   E. Mitnick

10. What group was responsible for conducting a series of denial of service attacks against organizations associated with copyright enforcement?

    A. Anonymous
    B. Syrian Electronic Army
    C. APT1
    D. WikiLeaks
    E. Mitnick

11. Who was responsible for creating the WikiLeaks website that published massive quantities of sensitive government information?

    A. Anonymous
    B. Chelsea Manning
    C. Edward Snowden
    D. Kevin Mitnick
    E. Julian Assange

12. Who was responsible for providing WikiLeaks with massive quantities of government documents related to the Iraq War and diplomatic cables?

    A. Anonymous
    B. Chelsea Manning
    C. Edward Snowden
    D. Kevin Mitnick
    E. Julian Assange

13. Who was responsible for leaking documents about the NSA to media outlets?

    A. Anonymous
    B. Bradley Manning
    C. Edward Snowden
    D. Kevin Mitnick
    E. Julian Assange

14. Which one of the following is *not* a common motivation for individual hackers?

    A. Greed
    B. Military objectives
    C. Ideology
    D. Ego
    E. Technical challenge

15. What kind of individual normally operates with permission from the target that is being attacked?

    A. Whistleblower
    B. White-hat hacker
    C. Leaker
    D. Black-hat hacker
    E. Gray-hat hacker

Defense-in-Depth Strategies

L IKE TRADITIONAL WARFARE, cyberwarfare is fought both offensively and defensively. Defending against cyberattacks is a complex task because of the broad range of attackers, the number of potential targets, and the huge variety of ways in which attacks are conducted. When you factor in that people also create vulnerabilities that can be leveraged to attack computers and networks, cyber defense can seem nearly impossible.

The adversaries that network and systems defenders face vary. Nation-states, corporations, insurrectionists, cybercriminals, and hacktivists each have their own goals in cyberwarfare. They may choose different targets and different methods, and of course they also bring different levels of capability. They might attack using targeted malware or massive brute-force network attacks. Or they might attack using subtler methods that leverage human factors in addition to technological means over weeks or months. If they succeed, they may quietly gather data; continue their attacks to gain greater access; or immediately use their access to damage systems, networks, or infrastructure.

In traditional information security operations, security professionals warn their employers that there is no way to be perfectly secure. If a system is usable and useful, it has the potential to be attacked—no matter how well defended the networks, systems, and other assets are. Worse, organizations have a finite amount of resources to spend on cyber defense, and cyber defense can often only defend effectively against threats that are known and understood. With technology's complexity and rate of progress, staying abreast of an organization's defensive needs and ensuring that legacy systems and software don't become an issue is an ongoing challenge. When you consider the potential to have far more attackers than defenders, and for those aggressors to have far greater resources than your own organization possesses, defense can feel like a losing battle.

Despite these challenges, you can use methods to effectively defend assets, to reduce the chances of compromise, to detect those attacks that do occur, and to provide a competent response. Computer network defense (CND) strategies attempt to first identify likely opponents, then to enumerate the threats and risks that an organization will face from those attackers. Once an organization has a good understanding of what it may face, it can design strategies to counter them using policies, procedures, technology, training, and a variety of other defensive options.

Since the creation of cyberwarfare as a concept, one of the key concepts for many CND strategies has been defense in depth. *Defense in depth* is the idea that defenses should have more than a single layer of protection between an attacker and the protected systems, data, or networks. Defense in depth in cyberwar is much like defense in depth in conventional military operations. It employs layers that use different methods to stop attackers so that a single attack or technology cannot succeed simply by penetrating a single system or layer of protection. In addition, it offers real advantages to those who are defending because they can use simpler, easier-to-understand, and sometimes less-expensive defenses in each layer. Network defense in depth often starts with a strong design that involves network security devices. These include firewalls, intrusion detection and prevention systems, antivirus, authentication, logging, response, and restoration capabilities. Network defense in depth can also include the policies, procedures, training, and knowledge of the staff who use and support computer networks and systems.

This chapter looks at how modern computer networks and systems implement defense in depth by using a variety of strategies and technologies. It explores U.S. Department of Defense, U.S. Department of Homeland Security, and the U.S. National Security Agency (NSA) strategies and concepts, as well as civilian know-how regarding the way in which people, technology, and operations influence defense strategies. You'll also learn where defense in depth can fail and why some experts have begun to claim that defense in depth is no longer the strong cyberwarfare defense strategy it once was.

Chapter 9 Topics

This chapter covers the following topics and concepts:

- What defense in depth is
- What the defense-in-depth strategies and concepts are
- Where and why defense in depth fails
- How zero trust fits into a defense-in-depth strategy
- What the design elements of a modern defense-in-depth strategy are

Defense in Depth

From ancient Roman fortifications to medieval castles, the concept of providing defense in depth by layering protective capabilities has been in use for thousands of years. The earliest motte and bailey fortifications used by the Norman invaders in England in the 11th century are recognizable as the predecessors of the mighty medieval castles you are probably familiar with. This design layered ditches, mounds of earth, and wooden palisades around a central multistory, defensible house (keep). The Normans ruled the recently conquered countryside from these very early castles. They relied on the multiple layers to keep them safe even if attackers successfully crossed the ditch and burned the palisade down.

Over the next 200 years, those early fortifications evolved as technology and strategies for attacking castles changed. Castles became increasingly more complex as attackers became more organized and the technologies used to attack them became more effective. Stone replaced wood to avoid fire, and layers of defenses became deeper and stronger to combat larger, more organized armies. By the 13th century, concentric castles like those shown in Figure 9-1 had layers of stone walls, strong towers, and heavily fortified gatehouses with drawbridges, strong doors, strong internal gates that could divide invading groups, and a myriad of ways to attack enemies trapped inside. These concentric castles are a common sight when describing defense in depth because they so clearly show the layered defenses available to a medieval lord, and thus are a useful metaphor for how to layer modern defenses.

The weapons and strategies used in warfare have never stood still for long, and changes were already beginning to occur even as these mighty stone castles were being built. By the middle of the 15th century in Europe, cannons and gunpowder had begun to change the balance of power in warfare. Traditional castles, keeps, and city walls with their tall stone construction were particularly vulnerable to this new form of warfare. For example, a cannon-equipped army could reduce the mighty fortifications to rubble from a distance.

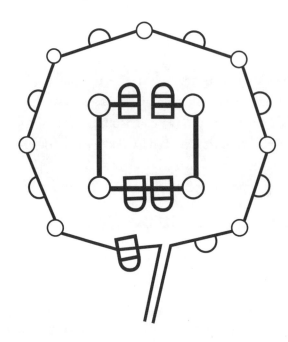

Concentric castles provided defense in depth using stone walls, moats, gates, and terrain features like hilltops and raised earthen mounds. Note the layered walls, strongpoints near entrances, and narrow pathway to the castle.

Designers realized this, and they developed an updated castle design that specifically addressed the new world of cannon and siege warfare. They recognized that traditional defenses were no longer relevant and that a new type of layered defense was necessary. Their fortification style, known as star forts (see Figure 9-2), changed how fortifications were designed and remained in use until the 19th century.

The constant change in both the weapons and technologies attackers use, and the ways in which defenders attempt to counter them, is the same challenge faced by information system defenders today. In fact, the experts assigned to defend modern computer networks and individual computers have often adopted similar strategies for the same reasons that fortress builders and defenders have throughout history: Enemies can often breach one layer of the defense. Layered defenses make it less likely that a single attack can completely compromise a network or system. They also allow for weaknesses and mistakes on the parts of both defenders and those who provide defenders' software, hardware, and devices.

Modern defense-in-depth strategies still use layers, but the layers are no longer stone walls and ditches. Figure 9-3 shows the U.S. Computer Emergency Readiness Team's defense in depth strategy elements for Industrial Control Systems as an example of the many layers of defenses that a modern environment may need. Technical, administrative, and human elements all play a role in building a secure environment, and successive layers implement a combination of technologies, as well as human knowledge, to prevent and detect attacks.

FIGURE 9-2

Star forts like the Italian fort in Nicosia, Cyprus, marked a major change in defensive strategies due to technological change. Note the multiple layers of low angled walls to defeat cannon fire, the dry moat and ditches to prevent foot-soldier assaults, and the angled projections that allowed defenders to fire sideways at attackers.

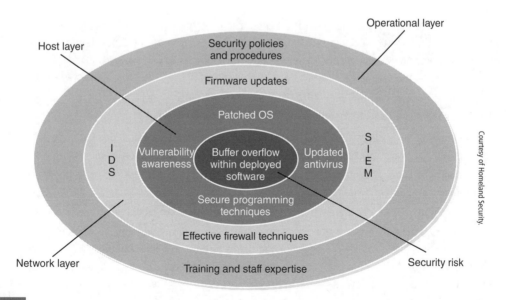

9

Defense-in-Depth Strategies

FIGURE 9-3

The US-CERT's defense-in-depth strategy elements include risk management, cybersecurity architecture, physical and host security, network architecture and perimeter security, monitoring, vendor management, and human elements.

FIGURE 9-4

The C-I-A triad shows the interaction between confidentiality, integrity, and availability when handling data and services.

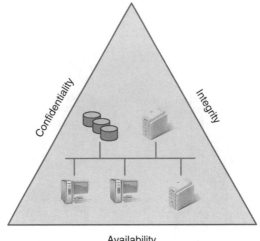

Availability

The C-I-A Triad

Security practitioners around the world use a common information security conceptual model known as the C-I-A triad as part of defense in depth designs. The **C-I-A triad** consists of confidentiality, integrity, and availability, as shown in Figure 9-4. It is often a key part of defense-in-depth designs, as well as throughout information security and cyberwarfare theory and practice. It is worth noting that some sources refer to this as the A-I-C triad to distinguish it from the U.S. Central Intelligence Agency (CIA), but the C-I-A triad is the most commonly used term and the term used in this book.

The C-I-A triad has three components:

- **Confidentiality** ensures that information is not accessible or disclosed to unauthorized systems or individuals.
- **Integrity** ensures that information has not been modified by unauthorized users or systems, and remains accurate and consistent.
- **Availability** ensures the system, data, network, or service is available and can be used or accessed.

Many security practitioners also commonly add *authentication* or **authenticity**, which is the ability to validate that the system or user is who he or she claims to be, and **nonrepudiation**, which means that the sender cannot claim to have not sent the data or messages received. Adding authentication is also a critical part of designs that validate the permissions or rights of individuals, systems, and services when they are used—a key part of more advanced modern techniques.

Defense-in-Depth Strategies

Many groups and organizations have published defense-in-depth strategies or strategies that rely on defense in depth as a core concept. All have emphasized elements that are specific to their organizational goals, the technologies that they rely on, and the attackers and attack techniques they expect to face. The following sections examine publicly available defense-in-depth strategies from the U.S. NSA and the Department of Homeland Security. You'll also examine how elements of the U.S. government's defensive strategies and priorities mirror those of the SANS Critical Security Controls, a popular list of network security design considerations that the business world often uses in network security implementations.

The NSA People, Technology, and Operations Defense Strategy

The NSA's Information Assurance–based defense-in-depth strategy is based on the idea that people, technology, and operational security must be provided to ensure end-to-end defense. The NSA points to availability, confidentiality, integrity, authentication, and nonrepudiation services as key parts of the ability to protect against, detect, react to, and recover from attacks.

As with most high-level conceptual security designs, the NSA's Information Assurance strategy model is typically explained with a simple diagram. (See Figure 9-5.) Note the emphasis on robust and integrated measures and actions. The NSA realizes the importance of a strong defensive design, with multiple supporting layers that cover both the individuals who use technology; the technology itself; and how the daily operations that support, monitor, and maintain them work. Although the diagram looks simple, the underlying implementation can be quite complex, as you'll see in looking at how these elements interact.

People

The NSA's people-based strategy relies on hiring talented staff, training and rewarding them, and penalizing unauthorized and unacceptable behavior. To do this, the NSA built a framework that includes policies and procedures, training and awareness, system administration, physical security, personnel security, and facilities countermeasures. Combining these elements provides depth by ensuring that (1) the staff knows the right thing to do based on policies and procedures, (2) they know how to do it because of their training, and (3) they are in an environment that helps to

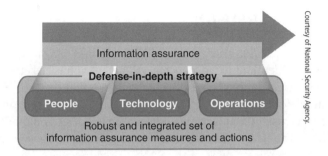

FIGURE 9-5

The NSA's Information Assurance and defense-in-depth conceptual model combines people, technology, and operations into a defense-in-depth strategy in support of information assurance.

enforce those requirements with effective system administration and physical security. These elements also help ensure security through personnel security. To do this, they use background checks and other review of their staff, and then ensure that staff work facilities can provide appropriate levels of security, oversight, and separation.

Technology

The technology portion of the NSA's recommendations focuses on how technology is designed, acquired, configured, managed, and maintained. Technology focus areas emphasize the need to defend in multiple places at once:

- Defending the network and infrastructure
- Defending the enclave boundary
- Defending the computing environment
- Supporting infrastructure like key management, public key infrastructure, detection, and response

The NSA's defense-in-depth strategy uses layered defenses that work together to ensure that the failure of one layer of protection will not expose the data, service, or system to attack. The NSA emphasizes the need for each layer of defense to create unique barriers to access, so that a single attack cannot bypass multiple layers simultaneously. It also focuses on the need to detect and respond when a breach of a defense layer occurs.

Operations

The operations leg of the three-part Information Assurance defense-in-depth strategy focuses on the following:

- Security policy
- Certification and accreditation

technical TIP

Key management and public key infrastructure are parts of an encryption strategy. Keys are the part of a cipher that is used with the algorithm to specify how the cipher's encoding will transform the original unencrypted data. Public key infrastructure is the set of systems and software that make public key encryption work and includes a certificate authority that issues and verifies certificates, registration authorities that verify whether the entity requesting a certificate is valid, directories of certificates, and the certificate-management system itself.

- Security management
- Key management
- Readiness assessments
- Attack sensing
- Warning
- Response
- Recovery and reconstitution

In essence, the daily activities of the defense-in-depth strategy occur here. Elements of this strategy include testing and validating configurations, systems, and software; ensuring that patching and updates occur; performing regular assessments; monitoring; and restoring normal functionality after a successful attack.

When taken together, the People, Technology, and Operations design philosophy reflects common practices for most mature information security operations. It should come as no surprise that the threats the NSA faces mirror those that are found elsewhere, even though the NSA may face them in the form of cyberwarfare activities.

9

Defense-in-Depth
Strategies

FYI

The Information Assurance Directorate (IAD) of the NSA provides standards and a technical framework at *https://apps.nsa.gov/iaarchive/*.

The Common Criteria Protection Profiles provide both configuration and testing certification information for systems, software, and other products tested to meet the Common Criteria. The **Common Criteria** make up an international standard for computer security certification and testing. You can find them at *http://www.commoncriteriaportal.org/pps/*.

The 20 CIS Controls

In the first edition of this book, we reviewed the IAD's *Confidence in Cyberspace* guide, intended to provide guidance on how to fight attacks throughout their life cycle. These controls remain relevant today, although the IAD no longer publishes the guide. Instead, industry professionals now commonly use industry standard references like the 20 CIS controls and resources found at https://www.cisecurity.org/controls/cis-controls-list/.

The CIS controls are broken down into three primary areas: basic, foundational, and organizational, as shown in Table 9-1.

TABLE 9-1 The 20 CIS Controls by Type.	
CIS CONTROLS	**TYPE**
1. Inventory and control of hardware assets	Basic
2. Inventory and control of software assets	Basic
3. Continuous vulnerability management	Basic
4. Controlled use of administrative privileges	Basic
5. Secure configuration for hardware and software on mobile devices, laptops, workstations, and servers	Basic
6. Maintenance, monitoring, and analysis of audit logs	Basic
7. Email and web browser protections	Foundational
8. Malware defenses	Foundational
9. Limitation and control of network ports, protocols, and services	Foundational
10. Data recovery capabilities	Foundational
11. Secure configuration for network devices such as firewalls, routers, and switches	Foundational
12. Boundary defense	Foundational
13. Data protection	Foundational
14. Controlled access based on need to know	Foundational

TABLE 9-1	The 20 CIS Controls by Type.	
CIS CONTROLS		**TYPE**
15. Wireless access control		Foundational
16. Account monitoring and control		Foundational
17. Implement a security awareness and training program		Organizational
18. Application software security		Organizational
19. Incident response and management		Organizational
20. Penetration tests and red team exercises		Organizational

While the entries themselves are short, the actions required to achieve them can be complex. Control 1, inventory and control of hardware assets, can include both administrative tasks like inventory control when devices are acquired, tracking throughout their life, and proper disposal. It also often includes automated discovery, using technical tools and management techniques to ensure that changes are detected and inventories are updated.

Use of controls like the CIS list can help in the design of a defense-in-depth environment but should not be used without understanding the context and requirements of the organization where they are deployed. Organizations that simply copy a list and implement it may be missing critical controls due to a different risk or threat model or may be spending resources on items that are less beneficial than others that they might choose if they did their own analysis.

The Department of Homeland Security and Defense in Depth

The U.S. Department of Defense (DoD) also provides public information about cybersecurity and DoD defensive design. DoD Instruction 8500.01E, the DoD instruction addressing cybersecurity responsibilities, does not mention defense in depth in so many words. It does, however, provide a significant amount of information about how the DoD expects cybersecurity defense in depth to be implemented. In addition, many of the same elements that are found throughout the NSA's People, Technology, and Operations strategy and the CIS 20 controls can be found in the DoD's instruction.

 NOTE

You can find publicly released DoD instructions related to DoD Chief Information Officer operations at https://dodcio.defense.gov/library/

9

Defense-in-Depth
Strategies

NOTE

The term *self-defending network* implies that the network is able to respond to attacks by changing rules, modifying how it is configured, and otherwise responding to problems. You'll learn more about the concept of a self-defending network later in this chapter.

Much like the designs you've already looked at, the DoD Instruction requires operational resilience, or availability. In fact, it requires the following:

- Information and services to be available to authorized users "whenever and wherever required according to mission needs..."
- Security posture to be monitored and available to those responsible for it at all times
- Technology components to be able to self-defend, reconfigure, and optimize with little or no human intervention

Similarly, the DoD's Cybersecurity Discipline Implementation plan focuses on four distinct areas:

1. Strong authentication—to degrade the adversaries' ability to maneuver on DoD information networks
2. Device hardening—to reduce internal and external attack vectors into DoD information networks
3. Reduce attack surface—to reduce external attack vectors into DoD information networks
4. Alignment to cybersecurity/computer network defense service providers—to improve detection of and response to adversary activity

While each of these documents has a slightly different approach, each shares a common theme: Defense in depth requires efforts throughout an organization, at each layer, and in a coordinated manner.

Computer Network Defense and Defense in Depth

At the beginning of this chapter, you learned that defense in depth is part of the defensive side of cyberwar operations known as CND. To best understand where defense in depth fits in a complete CND strategy, you need to understand all of the elements that CND typically entails and how current network defense strategies came about.

The U.S. DoD and NSA's early Information Assurance designs commonly cited defense in depth. However, the term *computer network defense* was first used in the late 1990s. The Joint Task Force for Computer Network Defense was created in 1998 as part of U.S. Space Command. This creation, and its later growth to cover computer network attacks, is one of the first highly visible uses of the term CND.

Over time, the U.S. government grew CND into a complete discipline with dedicated cyberwarfare support organizations. Those organizations now implement a top-to-bottom CND strategy with elements similar to those shown in Figure 9-6, which depicts the U.S.

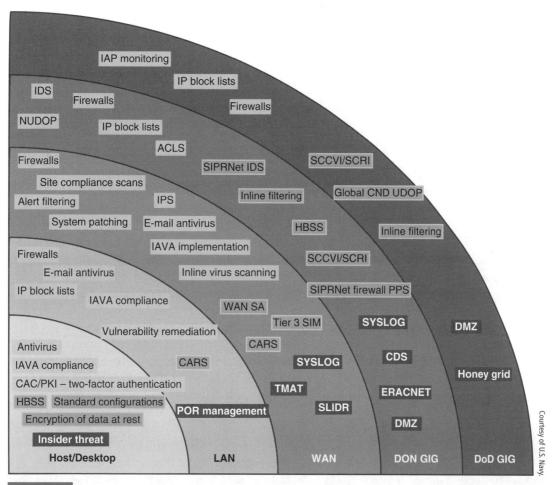

Host/Desktop

Antivirus
IAVA compliance
CAC/PKI – two-factor authentication
HBSS Standard configurations
Encryption of data at rest
Insider threat

Firewalls
E-mail antivirus
IP block lists
IAVA compliance
Vulnerability remediation

CARS

Firewalls
Site compliance scans
Alert filtering
System patching
IAVA implementation
Inline virus scanning

IPS
E-mail antivirus

Firewalls
IDS
NUDOP
IP block lists
ACLS

IAP monitoring
IP block lists
Firewalls

LAN

POR management
CARS
TMAT

WAN SA
Tier 3 SIM
CARS

SIPRNet IDS
Inline filtering

WAN

SLIDR
SYSLOG

HBSS
SCCVI/SCRI
SIPRNet firewall PPS

SCCVI/SCRI
Global CND UDOP
Inline filtering

DON GIG

DMZ
ERACNET
CDS
SYSLOG

DoD GIG

DMZ
Honey grid

Courtesy of U.S. Navy.

FIGURE 9-6

The Department of the Navy CND defense-in-depth strategy combined elements at the host, network, network edge, and policy layer.

Navy's CND strategy from 2009. Over time, this list has grown with the addition of new elements and additional technologies like artificial intelligence, machine learning, containerization, and cloud services have become commonplace.

As you dig into U.S. CND documentation like CJCSI 6510.01F, you will still find defense-in-depth language like "Implement a defense-in[-]depth strategy for ISs and supporting infrastructures through an incremental process of protecting critical assets or data first. The defense-in[-]depth strategy must establish protection and trust across various network layers (e.g., application, presentation, session, transport, network, data link, or physical)."

Where and Why Defense in Depth Fails

Despite multiple layers of protection, defense in depth can fail in a number of scenarios. Much like the castles that come to mind when you diagram defense-in-depth designs, changes in technology, flaws in design, and trusted insiders can all breach or betray even the strongest defenses. But attacks aren't the only reason that defense in depth can have problems. Some of the biggest problems with defense in depth result from trade-offs it creates simply because of the way it must be implemented.

Recall the three main concepts from the C-I-A triad: confidentiality, integrity, and availability. If you think about these three concepts in the context of defense in depth, it quickly becomes obvious that the more you protect confidentiality, the harder it becomes to provide provable integrity. This is because proving that something has not been changed requires access to it so you can compare it with a known good version of itself. Providing access means that you have another component that must be trusted. That addition of trust makes it harder to have confidentiality. This simple issue can drive significant costs, or it can create unexpected holes in your layers of security as you increase either confidentiality or integrity at the expense of the other.

Availability can be similarly affected. As environments grow increasingly complex, the infrastructure and overhead required to maintain high levels of availability typically increases. Thus, as you increase the number of security layers, or increase their complexity, availability often suffers. Modern cloud solutions attempt to fix this by creating multiple regional data centers with hundreds or thousands of systems. However, they face the same issue: If you increase the number of places your data can be, you face the issue of securing it in all of those places at the same level of confidentiality and integrity.

If you have assessed your organization's threats and risks, and have spent your time and resources well, it's worth exploring the common places that defense in depth fails: neglecting layers, trusted attackers, human factors, and changes in technology.

Neglecting Layers: Getting Past the Shell

Much like a medieval castle attacked through its postern gate (a secondary entrance used for messengers and others who needed to come and go without being noticed), modern networks are vulnerable to attacks that are allowed to bypass their layered defenses. One of the most common mistakes made in defense-in-depth strategies occurs when defenders build strong outer layers of firewalls and network defenses, and then neglect the systems that inhabit the core of their defensive rings. That mistake may involve trusted individuals or systems, reliance on the security of software or hardware, or simply leaving assets physically accessible.

A great example of an attack that did exactly this occurred in 2011 when attackers targeted RSA, a major security company (and part of EMC, a large technology company). The RSA attackers crafted an email and sent it to specific users in the company. The email appeared to come from Beyond.com, a job-hunting site. The email included an infected spreadsheet as an attachment, and that spreadsheet included an exploit that targeted the popular Flash player. When the user opened it, the exploit compromised the user's system and turned it into a remote-controlled gateway into the RSA network.

Once attackers had an entryway into RSA's network, they moved laterally, targeting accounts and systems with greater privileges and access. Eventually, they found the data they were looking for. In the case of RSA, that data were key information about RSA's core business: one-time passwords, also known as two-factor authentication software and hardware. In cyberwar, it's a huge win to successfully target the technology used to secure an enemy's authentication systems that prevent simple passwords from being targeted. The attackers who went after RSA had the keys to the kingdom for major companies and military users around the world.

NOTE

Interestingly, none of the RSA employees who received the email were high-profile employees, and the email was actually flagged as spam. This crosses over into human factors: The employees who opened the email had to retrieve it from their spam folder to open it!

In Figure 9-7, a strong defensive layer of intrusion prevention systems, email spam and antivirus filters, firewalls, monitoring systems, and patching existed at RSA. However, users were allowed to receive attachments from the outside world. Their systems used the Flash plug-in, and they may not have used application whitelisting, which would have helped prevent the malware from running. Once a user opened an Excel spreadsheet with its embedded Flash malware, his or her system was compromised and allowed access to other systems in that network segment and beyond. If internal defense isn't sufficient, the attackers can **pivot**, attacking systems inside the heavily defended outer shell.

2020's SolarWinds Orion breaches are another example of where defense in depth failed for many organizations. Orion is a monitoring and management platform that is used extensively throughout enterprise networks and is well known for the broad range of capabilities it provides. Advanced attackers were able to compromise the update functionality for Orion, resulting in malware being installed as part of the update

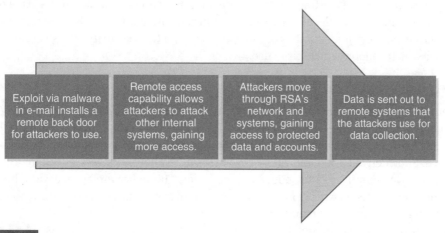

Exploit via malware in e-mail installs a remote back door for attackers to use.

Remote access capability allows attackers to attack other internal systems, gaining more access.

Attackers move through RSA's network and systems, gaining access to protected data and accounts.

Data is sent out to remote systems that the attackers use for data collection.

FIGURE 9-7

The attack process at RSA started because users had the Flash plug-in installed, allowing attackers to use a zero-day exploit delivered via email to compromise systems and to move through its network.

9

Defense-in-Depth Strategies

> ### Zero Trust: An Attempt to Close the Gap
>
> Defense in depth's frequent reliance on a hard outer shell has resulted in the idea of zero trust growing in popularity for network and systems design. In a **zero trust** environment, there is no default assumption that any system or user can be trusted—zero trust is often summarized as "Never trust, always verify." In fact, even authenticated users or services will have to reauthenticate and be revalidated to ensure that a currently trusted user or system still is!
>
> Zero trust sounds like an ideal design for safety, and the concept of zero trust has a lot to offer organizations. The challenge of zero trust is much like the challenges faced by defense-in-depth architectures before zero trust was conceived: Implementation can be complex and costly, can interfere with doing the actual work that must be accomplished, and is still prone to human error or intentional violations. That doesn't meant that zero trust doesn't have a significant role to play in modern defense-in-depth architectures. Instead, it means that zero trust designs are used in more places as capabilities and knowledge grow and as technologies and systems are more able to support it.

process. This bypassed many of the defenses organizations had in place—in fact, many relied on SolarWinds as part of their defensive tools to monitor traffic and identify unexpected behaviors. Once the malware was in place across thousands of organizations and infrastructures, attackers were able to pick and choose their targets for ongoing exploitation. Since they were using trusted back-end management tools, they had broader access than they might have through an attack from the outside to a vulnerable system.

The SolarWinds attack demonstrates a likely pattern for future attacks. Attackers with sufficient resources will continue to target the underlying infrastructure and trusted tools so that they can gain privileged access inside of organizations. This makes concepts like Zero Trust even more important as organizations design their security infrastructure and consider which vendors and tools to use.

System Administrators: Trusted Attackers

System administrators provide another potential way in for attackers. Edward Snowden was a trusted contractor. Snowden and his collection of NSA data provide a case study of a privileged user leveraging his access to gather significant amounts of data without being detected. Snowden did not compromise NSA systems to allow outsiders in, but he still left with a massive amount of data about NSA programs. This cyberintelligence was a huge coup for foreign intelligence agencies, but access to the systems he was trusted with would have been an even bigger win for them.

Publicly released information about Snowden's attack indicates that it was not an incredibly technical exploit or one that compromised systems to succeed. Instead, he used a commonly available tool used to download the content of websites for later viewing, as shown in Figure 9-8. Using the tool from a trusted position in the network allowed him to

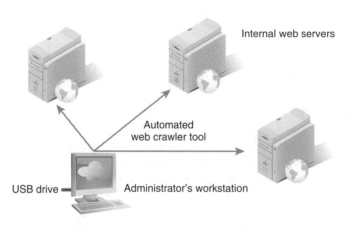

Trusted NSA network

Internal web servers

Automated
web crawler tool

USB drive — Administrator's workstation

FIGURE 9-8

The trusted administrator's system has access to systems inside of a secure network. Because the administrator's system is expected to contact those machines, the web crawler's data gathering is likely to go unnoticed, or at least be easily explained.

gather huge amounts of data and explain the network traffic as part of his normal system administration job. Once he had the data, it was a relatively simple process to find a way to leave with that data, whether it was via a USB thumb drive or some other method. System administrators often carry around drives or send large files as part of their jobs, and this too would likely go unnoticed in many cases.

For those who want to have truly secure networks, it is an unfortunate fact that security systems require at least one trusted staff member to administrate them. This fact creates a potential problem with trusted attackers. It also creates a target for attackers who want the administrator's privileges. Thus, system administrators are a dual threat: They could choose to become attackers themselves, and their accounts and privileges are targets for attackers, even if the administrators can be trusted.

In the defense-in-depth strategies discussed earlier in this chapter, system administrator privileges are one of the most heavily protected and monitored assets in any successful design. Knowing when those privileges are used, that the person using them is the authorized user, and what they are doing with those privileges is critical to preventing attacks and detecting any that succeed.

Attacking the User: Human Factors

Humans are in many ways the most vulnerable part of technological systems and are the hardest to fix. Human instinct drives people to be helpful. Further, people are trained to assist others, to trust, and to sympathize. The problem is that attackers gladly use these natural instincts to their advantage. People tend to behave in ways that attackers can manipulate to access systems and data. If an attacker can predict how people will act, he or she can target the attack to use them to bypass security.

9

Defense-in-Depth
Strategies

Using human factors to attack systems was an important part of the Stuxnet malware attack on Iranian nuclear facilities. The facilities used an air gap, or physical separation of systems that were network accessible from the control systems in the design of nuclear-enrichment facilities. This is a common feature in power plants and industrial facilities as well as in high-security military operations where Internet-accessible systems could lead to fatalities or infrastructure failure. Air gaps are designed to prevent users who have access to the protected area from transferring materials from an untrusted area across the gap without taking appropriate security measures.

In Figure 9-9, you can see how a critical part of the Stuxnet attack leveraged expert technical staff. The staff members unwittingly carried malware on USB thumb drives that they used to transfer files from Internet-connected laptops to the protected network. Once the air gap had been crossed with human help, the Stuxnet malware could directly target the hardware it was looking for.

This type of attack takes careful planning and required the attackers to cast a broad net. The attackers had to infect many systems belonging to potential technical advisors and workers who might work with the Iranian facilities. The Stuxnet malware had to be able to bypass their antivirus software. In addition, the attackers had to rely on the workers using removable drives to transfer files. The sequence of events that Stuxnet relied on was likely to happen eventually, but setting up the chain of infections to get it to the right place at the right time was also incredibly complex.

Changes in Technology

When designing defense in depth, technological change is one of the hardest challenges to defend against. Last week's secure design might be an outdated relic this week if a technological breakthrough is made or if a critical flaw is found in an old technology. Even before the modern idea of cyberwar existed, technological progress was responsible for success in breaking cryptographic systems. The German World War II Enigma device,

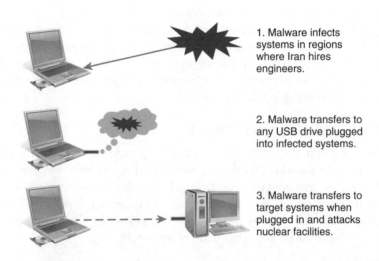

FIGURE 9-9

Targeted malware seeks to infect systems like those that frequently hired technical consultants to the Iranian nuclear program use. Once infected, those systems deliver malware via any USB drive plugged into them.

1. Malware infects systems in regions where Iran hires engineers.

2. Malware transfers to any USB drive plugged into infected systems.

3. Malware transfers to target systems when plugged in and attacks nuclear facilities.

once considered unbreakable, drove the development of computerized code-breaking systems, which made Enigma and similar systems obsolete and led to entirely new types of encryption systems.

Modern advances in code-breaking technology, known as rainbow tables, have made the use of MD5 and SHA1 hashes for password storage similarly obsolete. Rainbow tables allow attackers to quickly retrieve passwords if the attacker knows the password's hash. Unfortunately, some websites continue to store their passwords as MD5 or SHA1 hashes because both are commonly available in toolkits for developing web applications even after years of breaches and established best practices that point to more secure options. Thus, organizations and developers that are unaware of the ease with which rainbow tables can allow password recovery have lost the race to protect their users' data. If organizations' storage for hashed passwords is breached, attackers frequently discover the actual passwords.

Unfortunately, defenders are often unaware of the technological changes that have made their previously secure technologies obsolete. Some defenders unwittingly make this mistake because the vulnerable technology is embedded in their infrastructure, leaving them vulnerable without their knowledge. Thus, attackers frequently hold the advantage, and defenders need to stay vigilant and update their technology before it reaches that point. Defense in depth does offer some remedies for this by layering technologies with different weaknesses, but that defensive measure usually comes at a high cost.

One of the biggest problems for defenders is that changes in technology are typically far easier for attackers to exploit than they are for defenders to update. Defenders often have existing infrastructure and businesses they cannot disrupt without creating losses. Thus, they have to balance security against changes in technology and may end up on the losing end of the battle. The star forts mentioned at the beginning of this chapter are a perfect historical example of this. They were created because cannons were capable of shattering the huge existing stone castles of the day, leaving the defenders unable to protect

9

Defense-in-Depth Strategies

---

technical TIP

Hashes are cryptographic functions that take a block of data and perform operations on it to produce a fixed-length string of characters. If the same data are hashed, they will always return the same string. However, modifying the data will always change the hash. Hashes aren't intended to be reversible, which makes them a great way to compare data like passwords without exposing the data itself.

Rainbow tables are a prebuilt list of all the hashes that strings can hash to. Thus, if an attacker has the MD5 hash of a password that was created by simply hashing the password with no additional steps taken to protect it, the attacker can simply perform a lookup against the rainbow table and identify the original password.

Using rainbow tables to test against sample password lists available from previously hacked sites has resulted in exposing more than 50 percent—and in some cases more than 90 percent— of all the hashed passwords.

themselves. By the time that defenders built star forts in large numbers, technology had again advanced, and new techniques were in place to overcome them.

Designing a Modern CND Strategy

Modern CND strategies must be designed to defeat both nation-state level capabilities and those of asymmetric actors like hackers, organized crime, and even corporate employees seeking to engage in corporate espionage. Thus, a CND strategy makes use of many traditional information security design elements layered with government- and military-specific procedures, policies, and technologies. The additional considerations these designs take into account reflect the needs of military operations, such as chain of command, classification, and resistance to electronic warfare (as opposed to cyber-warfare). The hardware, operating systems, and software they are deployed on may be significantly different from their civilian counterparts, or they may simply be civilian versions configured to meet CND requirements. This part of the chapter explores the concepts of dynamic defense and some of the common elements of a secure network.

Dynamic Defense

A primary criticism of defense in depth is that adding layers makes using the defended assets more difficult. Each successive layer adds overhead to manage, monitor, and access the environment. Those layers can also cause failures if they themselves have problems. When organizations consider layer upon layer of fixed defenses, they often wonder if there is a way to have a reactive defense mechanism that uses situational knowledge, a range of capabilities, and an adaptive mechanism to handle unexpected events. Thus, the concept of *dynamic defense*, or defense that can change in reaction to threats and new risks, has become more popular. In fact, dynamic defense is often what organizations and authors are describing when they propose self-defending systems and networks that can react to threats and modify their defense schemes to meet them.

The idea of dynamic defense isn't new. J. R. Winkler, C. J. O'Shea, and M. C. Stokrp outlined the basic concept in a 1996 paper, noting that operational security analysis should allow dynamic defenses to do the following:

- Add additional countermeasures that can reconfigure to handle new threats
- Implement interoperability between defensive systems, allowing them to coordinate their defenses
- Provide both centralized and decentralized components
- Use a database of operational security information to support the components of the defense system

Despite almost two decades of conceptual development, well-integrated and organization-wide dynamic defenses largely remain a dream for most organizations. More advanced security systems, however, have begun to implement these ideas within their own bounds. Some defense systems are designed to allow custom-built modules to control other systems, but those are rarely built using a shared control standard.

The Self-Defending Network and Self-Defending Hosts

One concept for defense in depth is that of a *self-defending* network or host. These self-defending systems are intended to adapt to prevent attacks by monitoring systems, users, and network traffic. When they detect something that should not be occurring, or that does not fit their security posture, they block the traffic, disable the user or service, or otherwise take what is hopefully appropriate action. As with any automated system, a system that can modify itself to handle problems can also likely cause itself harm by responding inappropriately to those threats.

A few common elements show up in most literature about self-defending networks and hosts:

- They use large datasets, artificial intelligence, and machine learning capabilities to use information to monitor for attackers, known malicious sites, and malware. These capabilities combine an understanding of what is normal for a system or network, what behaviors are common or acceptable, and what malicious activity often looks like.

- Software-defined networks can redefine their boundaries as needed, moving systems into much more granular enclaves than a traditional network design typically provides.

- User- and system-contextual awareness can grant permissions to users based on behavior and their role in the network at the time they use it. Rather than giving users all of their rights all of the time, this contextual awareness focuses on giving users only the rights they need for the time they need it.

- Direct, secure interconnection between endpoints allows end-to-end security of data without systems and networks between them being able to see or modify the traffic between them.

- System posturing relies on more than just the user or system connected to the network. Often systems are granted rights based on the logged-on account or where in the network the host resides. In a fully self-defending network, the host would have a more complex posture based on an up-to-date profile of the system, the user's rights, and where the device is.

As you might imagine, self-defending networks and hosts are a challenge to pay for, build, configure, and maintain. In many cases, organizations are using some of these concepts and capabilities, but true end-to-end, self-defending networks that use all of these elements remain rare. As organizations build more software-defined infrastructure, and as container-based systems become commonplace, components of self-defending infrastructures are continuing to appear—even if they're usually not fully realized.

CND and Defense-in-Depth Design

System and network security design is an incredibly complex topic with a wide range of options and strategies available to defenders. Often the best place to start is with a strong policy and procedural background, as well as with awareness and training to support the design. Figure 9-10 shows a conceptual layered design that uses many of the elements and concepts discussed in this chapter, built on a foundation of policy and procedure. Keep this design in mind as you review the common elements of modern security designs.

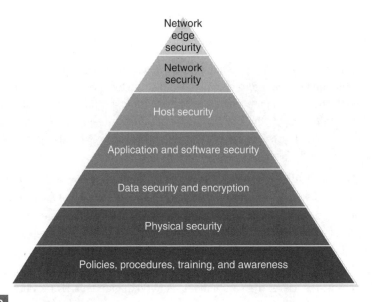

FIGURE 9-10

A sample CND design starts with policies and procedures. Physical security then follows the layers of the technical environment from the data resident on the hosts to the software and applications, hosts, network, and the border of the network.

Risk and Threats

CND designs must respond to the threats and risks they face. Standardized designs that do not address the realities of the attackers and defenders and the capabilities of both will fail quickly. Modern designs typically start by identifying the risks to the organization through some form of risk assessment methodology, which also identifies the threats and actors that the organization will face.

Once the risks the organization faces have been assessed and prioritized, best practices and standardized design concepts like those discussed earlier in this chapter are applied to counter them. Layered defenses are typically designed where possible to ensure that a single failure cannot bring down the organization's entire security architecture.

FYI

Risks are defined as the potential that a threat or actor will exploit vulnerabilities, causing harm to an organization. There are many different risk assessment methods, but the key element of each is that they identify what could happen, what the result would be, how likely it is, and how much harm it would cause.

Threats are possible dangers that could cause a security breach or exploit resulting in harm. Understanding the threats an organization faces helps to inform risk assessments and security design.

Secure Networks

The technology elements of computer network defense rely on having a secure network, including the systems on the network, the network devices, and the technologies used to connect them. Although this is a broad topic, this section briefly explains a few of the key technologies and concepts associated with current network security designs, including compartmentalization, defensive technologies, monitoring, cryptography, defense against malware, endpoint security, and physical security.

Network Enclaves and Properties

The concept of protected network enclaves has increased in popularity as networks have become increasingly complex, and the uses for them have expanded. A **network enclave** is a logically or physically separated portion of a network that isolates systems and devices from others based on rules, identity, location, or purpose. Enclaves are a particularly useful concept when organizations need to separate sensitive data or systems from other parts of their infrastructure.

Network enclaves are often built using internal firewalls using virtual network segmentation—a **virtual LAN (VLAN)**—or using a **virtual private network (VPN)** that allows creation of virtual networks of systems. Network enclaves often use a technology known as **network admissions control (NAC)** that requires systems to authenticate and verify that they match required settings and policies before they can join a protected network.

In Figure 9-11, Workstation A attempts to enter the protected network enclave by authenticating and performing tests required by the NAC system. It succeeds and is added to the protected network. Workstation B performs the same process, but it is not properly configured and fails the test, leaving it in an untrusted network.

Network Defense Technologies and Devices

Technical network defenses are typically selected based on the security requirements of the network. The defense-in-depth designs explored in this chapter combine many

9

FYI

The U.S. Department of Defense describes *enclaves* as follows:

> Enclaves provide standard cybersecurity, such as boundary defense, incident detection and response, and key management, and deliver common applications, such as office automation and electronic mail. Enclaves may be specific to an organization or a mission, and the computing environments may be organized by physical proximity or by function independent of location. Examples of enclaves include local area networks and the applications they host, backbone networks, and data processing centers.

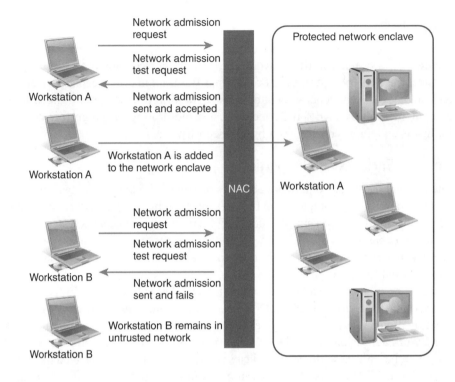

A sample network enclave admission process.

different network security technologies. Some of the most common defensive devices and systems are as follows:

- **Firewalls**—These are network security devices that allow or deny traffic based on rules. The rules can specify ports and protocols. More advanced systems apply additional filters intended to detect attacks and abuse of systems.

- **Intrusion detection systems (IDSs) and intrusion prevention systems (IPSs)**—Along with their local system sibling the **host intrusion prevention system (HIPS)**, IDSs, and IPSs are designed to detect and/or prevent attacks. They rely on either behavior-based detection, which looks for attacks based on behavior, or signature-based detection, which uses a fingerprint of known attacks.

- **Proxies and gateways**—These allow access either to or from secure networks, providing the ability to verify users and to filter what they send or receive.

- **Authentication and authorization systems**—These verify user identity and the permissions they are granted.

- **Virtual private networks (VPNs)**—VPNs provide encrypted secure networks virtually through existing networks.

- **Security Information and Event Management (SIEM) systems**—With their siblings, Security Information Management (SIM) and Security Event Management (SEM) systems, SIEM systems combine event logs, network traffic data, and other security and log information to detect and take action on security events, attacks, and even service issues.
- **Security Orchestration Automation and Response (SOAR) systems**—Another type of central security management system is a SOAR system, which focuses on taking data from sources like SIEM systems and using that data to manage security workflows and response procedures. The sheer volume of data gathered by a comprehensive SIEM means that doing something with it can be a challenge, leading to SOAR deployments.

Monitoring

Networks, devices, and software all typically offer some method of providing status information in the form of logs. In addition to logs, you can add additional instrumentation to validate and monitor what is occurring. This can help provide a layer of defense against both attackers and system errors.

Monitoring inside of a CND environment typically takes on a number of forms:

- Network traffic monitoring, such as network flows, bandwidth utilization, and other information about network traffic levels and types.
- System monitoring, such as error logs, administrative and privileged user access logs, authentication logs, and a host of other details about what systems are doing and what problems they encounter.
- Software and application logs, which provide information about how software and applications are performing and whether they have encountered errors.
- Behavior monitoring and anomaly detection capabilities, which increasingly are commonly implemented to detect threats as well as problems.
- A host of other items. Almost anything that happens on a computer or a network can be monitored.

Due to the broad variety of potentially monitored items, many organizations use automated log analysis and SIEM systems. It can be difficult to balance the desire to gather logs in case they are needed with the sheer amount of data that in-depth logging can create.

9

Defense-in-Depth
Strategies

technical TIP

Network flows are a useful type of network monitoring and can be thought of like a traditional phone bill's call log: They show which systems communicated, via which port and protocol, and how much traffic they sent. This can be a treasure trove of useful information when you are looking for a system compromise or attempting to identify attackers.

Cryptography

The ability to securely store and transmit data is important for CND. Encrypting data in transit can ensure that attackers cannot capture what is being sent. That same data must be protected when it is at rest, waiting to be accessed on a hard drive or other storage device. Thus, encryption capabilities and their implementation are part of the CND plan:

- In-place encryption is often based on full-drive encryption or encryption of specific files or data stores that are sensitive. In-place encryption is often stronger than in-transit encryption because the data are not transient and thus must resist potential long-term attacks.
- In-transit encryption includes the encryption used to protect and encapsulate VPNs, allowing them to operate inside of and through other networks without having their data exposed. It also includes the encryption used for secure web traffic (SSL/TLS) and other technologies that help to ensure that network and other transmissions remain secure. Because in-transit encryption is typically short lived, the encryption used is often weaker than that found in in-place encryption systems.

Smart attackers look for moments when data are exposed and target locations where they can capture that data. Often, this is when the data are transmitted or are momentarily unencrypted for access, such as when a credit card or password has to be processed by the system receiving it.

Defense against Malware

Malware is a major concern for organizations—particularly state-sponsored threats, such as the very complex and well-supported advanced persistent threats that have appeared in the past few years. Malware can provide access into your network, can allow control of your systems, and targeted malware like Flame and Stuxnet can attack specific systems and capabilities for cyberwarfare purposes. Even criminal malware like Ryuk and Zeus can cause significant problems for organizations even if they're not directly targeted by nation-state cyberwarfare capabilities. Thus, protection against malware is a major part of defense in depth.

Anti-malware packages provide a number of noteworthy abilities that should be considered when designing a defensive strategy:

- Whitelisting capabilities allow only trusted software to run, and **application whitelisting** checks to ensure that the software running matches a known good version of the software. Unfortunately, whitelisting also requires a lot of work to verify each package every time it is updated and limits flexibility by requiring every piece of software to be added to the list.
- Heuristic or behavior-based anti-malware capabilities look for behaviors associated with attackers, such as unexpected data transfers, scans of other systems, or access to memory or data that is atypical.

- Signature-based systems attempt to gather fingerprints of known malware and then compare files and applications with those fingerprints. Signature-based detection is becoming increasingly difficult as malware packages frequently change themselves using a variety of techniques every time they are copied or infect a new system. Thus, signatures don't match, and anti-malware vendors must track millions of similar packages.

Anti-malware technologies can be deployed on networks and on systems and devices. Unfortunately, modern malware concealment techniques can make advanced malware nearly impossible to detect. It has been shown that advanced persistent threat actors have been able to exploit the detection of their malware infections to upgrade them, using what they learn to defeat the tools that detected them. This results in an ongoing game of cat and mouse that pits defenders against attackers.

Endpoint Security Design

Security design and implementation for the endpoint is sometimes one of the hardest elements of a defense-in-depth plan. Endpoints are typically laptops, desktops, tablets, mobile devices, and other individual systems. They can also be network devices, cameras, and other parts of the technology infrastructure.

Endpoint security design is typically based on a few important elements:

- Securing the operating system or software that controls the device's basic functionality
- User accounts and privileges
- Controlling the software or other key functions of the device or system
- Configuration baselines that ensure that the system matches expected settings
- Monitoring and exception reporting
- Data protection, often in the form of encryption, backups, and integrity monitoring to ensure that data are not modified without permission

The need for advanced security capabilities has led to the appearance of a new endpoint focused set of technologies often called EDR, or Endpoint Detection and Response. These expand beyond anti-malware and monitoring capabilities to focus on behavior analysis and understanding the context that actions are occurring in on endpoint systems and devices. EDR systems may also block potentially malicious activity and can help with response activities including remediation and recovery.

Endpoint security can be incredibly complex. Modern operating systems have hundreds and sometimes thousands of possible security settings when loaded with applications and services. This is where configuration baselines come in to help organizations ensure that they have common settings properly in place.

Vendor and Solution Diversity

Another element of defense-in-depth design is vendor and solution selection to ensure that a vulnerability or problem with a single vendor, type of device, or solution is unlikely to cause an organization-wide issue. Zero day vulnerabilities that do not have an available patch, denial of service issues, or compromises like the SolarWinds Orion attack all argue for solution diversity where risks and capabilities can be matched to provide greater security. Unfortunately, significant trade-offs are also created by diverse implementations due to the additional cost of integration, support, and maintenance resulting in decision points for each solution based on the risks and threats organizations face. The risks of these trade-offs must be assessed against the risk of operating a homogenous infrastructure.

Common examples of diverse or differentiated defense-in-depth choices include using different border routing infrastructure than an organization's core routers and switches, ensuring that cloud infrastructure and third-party services do not all rely on the same underling vendors, and choosing security device platforms like firewalls, IPS, and SIEM systems from different vendors.

Selecting a Configuration Baseline

One of the key parts of a defense-in-depth strategy is **baseline configuration** of systems and devices. This can be a complex task, as modern operating systems often have hundreds of possible security and configuration options. Organizations that are attempting to build a security standard are faced with a massive amount of work to design and test a security baseline for each of the operating systems or devices they intend to deploy.

Fortunately, a variety of configuration baselines are available to start from, and the best of the baselines provide useful commentary and information about the settings themselves. This allows security administrators and system administrators to make informed decisions about which settings they may need to modify due to the requirements of their business or organization.

- Commonly used baseline standards are available from a variety of sources:
- Microsoft at *https://docs.microsoft.com/en-us/windows/security/threat-protection /windows-security-baseline*
- Apple at *http://www.apple.com/support/security/guides/*
- The NSA at *https://apps.nsa.gov/iaarchive/library/ia-guidance/security -configuration/*
- The Center for Internet Security at *https://benchmarks.cisecurity.org/*
- The DoD Cyber Exchange public site provides Security Technical Implementation Guides at *https://public.cyber.mil/stigs/*

Important decisions that your organization might face when adopting a baseline include whether you should create your own baseline or modify one or more of these, which changes you should make, and how you will manage systems to ensure that they meet the baseline standard.

Unfortunately, vendors often lag behind their own releases of new operating systems, devices, and software. It is not uncommon to have a new operating system available for months or even years before a baseline is released. This leaves administrators scrambling to build a new baseline using the knowledge they have from older versions.

Physical Security

Although it might seem odd in the context of CND, physical security is actually one of the most important parts of defending a computer network and its systems. If an attacker has physical control of a system, he or she can almost always eventually compromise it in some way. Thus, a well-designed defense-in-depth plan must account for how computers, network devices, and the physical network itself are protected.

Physical security designs typically start with an understanding of the environment to be protected and how that environment will be used. You can use the same methods you used to assess risks and threats for the computer-based environment to assess physical threats and risks. Unlike network defenses, physical security must also take into account nature itself, as the availability portion of physical security can be greatly harmed by events like earthquakes, hurricanes, floods, and tornadoes.

Physical security efforts also have to include policy and procedure. A heavily secured location that protects systems but allows travelers with laptops to have remote access can find itself leaking data through theft or because the laptop was physically bugged while it was out of the traveler's hands.

Travel Danger

The authors of this book have spoken with corporate security officers who have discovered physical and software bugs placed in and on laptops returning from overseas. The security staff said that they now issue disposable laptops and phones that are physically disassembled and inspected or that may be discarded when they return from overseas rather than allowing travelers to take their daily work systems with them. They also provide a clean operating system and limit remote access to corporate networks for the systems they send overseas so that corporate data cannot be copied off the laptops while they are out of the country.

CHAPTER SUMMARY

This chapter examined the concept of defense in depth in cyberwarfare. Defense in depth is a key component of published cyberwar computer network defense strategies provided by the U.S. NSA, DoD, and civilian organizations that specialize in information and network security. These defense-in-depth strategies focus on people, operations, and technological strategies that blend how networks and systems are designed, built, managed, and monitored, as well as the handling of responses to and recovery from cyberattacks.

The concept of zero trust in defense-in-depth strategies as well as the increasing adoption of code-defined systems and infrastructure has enabled new designs for defense-in-depth environments. While complete zero trust environments remain rare, elements of zero trust are increasingly appearing in architecture designs where they can make a significant difference. Security practitioners must now consider how to combat advanced persistent threat actors with long-term, high-level access to systems, networks, and accounts, and zero trust offers a greater chance of preventing compromises from becoming long-term, ongoing issues.

Although a key concept in CND strategy, defense in depth can create problems. In fact, defense in depth involves trade-offs among confidentiality, integrity, and availability, the three common precepts in information security designs. Defense in depth is also vulnerable to design issues, trusted administrators and users acting in bad faith, and technological change. These issues don't mean that defense in depth doesn't make sense. But they do require you to think clearly about which defenses you use, and both where and why you use added layers—because these have costs and potential flaws themselves.

KEY CONCEPTS AND TERMS

Administrative privileges
Application whitelisting
Authenticity
Availability
Baseline configuration
C-I-A triad

Common Criteria
Confidentiality
Host intrusion prevention
  system (HIPS)
Integrity
Network admissions control
  (NAC)

Network enclave
Nonrepudiation
Pivot
Virtual LAN (VLAN)
Virtual private network (VPN)
Zero trust

CHAPTER 9 ASSESSMENT

1. A segment of a network that is protected from other parts of the network is commonly called a(n) _____.

2. The NSA designs its defensive strategy around which three key areas? (Select three.)

   A. People
   B. Security
   C. Technology
   D. Operations
   E. Defense

3. Which of the following is frequently used to identify security settings all systems should have?

   A. Manual
   B. Configuration baseline
   C. NSA/DID STIG
   D. Vulnerability scan

4. Self-defending network technologies are widely available and standardized.

   A. True
   B. False

5. The _____ Criteria provide an international standard for computer security certification and testing.

6. Defense in depth becomes less likely to cause availability issues as more layers are added.

   A. True
   B. False

7. Attackers who have gained access to a system at a desk have overcome which type of security?

   A. Physical
   B. Network
   C. DoD
   D. Departmental

8. The weakest part of most defense-in-depth strategies is the _____.

9. Systems that monitor for behavior to identify attacks are using _____ detection.

   A. signature
   B. heuristic
   C. malware
   D. fingerprint

10. When do zero trust designs require reauthorization to perform tasks?

    A. The first time the action is performed
    B. When changing roles or activities
    C. Whenever the right or capability is exercise
    D. At random intervals to identify compromised accounts

9

Defense-in-Depth Strategies

Cryptography and Cyberwar

THROUGHOUT HISTORY, military commanders and national leaders have needed to communicate securely. Since ancient times, they have used various techniques to help ensure that enemies could not read intercepted communications. Thus, cryptography was born. Cryptography, or secure communication in the presence of those who should not be able to receive the communication, typically relies on ciphers. These methods or algorithms make the information hidden or keep it secret. The processes used have ranged from simple letter substitution like that used in children's secret codes to complex encryption systems that required prior knowledge of a secret key to decode the messages.

In modern cyberwar, cryptography is still used for the many of the same purposes as those ancient ciphers, but on a far greater scale and with significantly more resistance to being broken. Modern cryptographic systems are far more complex, involving fast computers, huge prime numbers, advanced algorithms, and techniques that allow individuals and systems that have had no prior contact with each other to create secure connections. The broad use of cryptography in its many forms is a key part of cyberwarfare. Understanding how it works, where it can be used, and how it can be attacked is critical to both computer network attack and computer network defense strategies.

Cyberwarfare now involves both attackers and defenders using cryptographic systems and attacks aimed at defeating them. Attackers in cyberwarfare frequently target data stored on systems and in transit across networks. This means that defenders must try to protect that data from being acquired or modified throughout its life cycle. Defenders also must understand the encryption systems they put in place and how to defend the systems themselves. To do this, they must ensure that they are using secure cryptographic systems, which has proven to be a challenge despite their best efforts.

That challenge exists because attackers now target the underlying cryptographic algorithms, software, and even the hardware used for encryption.

Publicized information indicates that organizations at the nation-state level have actively worked to weaken the ways that important parts of widely used encryption algorithms function—to make them easier to break. Other attackers have focused on gaining access to the software used to protect data so they can reverse engineer it and find areas where it may be weak.

Attackers have also begun to use encryption as a weapon in its own right. Encrypting data in a way that prevents legitimate users from accessing it allows attackers to negotiate for the victims to perform an action or to choose when and where the data are available. The world of connected computers and networks now allows attackers to seize what would have been the contents of file cabinets and vaults and put that data in a virtual, unbreakable safe that only they can provide access to.

This chapter introduces important cryptographic concepts and some important historical cryptosystems as well as modern encryption algorithms that are currently widely used. It then explores how cryptography is used by both attackers and defenders in cyberwar—including how nation-states seek to influence for their own advantage the way in which cryptographic systems are created. Next, the chapter discusses how to defeat attacks on cryptographic systems and how malware writers and cyberwarriors have weaponized them. Finally, it looks at the future of cryptography in cyberwar as combatants attack and defend, and as next-generation technologies and techniques become available.

Chapter 10 Topics

This chapter covers the following topics and concepts:

- What the basics of cryptography are
- How cryptography is used in cyberwar
- How cryptanalysis attacks cryptography
- How to defeat attacks on cryptographic systems
- How cryptography is weaponized
- What the future of cryptography in cyberwarfare is

An Introduction to Cryptography

Cryptography is the study and practice of techniques to secure communication and protect it from adversaries. Cryptography in cyberwarfare often involves **encryption**, which is a process that encodes information in a way that prevents unauthorized parties from reading the data. Of course, data that are encrypted need to go through the **decryption** process, which removes the encoding, allowing authorized parties to read the original text.

Cryptographic systems have existed for thousands of years. In fact, one of the earliest documented ciphers dates back to Roman times. Julius Caesar used what is now called the Caesar cipher, a simple cipher that encrypted messages by simply shifting letters to the right by the same number of letters every time. Thus, if adversaries captured his communications, they appeared nonsensical. Over time, adversaries figured out the encryption methods in use, so more advanced techniques were developed and used.

Of course, cryptographic techniques include more than encryption. One early technique for hiding messages reportedly required the messenger to shave his head. Once the messenger's head was bald, the message would be tattooed onto his scalp. When his hair regrew, he would be sent with the message. Recovering the message was as simple as shaving the messenger's head and reading it. Obviously, this wasn't a great method to transport messages in a timely fashion—as you would have to wait for the messenger's hair to grow back before sending the message—but it did conceal them well. Few opponents think to shave a captured enemy's head, and messages on the back of your head underneath your hair would be very hard to read.

> **NOTE**
> Cryptography includes more than just encryption. Most modern cyberwarfare uses involve encryption, so encryption is a main focus throughout most of this chapter. Despite that, as you read this chapter, remember that encryption is but one cryptographic technique—not the only one.

10

Cryptography and Cyberwar

Advances from simple ciphers and other cryptographic techniques like these have continued since those early developments. The need to send secure messages has been a constant, and the ability to conceal meaning in written messages was necessary for communications in warfare and in trade. In fact, the cryptographic arms race has been a fact of life for military, political, and commercial reasons for thousands of years.

Modern cryptographic systems are far more advanced than the systems used in Julius Caesar's time or even those used in World War II by either the Axis or the Allies. Computer systems and advanced mathematical algorithms use very large prime numbers, fast processors, and combinations of complex techniques to secure data from prying eyes. The Internet has created a need for very fast, open, and secure encryption systems that people can use in their daily lives. In fact, most Internet users use encryption every day without realizing it as they log on to websites and as their traffic is carried from service provider to service provider across the Internet.

Both computer network attacks and computer network defense involve cryptographic systems on a regular basis. Attackers may try to break through encryption, weaken encryption systems or inputs into the encryption algorithms that defenders use, or even use encryption itself as a means of attack. Defenders use encryption to protect data at rest and during transit. Further, defenders choose from a host of available options when selecting how strong their encryption is, how and when they use it, and how users interface with it.

Cryptographic Concepts

The term *cryptography* covers a range of techniques for secure communication, all with the same goal: to protect communication from adversaries. It relies on four major concepts:

- *Confidentiality*, or the ability to ensure that data are not exposed to those who should not see it. Encryption systems that cannot be broken by opponents in a timely manner can ensure confidentiality—at least in a reasonable time frame!

- *Data integrity*, which is the ability to ensure that data have not been modified, either by addition or removal of data, or by modifying the data's meaning. Message integrity is enforced by values known as message digests or checksums, which prove that the message has not changed by using algorithms to compare their original and received versions.

- *Authentication*, or the ability to verify the identity of a sender.

- *Nonrepudiation*, which requires that it be possible to prove that the sender did send the file or message. Nonrepudiation means that the sender cannot claim that someone falsely sent the message posing as the sender.

> **NOTE**
> The goals of cryptography are very similar to the C-I-A triad of confidentiality, integrity, and availability, but they aren't the same. Be careful as you review the goals because authentication and availability can be very easy to confuse.

Many modern cryptographic systems offer all of these abilities by providing a means to encrypt a message; the ability to include a digital signature in communications that proves who the sender is; and a means to check that the data are intact, have not been seen by unintended recipients, and that it can be proved that the sender actually sent it.

Ciphers and Encryption

In cryptography, encryption is the process for encoding information, such as a message or file, so that only authorized parties can read it. This is accomplished by using an encryption algorithm or process to encode the original message and to decode the encrypted message when it arrives. This section examines encryption concepts and terms and then explores the basics of the two major types of encryption algorithms: symmetric and asymmetric ciphers. It also examines the concept of cryptographic hashing, a technique used to ensure that data have not been changed by unauthorized parties.

Key Encryption Terms

As you discover how encryption is used in cyberwar, you must become familiar with some common terms. Keep the following terms in mind as you read about encryption systems throughout this chapter:

- Codes are often confused with ciphers, but aren't ciphers. A **code** is, however, a cryptographic system that substitutes values or words for other words. Morse code is an excellent example of a code that substitutes a series of dots and dashes for letters, and most people are familiar with the police use of codes like "10-4" from movies and television. Both codes are commonly available and don't provide any confidentiality, message integrity, or authentication.
- **Plaintext** is the original data prior to encryption.
- **Ciphers** are encryption algorithms that are applied to plaintext. Unlike codes, ciphers are intended to provide confidentiality and message integrity, and some provide capabilities for authentication and nonrepudiation.
- **Ciphertext** is the output of a cryptographic algorithm.
- A **key** is the secret variable in an encryption algorithm. In modern encryption, keys are typically very large prime numbers. Keys exist in a *key space*, which is the range of all possible values that a key can have and remain valid for the purposes of the encryption algorithm. In modern digital encryption systems, a key space is typically measured in bits, or the number of binary 1s and 0s that it contains. A key doubles in complexity for every additional bit larger it becomes.
- A **cryptosystem** is the hardware and software implementation of a cipher.
- **Checksums** are mathematical values calculated to check for common errors in data and are quite fast.
- **Message digests** or **hashes** are cryptographic functions that protect the integrity of a message by allowing you to check that the message has not changed. Hashes are slower than checksums but provide additional capabilities and features that checksums do not.

 NOTE
You will explore hashes and checksums in more depth later in this chapter.

10

Cryptography and Cyberwar

Symmetric Ciphers

Symmetric ciphers rely on a shared key. When a sender and recipient want to communicate, they use the same key for both encryption and decryption. This means that **symmetric encryption** can't provide authentication or nonrepudiation because you cannot identify who the sender is based on the encryption method. In addition, symmetric encryption between many users typically uses the same key for all of them to avoid having to encode the message many times. If one of those users were to no longer be trusted, the symmetric key for all users would need to be changed for future communications.

> **NOTE**
>
> Because a symmetric cipher may be in use by many users (or devices), the security of the system is only as good as the practices of the least careful user or the security of the least secure device. This can be a major issue for symmetric ciphers because key distribution is difficult and knowing if a key has been lost can be almost impossible if an attacker can intercept messages but only reads conversations without taking other action.

Symmetric encryption can be very hard to break when sender and recipient are using a strong key that is handled responsibly. Perhaps more important, symmetric encryption is relatively fast compared with asymmetric encryption. Thus, symmetric encryption can be very useful in ongoing communications when a strong cipher requiring less processor time is desired.

Symmetric cryptography has a number of problems that influence its use:

- The algorithm isn't scalable. Because any user with the key can decrypt any communication sent with it, security between individuals in large groups can only be maintained by having a key for each pair of individuals. As groups grow, this number grows to unsupportable numbers, as shown in Table 10-1. The formula to calculate the number of unique keys is keys $= n(n-1)/2$ where n is the number of users.

Table 10-1 Symmetric cipher scaling.

NUMBER OF USERS	UNIQUE KEYS
1,000,002	499,999,500,001
1,000,003	499,999,500,003
1,000,004	499,999,500,006
1,000,005	499,999,500,010
1,000,010	499,999,500,045
1,000,100	499,999,504,950
1,001,000	499,999,499,500
1,100,000	494,999,950,000
1,000,000	499,999,500,000

- **Key distribution** is a challenge because keys need to be distributed securely. Thus, if you don't already have a secure way to exchange keys, you can't ensure that the shared key won't be compromised when it's transferred. This is a chicken-and-egg problem, with the inability to create a secure transfer process also meaning that you can't transfer the means to build one safely.

- Symmetric encryption doesn't handle changes in trust relationships well. When one person who has a shared symmetric key leaves a group, that entire group of users must generate a new key. If the key is exposed, the same thing must occur, potentially resulting in the creation of as many new keys as there are users who use them to communicate with the person whose key was exposed.

- Symmetric encryption doesn't provide nonrepudiation. You cannot prove that a symmetric key was used by the expected owner. In fact, every communication has at least two users who could have sent the message. Thus, although symmetric encryption can provide useful advantages, it misses an important use case for many cryptographic exchanges.

Figure 10-1 shows how a typical message is sent between User A and User B. User A encrypts the message using his or her shared secret key and transmits the message to User B. User B then uses the same shared secret to decrypt the message. Bear in mind

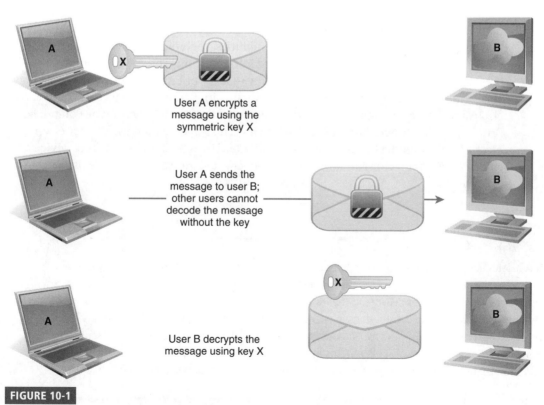

User A encrypts a message using the symmetric key X

User A sends the message to user B; other users cannot decode the message without the key

User B decrypts the message using key X

FIGURE 10-1

Symmetric ciphers use the same key to encode and decode a message.

that User B has no way to prove that User A is actually the person who sent the message and is forced to rely on User A's secure handling of the shared key.

Enigma: Using and Breaking Wartime Symmetric Encryption

During World War II, Germany used an electromechanical symmetric encryption device called the Enigma device. The Enigma devices originally used by the Germans used three wheels to perform complex substitutions of letters very much like a series of Caesar ciphers. Each day, a codebook was used to set the initial settings of the rotors, and then various changes were made for each message. The Enigma actually used techniques similar to many cryptosystems today, including the concept of an initialization vector, or starting position for the encryption, and basic error detection, which was accomplished by sending the initial message settings twice.

Decrypting the ciphertext based on this cipher required the recipient to know how the wheels were set when the message was encrypted. Thus, protecting the codebooks for Enigma devices was a critical part of the security of the Enigma encryption scheme. As the war went on, Germany added wheels and would issue Enigma devices with replacement wheels, which could be inserted into the devices, increasing the number of potential combinations. Some late-war versions of German Navy Enigma devices used as many as eight wheels!

Cracking Enigma was a complex task because the number of potential combinations for each encrypted message was very large. Fortunately for the Allies, the way in which Enigma was used did have vulnerabilities, including the double sending of the initial setting and the use of a standard initial setting. The Allies leveraged initial breakthroughs in cryptanalysis to learn how to decrypt Enigma, but they needed to be able to more quickly crack messages as they came in. Poland attacked this using early electromechanical computing devices called *bomba*. After the invasion of Poland, both Britain and the United States developed their own *bombes*, which were faster electromechanical computers that could crack Enigma. The U.S. Enigma-cracking device could run a three-rotor run for an Enigma-encrypted message in as little as 10 minutes.

FYI

The substitution of one letter for another is known as a *substitution cipher*, which is a very basic symmetric cipher. Substitution ciphers that perform only a single substitution have been included in familiar children's toys for decades in the form of secret decoder rings and other cereal box toys. Enigma's multi-round substitution cipher—and additional variables due to other settings—increased the potential substitutions. Each additional sequential wheel resulted in 26 times more possibilities. This increase in possible outputs remains a goal when attempting to make encryption stronger, but modern cryptosystems typically achieve it by increasing key length in bits.

In the end, the intelligence gathered from decrypted Enigma-encrypted messages allowed the Allies to sink German ships, to monitor German military research developments, and to better understand whether their operations against Germany and the Axis were successful. This use of early electromechanical computers was the predecessor for today's computer-based cracking of cryptosystems. It also demonstrated the power of cracking your enemy's encryption without your enemy being aware of it.

Asymmetric Ciphers

The problems with symmetric ciphers resulted in a desire to create a solution that would allow for a scalable key distribution system that could distribute keys safely and that could prove who senders were. Thus asymmetric key algorithms, also known as *public key algorithms*, were created. **Asymmetric encryption** relies on key pairs with a public and a private key used in the encryption algorithm. Each key is typically a very large prime number. The encryption algorithm's strength relies on the fact that determining which very large prime numbers were used in the algorithm is very difficult.

> **technical TIP**
>
> Identifying the two large prime numbers that were multiplied to create a final number is known as *factoring*. Some of the basic strength of current asymmetric encryption schemes comes from the difficulty of factoring the product of two large prime numbers.

If you compare the features of asymmetric cryptography with the issues of symmetric cryptography, you'll see that asymmetric cryptography handles them quite nicely:

- **Public key encryption** can provide all four critical requirements: confidentiality, message integrity, authentication, and nonrepudiation.
- Key distribution can be handled by simply distributing the public key for each user. Because the public key is intended to be distributed and cannot be used except to encrypt a message to that user, it's safe to send or even to publicly distribute online.
- Unlike symmetric keys, asymmetric keys only require twice the number of keys as the number of users in the system, because each user needs a public and a private key. New users only have to publish their public key. As shown in Table 10-2, asymmetric ciphers scale well.
- Users can be removed easily because they have no knowledge of a shared key. Removing them simply requires revoking their keys.
- Compromise of a private key requires generating a new key pair and distributing the public key again. Because it's safe to distribute, the only problem is getting it into the hands of all users who need it.
- Distribution of keys doesn't require some form of existing secure communication. Because you send a public key, you can exchange keys safely without pre-existing

Table 10-2 Asymmetric cipher scaling.

NUMBER OF USERS	UNIQUE KEYS
1,000,002	1,000,004
1,000,003	1,000,006
1,000,004	1,000,008
1,000,005	1,000,010
1,000,010	1,000,020
1,000,100	1,000,200
1,001,000	1,002,000
1,100,000	1,200,000
1,000,000	2,000,000

knowledge. This makes public key encryption ideal for online encryption, such as during online shopping, where the participants have likely not encountered each other before and have no pre-existing way to securely communicate.

Figure 10-2 shows the basic flow of an asymmetric encryption process. User A, the sender, encrypts a message using User B's public key. User A sends it, confident that only User B can decrypt the message. Upon receiving the message, User B uses his or her private key to decrypt the message. User B can then read the message.

The question of how to prove that a given public key is owned by the individual who claims to own it can appear challenging. Fortunately, solutions to this problem have been designed, too. Users sign a public key using their private key. The signatures are verified using those individuals' public keys, just as with messages. This means that trust is created by verifying that users whose keys are trusted have signed a public key. Therefore, the public key likely does belong to the person whom it appears to belong to.

In Figure 10-3, note the difference in the encryption process used by User A. User A uses his or her private key to digitally sign the message. This is something only User A can do because User A is the holder of the private key. Once User B receives the message, he or she can use User A's public key to verify the signature. This is done using a process similar to what was used to decrypt the message with his or her own private key. This process results in a message that meets all four of the desired capabilities of a cryptographic system: confidentiality, integrity, authentication, and nonrepudiation:

- Confidentiality is ensured because the message is encrypted.
- Integrity is provided by hashing the message and providing a means of proving that the hash hasn't changed.

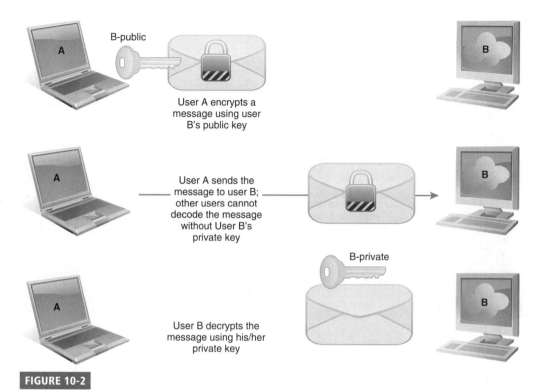

FIGURE 10-2

Asymmetric encryption systems use a recipient's public key to encode messages and the recipient's private key to decode messages.

- Authentication is provided by the digital signature because only User A has the private key.
- Nonrepudiation is provided by the same method: No other user could have sent a message as User A because no other user should have User A's private key.

Figure 10-3 shows how nonrepudiation can be provided in an asymmetric cryptosystem by using digital signatures. User A's private key provides the proof that User A was the one who sent the message.

FYI

Key-handling practices for asymmetric encryption systems are critically important. Often users who are unfamiliar with asymmetric key systems will accidentally send the private key, or both keys, to someone they want to communicate with. Once that happens, the private key should be considered compromised, and a new **key pair** must be generated. This means that the way you store a private key is really important, too. If attackers get access to private keys, they can send whatever messages they want as that user or service. Inadvertent release of keys allowing attackers to take control of accounts or services is a common problem for many organizations with cloud infrastructure, emphasizing the need for good key management practices.

10

Cryptography and Cyberwar

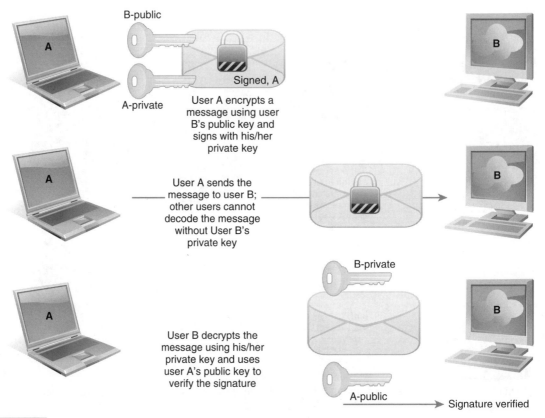

FIGURE 10-3

Users can use their private key to digitally sign a message, allowing recipients to verify their identity using the sender's public key.

This entire system relies on the safety of the private keys of the individuals involved. It's normal in cryptographic systems like this to treat private keys with great care and to require a strong passphrase to use the private key. If keys are lost or are exposed, trust in the system is immediately lost. In cyberwar, the obvious strategy is to acquire private keys if possible and to ensure that the target does not know that it has lost control of the private keys.

Steganography

Steganography, or the practice of concealing a message inside another message, picture, or file, is a topic that consistently comes up when discussing terrorists and insurrectionists who are attempting covert communication. Encrypted messages themselves are difficult to conceal, as ciphertext can be easily distinguished from normal plaintext. Thus, the ability to hide a message inside normally innocuous traffic is desirable.

Much of the focus on modern digital steganographic techniques has involved concealing data inside images and other digital data. Specialized steganographic tools have been created that can add data to images, and of course, detection tools known as *steganalysis tools* were created that can detect that same data when they have been added. For years, rumors abounded that terrorist groups were hiding messages and attack plans in images posted to their websites and sent via email.

Steganography isn't limited to just hiding text in images, however. Techniques for hiding encrypted messages in comments on websites, inside sound files and voice-over-IP conversations, and even simple ideas like hiding single frames in video or using text the same color as the background can all be considered a form of steganography. ISIS and al-Qaeda members were found to be using steganographic techniques in 2015 to hide messages in eBay listings and Reddit posts, and the use of steganography continues elsewhere today.

Identifying files with steganographic data concealed in them can be difficult without access to the original file. Well-done steganographic techniques can appear like random noise or corruption in a file, and because the data are encrypted, it can appear truly random.

Steganographic techniques aren't limited to cyberwarriors and information security practitioners. In 2009, *World of Warcraft* (WoW) players discovered that screenshots created by the WoW client embedded data, including the user's account name, server IP address, and a timestamp. Players were concerned, and rumors spread that Blizzard, the company that runs WoW, would use the data hidden in screenshots to punish players who took and published screenshots of in-game exploits.

Modern Cryptosystems

A number of cryptosystems are in widespread use today, and each of them has a role in modern cyberwarfare:

- DES, one of the earliest broadly adopted electronic encryption systems
- 3DES, its later upgrade
- AES
- RSA
- A number of other competitors that are available for programmers and system builders

Although the technical details of these algorithms' implementation are beyond the scope of this text, a basic understanding of the algorithms that are commonly used is helpful when exploring their use in cyberwar.

Data Encryption Standard

The Data Encryption Standard (DES) is a symmetric key cipher that was submitted in response to a U.S. government search for an algorithm to protect sensitive but unclassified electronic data. It was adopted in 1977 with U.S. National Security Agency (NSA) approval and was quickly adopted for systems worldwide. Unfortunately, DES used a relatively short 56-bit key and contained classified design components, leading to suspicion that the NSA was involved in its creation.

DES was broken in 1999 when groups working together were able to break a DES-encrypted message's encryption in less than 24 hours. DES has been considered outmoded for years, but it remains widely available and continues to be used in some software and systems.

Triple DES

Triple DES (3DES) was created using the pre-existing DES algorithm as a basis. 3DES works around the short key length of DES by using multiple rounds of encryption using the DES algorithm. This increases its key length to a maximum of 168 bits if run with three rounds using different keys. Unfortunately, due to the manner in which attacks against it are conducted, it's effectively only a 112-bit key.

---

technical TIP

You might expect that running DES three times would result in an effective 168 bits of security, but an attack known as the meet-in-the-middle attack means that attacks require less work to complete an attack against the full list of potential keys. Attacks like these can mean that the expected protection provided by a cryptosystem is far less than expected. For 3DES, this means that it effectively has 2^{56} fewer possible keys for an attacker to try.

Because DES was already commonly available and broadly implemented on a variety of systems, 3DES was a relatively simple upgrade for most software and hardware developers. Obviously, using 3DES instead of DES requires three times the amount of computational work, but DES is a relatively fast algorithm, which allowed 3DES to be a practical solution to the issues with DES at the time.

FYI

The U.S. federal government defines how cryptographic modules are accredited for modern cryptosystems in the Federal Information Processing Standard (FIPS), and comply with either the older FIPS 140-2 standard, or the newer FIPS 140-3. FIPS defines four levels of security from Level 1, which provides basic encryption capabilities like an encryption chip in a personal computer, up to Level 4, which has enhanced security features and strong security mechanisms that detect and prevent physical and other attacks. FIPS testing is provided by the U.S. National Institute of Standards and Technology (NIST) and the Canadian Communications Security Establishment.

Advanced Encryption Standard

In October 2000, with the highly publicized cracking of DES still fresh in the minds of security practitioners, NIST announced approval of a replacement for DES. The replacement, known as the Advanced Encryption Standard (AES), uses the Rijndael cipher and was approved for use for all sensitive but unclassified data, much like its predecessor DES had been.

AES allows three key strengths, 128 bits, 192 bits, and 256 bits, allowing a balance between the amount of processor time required to encrypt a file and the security desired from the cipher. Like DES and 3DES, AES is also a symmetric cipher. AES is now commonly used in many areas and has largely but not completely supplanted DES and 3DES.

RSA

Unlike DES, 3DES, and AES, the RSA algorithm created by Ron Rivest, Adi Shamir, and Leonard Adleman in 1977 is an asymmetric encryption algorithm. As you learned earlier, asymmetric algorithms provide a number of distinct advantages over symmetric algorithms, including authentication, nonrepudiation, and the ability to safely distribute keys without a pre-existing secure channel. RSA also has the advantage of allowing arbitrarily large key lengths, thus allowing users of RSA encryption to increase key length to keep encrypted data secure for longer against possible brute-force attacks. RSA, much like DES, 3DES, and AES, has been implemented on a wide variety of platforms and systems. RSA is the most common asymmetric encryption algorithm in use at the time of writing.

FYI

One important concept for cryptographic systems is Kerckhoffs's principle, which states that a cryptographic system should remain secure even if the enemy knows everything about the system except the key. Although not all cryptographers agree, exposing a system to third-party review can help ensure that a malicious actor hasn't compromised your cryptosystem by inserting a critical flaw in its design. All four of these algorithms have been broadly studied, although as noted earlier, DES was criticized for its NSA ties and secret elements when it was released.

Hashing and Message Digests

Earlier in this chapter, you learned about the need to ensure that messages haven't been modified when using a cryptosystem. The function typically used to do that verification is a *cryptographic hash*. Hashes take a message and generate a unique output value based on the message. The output of a hash is often known as a *message digest*.

Senders use message digests when they want to prove that their message has not been modified. To do so, they hash their original message and then send the hash with the message. If the hash sent with the message matches the hash calculated by the recipient, the message has not been modified.

10

Cryptography
and Cyberwar

Hashes have five basic requirements:

1. They must accept inputs of any length.
2. Their output must be a fixed length.
3. They should be easy to compute for any input data.
4. They should not be reversible. Thus, you should not be able to calculate an original message from a hash.
5. They should not have collisions, meaning that they should not generate the same hash for different inputs.

A number of hashing algorithms are commonly used. Two of the most common are MD5 and SHA. Both are in common use; however, MD5 has been shown to have a number of significant flaws, including the potential to create arbitrary collisions, or inputs that hash to the same value as other inputs. SHA-1, the original version of the SHA hash, has also been shown to have problems. Due to these vulnerabilities, a newer version of SHA known as SHA-2 is the preferred solution for most secure purposes today.

In addition to their uses for verifying message integrity, hashes are also commonly used for digital signatures in asymmetric cryptosystems. To digitally sign a message, the sender simply hashes the message and then encrypts the message digest with his or her private key. The sender then adds the encrypted message digest to the plaintext message and sends it. This proves that the sender truly is the sender and that the message has not been modified, as it will hash to the correct value once it's decrypted with the sender's public key.

technical TIP

A common technique to make applications that use hashes to store passwords more secure is *salting*. A *salt* is a value added to the data that are hashed. The salt is kept secret and stored securely. This means that if the hashed password list is stolen, attackers can't simply compare the hashes with a list of known password hashes. The salted hashes will be different than those of the same text simply run through the algorithm.

A second salting technique that is sometimes used is a *pepper*. A pepper is a secret like a password added to a salt to increase its resistance to brute force attacks. Peppers are typically kept in a separate secure location to reduce the likelihood of an attacker capturing both at once. If you had to choose just one, salts are the better option, but a pepper can also help.

The Hash Life Cycle

Hashing algorithms are vulnerable to attacks that create collisions. A collision occurs when two strings of text hash to the same hash value. If attackers can manage to do this with arbitrary strings they have chosen, the hash is broken and cannot be trusted. Because hashes are often used to verify that files haven't changed, this means that attackers who can create an arbitrary hash can replace trusted files with malware without any warning to users who responsibly check the file's hash. Obviously, if a hash can be replaced with data that an attacker chooses, the hashing algorithm should be replaced.

Research into how hashing algorithms can be defeated is often made public, and thus you typically have a reasonable idea when a hashing algorithm is first starting to be vulnerable to collisions or other attacks. Once you know that an algorithm has problems, it would make sense to start to retire the algorithm and replace it with a new, secure algorithm. That doesn't always happen due to the cost of replacing infrastructure or simply because of bad habits or lack of knowledge on developers' part.

Even with problems like long-term replacement costs, you might think that organizations would replace an algorithm that has had known problems for almost 20 years, like MD5. Unfortunately, not every organization makes wise choices about which algorithms they choose to use or how they use them. The MD5 hashing algorithm has known flaws that allow the creation of arbitrary collisions. Perhaps worse, tools called *rainbow tables* exist that allow a speedy lookup of the original value of text that generates many hashes. Unfortunately, because MD5 is widely available, fast, and easy to implement, many organizations have continued to use MD5 to store hashed passwords. When attackers capture their databases and discover MD5-hashed passwords, they are able to quickly determine the original text of those passwords and then use the email addresses and passwords they have stolen to attempt to log on to other sites.

Fortunately, stronger alternatives exist, including hashing algorithms specifically designed for passwords. They are designed to be resistant to attacks, including having higher computational costs, which are problematic when hashing large numbers of values like a password attack, but with reasonable time frames, are good at the speed at which passwords are typically used. Using a password hash, salts, and other techniques are published best practices and easily accessible, but organizations continue to experience breaches due to using poor practices.

Cryptography in Cyberwar

Now that you have a basic understanding of the underlying concepts and technologies used in cryptographic systems, you can look at their uses in computer network defense and computer network attack scenarios.

Computer Network Defense and Cryptographic Systems

The strategies and technologies used in computer network defense are typically the same types of technologies used in a traditional information security defense-in-depth plan. Layers of defensive encryption-based technologies are deployed to ensure that data are exposed as little as possible and to make sure that keys are properly protected. Among these technologies are the following:

- **Drive encryption** is encryption either for full drives or for volumes (drive partitions). Drive encryption has become increasingly important and increasingly common, not only for computers but also for mobile devices like phones, tables, and even wearable devices, because it helps to protect against attackers who manage to access the machine while it's out of its owner's control. A number of strategies are available to implement drive encryption—including those based on hardware devices built into the drive or the computer, as well as software-based encryption capabilities.

- **File encryption** is encrypting single files or groups of files. This common solution is used particularly when files need to be transferred or stored on unencrypted media. File encryption is built in to a variety of products, including AES encryption in modern versions of Microsoft Office.

- **Digital certificates** are electronic documents that use a digital signature to link a public key to an individual, organization, or system. Digital certificates are widely used and trusted based on the processes used to verify the certificate holder. The authenticity of this relationship is checked by a **registration authority (RA)**, the part of a public key infrastructure that registers new users. Certificate-based authentication is also commonly used for cloud services as well as for local environments and is supported by security tokens like Yubikeys, Titan keys, and other hardware solutions. These devices provide secure storage and use of certificates to identify individuals and make it harder for attackers to rely on a simple credential compromise to gain access to infrastructure and services.

- A **public key infrastructure (PKI)** is a set of hardware, software, practices, and people that creates, distributes, stores, manages, and revokes digital certificates. PKI is used to ensure that individuals and systems who request certificates are who they claim to be. PKI systems then provide a chain of trust between a certificate issuer known as a **certificate authority (CA)**, a registration authority, certificate requestors, and those who must accept the certificates. PKI also includes the ability to revoke a certificate, which ensures that a stolen or compromised certificate can have its trust removed.

- Transport Layer Security (TLS), which is the replacement for Secure Sockets Layer (SSL), is the encryption that is typically used for secure web transactions. TLS uses asymmetric cryptography to perform a secure exchange of symmetric keys, combining the abilities of an asymmetric cryptosystem with the speed of a symmetric cryptosystem. It also relies on PKI and x.509 certificates to verify trust relationships between certificates used for websites and their owners.

- Virtual private networks (VPNs) rely on encryption when they want to protect the data they are sending in addition to simply creating a private network inside of another network. VPNs allow users on untrusted networks to become part of a trusted network and can allow users and systems to safely transfer data through those untrusted networks in a secure way.

- Hashing, which is commonly used to verify that files remain unchanged, is part of cryptographic systems for digital signatures and used to ensure integrity.

CAC Cards

The U.S. Department of Defense (DOD) uses a Common Access Card, or CAC card, for identification of both military and civilian staff. The card combines a picture ID with a chip that provides an encrypted certificate, which is issued by a federal public key infrastructure.

When a CAC card user wants to log on to a system or access a resource that supports use of the card, the user inserts the CAC card and then enters a personal identification number (PIN) number. The PIN information is stored on the CAC card, allowing it to be verified. If the PIN matches, it checks the user's identification number against a central directory, which matches the user with his or her rights.

CAC cards combine encryption technology with other authentication and user identification capabilities. This means they're an important part of the DOD's security design. It also means that stealing a CAC card and acquiring the PIN that goes with it could provide access to attackers.

All of these technologies rely on the algorithms, software, and hardware they are based on not being compromised. If attackers can successfully insert themselves earlier in the design process, they can attack in ways that most defenders cannot reasonably defend against because they don't have the resources or technology to assess underlying systems and algorithms.

Computer Network Attack and Cryptographic Systems

Many of the techniques used in computer network attacks focus on defeating the cryptographic systems deployed by defenders. Because of this, you can review the same technologies used for defense when looking at how to conduct attacks:

- Drive encryption is vulnerable to a number of attacks. Vulnerabilities include social engineering to persuade users to provide their passwords, acquiring passwords through the use of hardware keyboard capture tools, and attacking while the user is logged on with data unencrypted. In addition to these attacks, more-advanced attacks require physical access to the computer but can retrieve passwords from live memory even if the machine is encrypted. As with most systems, social engineering and human attacks are often the best way to bypass technical controls like encryption.

- File encryption is vulnerable to the same attacks as drive encryption, but files can be easier to acquire as they are often sent via email or other file transfer services and can be accessed from unencrypted drives.

- Digital certificates can be attacked by falsifying credentials or persuading a certificate issuer that the attacker is another person or organization. This has been leveraged multiple times by attackers, including well-publicized attacks that have created Google.com and Microsoft certificates. In each instance, the certificates were real certificates but had been issued to parties who were not the actual claimed owners.

10

Cryptography and Cyberwar

- PKI has many parts that must work together successfully to remain secure. One successful attack against a Dutch certificate authority called DigiNotar offers an excellent example of a successful attack against PKI. Attacks against certificate authorities as well as flaws in their technology and practices are relatively rare but are also not unheard of. In 2011, DigiNotar discovered that hackers had been able to issue more than 500 fraudulent certificates through its service. This scale of compromise resulted in DigiNotar being removed from the list of trusted certificate authorities by major browser vendors. As a result, DigiNotar went bankrupt because it was no longer able to function as a business. The threat was especially severe for the Dutch government, as DigiNotar was trusted with the ability to issue certificates for the Dutch government's PKI. In 2015, Symantec was discovered to have issued more than 100 test certificates for 76 different domains without the approval of the domain owners. It was eventually disclosed that Symantec was not following industry best practices for issuing trusted certificates, and Symantec was dropped from many major platforms due to this and other behaviors.

- TLS attacks are typically conducted by using fake digital certificates. The DigiNotar attack was used to create a rogue Google.com certificate, issued under false pretenses, that was trusted by users. That certificate was used for a large-scale attack in which the intruder pretended to be Google, allowing the intruder to intercept secure communications meant for Google.

- VPNs are most often attacked by intercepting data before they are encrypted or by compromising the systems that host the VPN service. As with most encryption, getting to the data before they are encrypted is often the attacker's best solution.

- Hashing attacks are typically done through attacks on the hashing algorithm. In 2005, attackers launched attacks that allowed fake certificates to be created that had the same MD5 hash as the original certificates. Defenders must ensure that they don't use outdated cryptosystems that allow attackers to easily defeat their defenses.

Attacking Cryptography

Attacks against cryptographic systems are conducted in a variety of ways ranging from brute force attacks to stealing or acquiring keys or performing cryptanalysis—the process of finding a weakness in the protocol or key management scheme used for the algorithm or service. Cryptanalysis techniques include analyzing encrypted data to attempt to find patterns, finding vulnerabilities in the underlying cryptographic systems, and even simply trying every possible combination in a cryptosystem to decrypt the message.

The attack type chosen is often based on how much information is available to the attacker. That information is normally classified in one of five ways:

- Ciphertext only, in which the attacker has only one or more encrypted messages
- Known plaintext, where the attacker knows what the encrypted message was, and knows which ciphertext it matches to

- Chosen plaintext, where the attacker is able to choose the message that is encrypted, and thus can see the effect on the output in the ciphertext

- Adaptive chosen plaintext, in which the attacker can choose each successive plaintext, allowing the attacker to study changes in plaintext as messages are modified

- Related key attacks, which rely on analyzing messages sent with different, but related keys and similar to those used in Enigma intercepts, where the daily setting remained the same, but the operator setting changed for each message

The availability of strong encryption using publicly available algorithms like RSA and AES that have few, if any, currently known critical flaws means that attacks against them tend to occur before encryption is used. Or attackers have to target the keys used to encrypt the data to recover it. However, that doesn't mean that attackers don't attempt assaults against encryption. Both attackers and defenders in cyberwar must consider the common attack techniques that are used against cryptosystems. The following sections look at some of the most common attacks against encryption when only ciphertext is available.

Brute Force

A brute-force attack is in many ways the simplest attack that can be conducted against an encrypted message. Brute-force attacks simply attempt to use every possible key against an encrypted message until the original plaintext is recovered. Thus, brute-force attacks are sometimes called *exhaustive key searches*.

The problem with brute-force attacks is that brute forcing strong encryption requires an incredibly large number of tries to be successful. Tries can be made faster by using faster computers or specialized techniques, but a brute-force attack against a 4,096-bit, RSA-encrypted message could take years to succeed, even with supercomputers available to attack it.

> **NOTE**
>
> In World War II, the Japanese Purple Machine, which was very similar to the Enigma device, was targeted by the United States. The attack on the Purple Machine was made significantly easier by the Japanese custom of using the same message format consistently, allowing a known plaintext attack.

Acquiring the Keys

One of the best ways to attack encryption is to acquire the keys. You can acquire keys in a number of ways, but following are the most common:

- Social engineering the key owner into providing the key or using force or threats to convince the holder of the key to surrender it

- Attacking the system that holds the keys

- Using attacks against the key-generation system, producing less-random keys or revealing the keys as they are generated

- Exploiting a flaw in the encryption software

Once a system is compromised, any keys that it contains should be considered compromised. Many organizations make the mistake of storing keys on systems that are connected to the Internet or leaving keys in locations that are more accessible than intended.

10

Cryptography
and Cyberwar

Revealed in April 2014, the Heartbleed bug provided an excellent example of a way to acquire keys by exploiting a software flaw. Researchers at Codenomicon and Neel Mehta of Google Security discovered the flaw that came to be known as the Heartbleed bug in OpenSSL, a widely used open-source implementation of TLS. When exploited, the bug allowed attackers to ask the server for a "heartbeat" response and permitted the attackers to ask for a larger response than the query they sent. The response that was sent back included the contents of the memory allocated to the OpenSSL application. In addition, the response could include the secret keys used for encryption between systems as well as the decrypted traffic sent between users and servers.

The flaw itself was inadvertently added to the underlying code that makes up OpenSSL two years prior to its discovery, and OpenSSL was used by many web servers around the world, including by Google, Amazon, and other major cloud service providers. Any organization that was able to identify and exploit the Heartbleed bug before it became public knowledge could have gained access to those sites using captured usernames and passwords or could have pretended to be the sites themselves by using its secret keys.

Attacking the Algorithm

Encryption algorithms can be seriously weakened if parts of the cryptosystem have flaws. In a typical modern cryptosystem, you will often find a random number generator and a number of distinct parts of the algorithm. In addition, the encryption system is typically implemented as either hardware, software, or a combination of both. This provides attackers with a number of places to insert flaws. This type of attack isn't just theoretical, as you will read in the following section.

NSA and RSA

Attacking the cryptosystems after they already exist isn't the only way to succeed. If an attacker can ensure that the cryptosystem is flawed before it enters general use, the attacker can exploit it without the huge amounts of effort that it usually takes to find flaws in a cipher. This is a significant danger for cryptosystems and one reason that the most trusted encryption algorithms are often those that have had broad public review to help ensure that they don't have flaws lurking inside them. Unfortunately, cryptographic systems are also incredibly complex, and the number of individuals who can fully understand them is relatively small. Minor changes in how an algorithm is constructed or how it's implemented can have a major impact on how secure it is.

In 2013, Reuters reported that the NSA had paid RSA, a security company that specializes in encryption technologies, $10 million to create a cryptographic system that included a deliberate flaw. The flawed Dual Elliptic Curve cryptographic system was part of RSA's BSafe security kit, and the Dual Elliptic Curve system's vulnerability is reported to have made it possible for the NSA to crack it. Later reports suggested that that vulnerability was not the only one inserted into the software. In fact, in 2014, researchers alleged that the Extended Random extension, which was proposed by the NSA, would have simplified attacks on Dual Elliptic Curve. The NSA didn't find success with Extended Random, however, as it was removed from RSA's software.

Of course, ensuring that the cipher itself has a weakness isn't the only way to defeat a cryptosystem, and the NSA isn't the only organization to have successfully provided a weak cryptosystem to organizations it wanted to monitor. After World War II, the Allies sold captured Enigma machines to developing countries. The fact that the Allies had the ability to decrypt Enigma messages was not public knowledge, and thus customers in those countries, including national governments, unwittingly used a broken cipher to protect their secrets.

 NOTE

Cryptosystems can be attacked many other ways. This section has explored just a few to provide a basic understanding of the threats defenders face in computer network defense and that attackers may try in computer network attacks.

Defeating Attacks on Cryptographic Systems

Attacks on cryptographic systems occur during the creation of the systems, during their implementation, and most often, when the system is in broad use. There are as many ways to defend cryptographic systems as there are possible attacks, but most defenses focus on ensuring that the system is well designed and that its component parts themselves aren't vulnerable. A well-designed, strong cryptosystem that has been carefully tested and whose supporting infrastructure has not been compromised can usually be considered safe.

Defenses

Defending against attacks on cryptographic systems requires that defenders think like the attackers they might face. Attackers will attempt to make the cryptosystem less effective, they will search it for flaws that will make it easier to crack, and if they have no other options, they may simply attack using a brute-force attack.

Organizations that rely on widely available ciphers and hashes have to watch for news about problems with them. Those organizations must have plans in place to replace or upgrade compromised ciphers and hashes when they begin to have problems. The issues that MD5 and DES have encountered over the past 20 years have shown that many organizations don't handle the transition from one technology to another well and that they often leave themselves vulnerable.

In addition to building a plan to handle the ciphers themselves, organizations that engage in computer network defense need to carefully consider where they use encryption.

Defense in Depth Using Cryptographic Systems

Cryptographic systems are often deployed as part of a defense-in-depth strategy. Figure 10-4 shows common uses of encryption in a typical computer network defense scenario. Note that the workstation uses full disk encryption, files sent via email are encrypted, and web traffic sent from the workstation is encrypted in transit over the wire by TLS.

The use of encryption in defense-in-depth systems provides defenders with the ability to pick appropriate levels of encryption technology to match the sensitivity of data, the resources required to provide the encryption, and the risk to the data and systems. Thus,

10

Cryptography
and Cyberwar

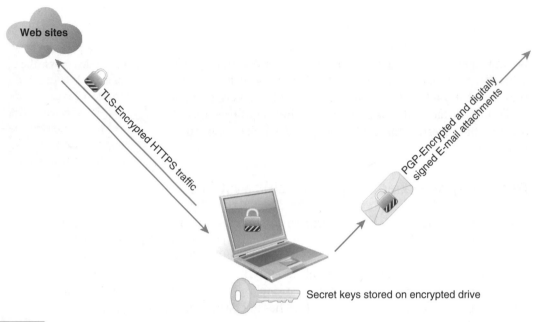

Web sites

TLS-Encrypted HTTPS traffic

PGP-Encrypted and digitally signed E-mail attachments

Secret keys stored on encrypted drive

FIGURE 10-4

Encryption uses for a typical modern laptop, including secure web access, e-mail, and full drive encryption.

a highly sensitive file might be kept encrypted, stored on an encrypted drive, and only transferred via a trusted, encrypted virtual private network connection. A lower-sensitivity file might be left unencrypted on the drive, might only be protected at rest by full disk encryption, and might be protected in transit by normal TLS used via a web browser.

Weaponizing Cryptography

The ways in which encryption is used in modern malware is a useful model to examine when considering how encryption techniques can be used in both offensive and defensive cyberwar activities. The following sections look at three major examples of malware that uses encryption. First, you'll examine the long history of using encryption to evade detection by defensive technologies. Second, you'll learn about the Zeus banking Trojan, which uses encryption to avoid detection and to keep its configurations secret. Finally, you'll read about Cryptolocker, an infection that encrypts data using modern asymmetric encryption and then extorts a ransom from users who need their data back.

Defensive Cryptography: Malware Encryption

Malware is one of the most prevalent threats in modern cyberwarfare. Stuxnet, Flame, and a host of other targeted threats have infected systems and provided backdoor Trojan access to systems. Malware has provided intelligence by leaking data, allowed infrastructure attacks, and acted as a remote-control system to take over other more critical systems.

Of course, malware wouldn't be effective if it could be easily detected and removed. For years, viruses and other malware were relatively easily detected by antivirus systems that could keep a known fingerprint of the malware and then remove it. Malware authors responded by making malware that changes its fingerprint every time it lands on a new system, but they have also made use of encryption to protect their malware.

Typically, encryption is used in three ways in modern malware:

- To encrypt the malware itself when on disk and running on systems. This helps to ensure that defenders cannot easily reverse engineer the malware to determine what it's doing or who built it by looking at its code.

- To protect the configuration files that the malware uses. This means that defenders cannot verify which targets, control systems, and other settings the malware is configured with. When defenders get access to that information, it can help them determine who the attacker is or which version of the malware they are fighting against.

- To protect the network communication of the malware. This provides concealment, as defenders theoretically cannot read encrypted traffic. This allows the malware to appear like any other encrypted web traffic. It also makes sure that defenders don't know what the malware is doing. If they could view the traffic, they could see command-and-control data in real time, possibly letting them know of attacks or other activities.

Cryptography isn't used only for defensive purposes. It can also be used as a weapon by applying the same technologies in ways that cause harm.

Offensive Cryptography

Encryption is increasingly used as part of the weapons packages deployed in cyberwarfare: viruses, Trojans, and other malware. Advanced malware packages now commonly include strong, openly available encryption capabilities that allow them to conceal their communications, protect their configurations, and avoid malware scanners by encrypting their internal workings. The following sections examine how malware packages that use encryption defensively and offensively have evolved. First, you'll look at Zeus, a malware package originally aimed at banking and other sensitive data but was updated and made broadly available as a configurable malware package, allowing attackers to easily deploy it without significant technical know-how. Next, you'll look at how Cryptolocker used strong RSA encryption to extort money from infected users, using encryption as a weapon and setting the stage for modern ransomware like Maze, Ryuk, Tycoon, and NetWalker.

Zeus

Cyberwarfare technologies and techniques often cross over into use by criminals, and the techniques developed by malware writers are often of interest to computer network attack developers. The Zeus Trojan is an excellent example of malware demonstrating new techniques that are of interest to both attackers and defenders while continuing to evolve over a long period of time. Zeus and similar Trojans are also of interest to governments due to their need to provide computer network defense and due to their broad deployment and usage as a remote access and data gathering tool. In 2007, when Zeus was first discovered, it was used to target data at the U.S. Department of Transportation as well as multiple commercial entities.

FYI

Viruses and other malware often have a variety of different names, as each antivirus vendor tends to name them differently based on information they discover about the malware or based on their own naming conventions. Thus, Zeus is has also been called Trojan.zbot, ZeuS, and Zbot, and various Zeus variants have other names like Zeus Sphinx that we discuss next. If you encounter malware, remember that simply searching for one name may not provide all of the information you need. Amusingly, even criminal malware uses aliases, even if inadvertently.

---

technical TIP

In addition to the Zeus Trojan, the Citadel, Carberp, and SpyEye Trojans were widely deployed. Each of these packages targeted user data, such as credentials, credit card information, or other similarly important information. Each of these packages also evolved over time, with features like plug-in architectures, which allowed the malware to adapt to the system it's on, built-in anti-detection technologies to avoid being noticed by antivirus software, and the ability to attack multiple operating systems and platforms. Carberp's claim to fame is that it was the first malware to use a randomly generated key rather than a static key that the creators of the software kept centrally.

Zeus started as a sensitive data theft tool specializing in the gathering of stored usernames and passwords as well as those that users type into systems on which it resides. Zeus and other banking Trojans like it are specially designed to recognize high-security sites that require multiple factors to log on and to use plug-ins designed for that site to capture the logon information. Zeus, however, is highly modular and has evolved significantly since its release. In fact, the creator of the Zeus Trojan treated it like commercial software, making licenses available for purchase for a few thousand dollars with a recurring subscription rate for updates before their retirement.

All of this isn't atypical for malware. However, the ways in which Trojans and other malware use encryption as both a defensive and an offensive tool means that the techniques previously used to protect data are being used to exploit systems and to protect malware. Because cyberwar tools are often similar to those used for network security exploit by non-cyberwarfare participants, it is reasonable to presume that the same techniques will show up in any reasonably competent cyberwarfare attack. Obviously, defenders must attempt to defend against these techniques.

FYI

We wrote about Zeus in the first edition of this book and expected it to have been replaced by other malware by the time this edition was published. Instead, Zeus remains active, remaining in the top 10 malware packages in the world. That's an incredibly long life span for a malware package and points to ongoing development, deployment, and usage.

Figure 10-5 shows an example of a Zeus infection and how it uses encryption during the infection process to help ensure that its configuration information remains secure from defenders. Zeus uses two primary infection methods: drive-by downloads, which use operating system and browser vulnerabilities to install Zeus when unsuspecting users visit sites where Zeus is active, and via spam and social media efforts that start with an email attachment or link to .zip file. Users click on the .zip file, unwittingly activating a downloader, which downloads and decrypts the Zeus executable. Once Zeus is live, it then reaches out to command-and-control servers for an encrypted configuration file.

Perhaps one of the most interesting turns in the story of Zeus is the alleged retirement of its author in 2010. After the announcement, the author released the Zeus source code

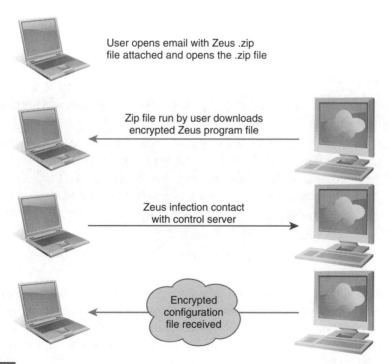

User opens email with Zeus .zip
file attached and opens the .zip file

Zip file run by user downloads
encrypted Zeus program file

Zeus infection contact
with control server

Encrypted
configuration
file received

FIGURE 10-5

A sample Zeus infection process starts when an unsuspecting user opens a .zip file sent by an attacker, infecting his or her workstation. The Trojan uses encryption to hide its installer and its configuration files.

and transferred the rights to sell Zeus to the creator of the SpyEye Trojan. Zeus source code became commonly available via many sites, allowing would-be attackers to gain easy access to a capable Trojan. In cyberwarfare terms, this means that the Zeus could be and has been weaponized by anybody with the ability to download and configure it from an easily accessible software repository, and the ongoing use of Zeus for more than a decade shows how potent that capability is. At the time of the writing of the second edition of this book, the Zeus Sphinx variant was being used to take advantage of unsuspecting users by leveraging COVID-19 worries—yet another new way to use a very old Trojan!

> **NOTE**
>
> Cryptolocker relied on 2,048-bit RSA encryption as well as AES-256 encryption keys for each file. Further, Cryptolocker actually informed the infected user that it uses RSA, ensuring that the user knows he or she cannot reasonably expect to decrypt the files on the drive. Even users who contact an expert will find out that the only way to get their information back without backups is to pay the ransom.

Cryptolocker and Other Ransomware Malware

Malware that encrypts files on a victim's hard drive isn't new, but the Cryptolocker malware was likely the first highly successful encryption-based malware. Cryptolocker used asymmetric encryption, generating a public/private key pair and sending the private key to a central command-and-control server where it's stored. It then encrypted all of the important documents, such as Microsoft Word, Excel, and other files on the victim's computer, and presents the user with a message demanding payment of 300 U.S. dollars or 300 euro within 72 hours.

Cryptolocker and Bitcoin

The Cryptolocker virus typically asked its victims to pay via common international money transfer methods within 72 hours.

As the price of Bitcoin went up, Cryptolocker's authors changed their payment methods to allow payment with Bitcoin. They relied on the relative anonymity of Bitcoin and demanded a half a Bitcoin in payment. At the time, Bitcoin was worth between $200 and $400, a significant increase from the previous value of less than $100, making it an attractive currency to request. Modern cryptographic malware continues to ask for Bitcoin; some versions also leverage payment gateways or other means of transferring funds.

If victims did not pay up within 72 hours, Cryptolocker deleted itself and the public key, leaving the encrypted files unusable. In fact, Cryptolocker's infection methods meant that when Cryptolocker servers were taken offline, victims could no longer recover their files. Reinfection of the system generated a new public/private key pair, meaning that the data were unrecoverable.

Ransomware-style attacks in cyberwar could be used to make critical systems unusable or as leverage in a negotiation. The same techniques used to encrypt sensitive data could be used for any valuable data, research, or communications. Once again, the techniques used in criminal malware attacks are likely to inform future cyberwar activities.

The Future of Cryptography in Cyberwar

Cryptography will continue to be a critical part of cyberwar. Encryption is the preferred way to protect live data at rest and in transit, and encryption capabilities are now built in to many products from the ground up. The increased use of encryption throughout data's life cycle means that attacks against encryption are necessary for attackers in cyberwar to succeed. It also means that defenders must account for changes in their use of encryption throughout data's life cycle. Physically securing a filing cabinet is no longer sufficient to preserve the security of data, and strong encryption placed on a file when it's created may be easily broken a decade or two later.

Attacks

The modes of attack on cryptosystems have changed as encryption has become increasingly strong. Modern encryption algorithms like RSA with a 2,048-bit key can reasonably be expected to remain secure for years, despite high-speed cracking efforts with massive supercomputers. Regardless of advances in technology, such as faster processors and graphics card–based cryptographic attacks, these algorithms remain strong.

Because strong and secure encryption algorithms exist, attacks continue to move to other locations in the encryption process. If data can be obtained while they are unencrypted, that is far easier than trying brute force to break strong encryption. Future attacks are likely to occur in two major places. Those areas are already the focus of attacks by both nation-state actors and others engaging in both cyberwar and traditional cyberattacks.

The first place that attacks occur against the modern encryption system is while the data are in use and unencrypted. Modern consumer-grade operating systems and devices often encrypt user data by default, making the data less likely to be accessed if the device is stolen, but they cannot leave the data encrypted while the user is accessing it. This means that attacks must target the user's processes and programs and active use of the device when the data must be unencrypted and accessible to the user. This drives a growth in targeted malware that obtains and sends unencrypted data using the user's rights.

The same large-scale adoption of encryption at rest and in transit makes attacks on the underlying algorithms and hardware that perform encryption more attractive. If a government or other organization can introduce a weakness into an algorithm or a hardware device or chip, it can use that vulnerability to more easily crack or possibly entirely bypass the encryption. Nation-state actors that do this create the potential for a widely adopted technology to have a critical flaw that they introduced. If that flaw becomes public knowledge, the impact is likely to be significant, as known flaws often result in broad exploit of the issue.

Defenses

Future defenses against attacks on encryption will continue to involve a few common solutions:

- Increased key length for existing strong encryption algorithms will be important. Algorithms without known vulnerabilities that can simply increase their key length can continue to outpace the speed of new cracking tools until a vulnerability is found. Even if a vulnerability is found, sometimes a modification of the system can be used to extend its life span, much like the evolution of DES to 3DES discussed earlier in the chapter.

- Open access to encryption systems for review will continue to be important for those who subscribe to Kerckhoffs's principle. Organizations are likely to continue to use internally developed encryption algorithms, but publicly available strong encryption will continue to benefit from exposure to the community.

- Increased oversight of the underlying design of hardware and software will be important as the NSA's alleged influence over the design of various RSA products as well as accepted cryptosystems demonstrates. Countries around the world will likely require greater oversight of new encryption standards to avoid a single country having access to a known flaw in an encryption system or one of its component parts.

- More integration of end-to-end encryption that makes accessing data more difficult for untrusted applications will be necessary. Malware that targets data that are unencrypted when viewed by the user of a system will mean that more data will be encrypted even when the user is logged on. There is a fine line between usability and security that will continue to move in sensitive systems.

Overall defense will remain a challenge due to the need to use data. Usability will almost always mean that there will be a moment when data are unencrypted and thus can be attacked.

Quantum Cryptography

One of the most anticipated and one of the most feared changes in encryption technology is quantum cryptography. **Quantum cryptography** uses the theories of quantum mechanics to perform cryptographic tasks like decryption and encryption. In theory, this allows a variety of possible advantages:

- Decrypting traditional encryption methods may no longer require huge time periods to accomplish using a quantum computer.

- Quantum encryption may be nearly impossible to break.

- Eavesdropping on key distribution would be detectable.

- Location-based cryptography may be possible, with the physical position being the only required key.

No quantum computers currently exist that can perform these functions. In fact, one estimate of how many qubits of quantum computational power are required to break modern RSA 2,048 bit encryption within a few hours places that number at 20 million qubits (or as few as 4,099 using Shor's algorithm, if the qbits were perfectly stable, which modern technologies can't manage). IBM announced that a 1,000 qubit quantum computer is expected to be ready by 2023, and D-Wave's specialized processors at the time of the publication of this book only have 1,100 qubits, which means that worrying about quantum attacks against encryption systems shouldn't be a high priority for some time to come. Despite this, ideas that quantum cryptography relies on will completely change cryptography as we know it if and when they become a reality. That change would mean that the role of encryption and cryptanalysis in cyberwar would also completely change, and traditional cryptosystems would become essentially useless.

CHAPTER SUMMARY

In this chapter, you learned about cryptography, the study of how to conceal messages from unwanted readers. You read about symmetric ciphers, which use the same key for both encryption and decryption, asymmetric ciphers, which allow safe exchange of public keys, nonrepudiation, and authentication, and signatures using hashes. You examined commonly used encryption and hashing algorithms and learned about how they are attacked and what drives their replacement.

You then used your understanding of cryptosystems to examine current cyberwarfare techniques for attacking and defending networks, systems, and data using encryption. The Zeus malware introduced you to the use of encryption to conceal malware downloads, communication, and configurations, and Cryptolocker provided a glimpse into how encrypting files can be used as an attack. With these attacks in mind, you looked at the future of cryptography and its likely effects on both computer network attack and defense.

KEY CONCEPTS AND TERMS

Asymmetric encryption	Digital certificates	Message digests
Certificate authority (CA)	Drive encryption	Plaintext
Checksums	Encryption	Public key encryption
Ciphers	File encryption	Public key infrastructure (PKI)
Ciphertext	Hashes	Quantum cryptography
Code	Key	Registration authority (RA)
Cryptosystem	Key distribution	Symmetric encryption
Decryption	Key pair	

CHAPTER 10 ASSESSMENT

1. Symmetric encryption provides nonrepudiation.

A. True

B. False

2. Which key do users of asymmetric encryption systems send to others to use?

A. Their private key

B. Their cipher key

C. Their public key

D. Their secret key

3. How many effective bits of security does 3DES provide?

A. 168 bits

B. 112 bits

C. 56 bits

D. 1,024 bits

4. Which of the following techniques is *not* a steganographic technique?

A. Hiding data in plain sight

B. Hiding data in an image

C. Inserting delays into communications to encode data

D. Concealing information in seemingly random data

E. Changing the font color of text to match the background color

5. What unique value is added to a value before hashing it to make it harder to brute force?

A. A salt

B. A spice

C. A qubit

D. A bacon

6. Asymmetric encryption is more secure than symmetric encryption.

A. True

B. False

7. AES is a(n) _____ encryption system.

A. asymmetric

B. symmetric

C. Caesar

D. Block Ring Product

8. Each bit of additional key increases the key space by a factor of _____.

A. 100

B. 12

C. 1

D. 2

9. A hash should be abandoned when a method to create _____ is discovered.

A. keys

B. collisions

C. MD5

D. a reversed hash

10. A successful attack on a(n) _____ in a PKI can result in all certificates issued by the certificate authority being revoked.

    A. registration authority
    B. certificate
    C. certificate revocation list
    D. X.509 array

11. Message digests are used to _____. (Select two.)

    A. check that messages haven't been changed
    B. calculate the value of a message
    C. allow files to be compressed via hashing
    D. verify digital signatures

12. How are data at rest protected from attackers who gain physical access to a computer?

    A. Full disk encryption
    B. VPNs
    C. TLS
    D. Hashes

13. Which attack is being used if the attacker can select what is encrypted?

    A. Chosen ciphertext
    B. Brute force
    C. Chosen plaintext
    D. Known ciphertext

Defending Endpoints

MANY PEOPLE THINK OF traditional information security designs from the outside in—as providing protection at a network edge or some other logical separation between organizations and the rest of the world. Each layer deeper inside the organization provides additional defenses until the workstations, laptops, and other endpoint devices are reached. Those individual systems and devices can be difficult to secure. Each system may be slightly different due to functional requirements, unique software or hardware configurations, or even individual user behavior. The large number of individual endpoints in an organization, as well as the endless variety of systems and devices that can be added to a network, makes protecting them a constant struggle.

Those individual systems and devices provide an excellent target for attackers, who typically seek to exploit vulnerabilities in the systems, their software, and the people who use them. Once an attacker succeeds with a single system or device, he or she can then use that system to explore the network around it, to steal data using the system or user's rights, or to launch further attacks against other parts of the network that may trust the compromised system. Although successfully attacking a single computer may not provide attackers with everything they need to successfully take over an organization's infrastructure, a single vulnerable computer is often their starting point. Like a breach in a medieval castle's wall, a single gap in the defenses of an organization can allow attackers in.

The past decade has offered many examples of attacks against endpoints using sophisticated malware and a deep understanding of the infrastructure and systems that their targets used. This evidence shows that attackers are talented, well supported, and patient. As nation-states continue to develop advanced cyberwarfare capabilities, the complexity of threats against individual systems and networks will continue to increase. Targeting endpoint systems is an attractive option for attackers who know that there are hundreds, and sometimes thousands, of opportunities for them to find a system that is weaker than the others around it—or a user who will fall for an attack and allow the attacker into the network through the user's own actions.

In addition to traditional cyberattacks, pairing advanced intelligence-gathering capabilities with nation-state-level resources creates an environment where a single flaw open to exploit can provide the way in for an attacker. This is true even if traditional security measures have been put in place. Intelligence operations in cyberwar are increasingly focused on acquiring digital data. Penetrating computer systems or other network-connected devices can provide intelligence operatives with an amazing amount of information. Networked devices not only store digital documents, designs, blueprints, and photos, but they also monitor the environment, run utility grids, control factories, and even provide direct medical care and monitoring for individuals.

The Stuxnet attack demonstrated that even systems that are not connected directly to the Internet can be reached by a clever and determined attacker. Stuxnet's use of a human factor to transfer the malware from infected laptops carried by engineers to the protected systems that it was meant to finally attack provides another reason to carefully consider how endpoint systems and networks can be protected. A single gap in an organization's defensive design can be exploited by attackers and can result in exposure of data or significant damage to systems and the infrastructure they control.

In cyberwar, the endpoint system is the final line of defense between attackers and defenders. Defenders must understand the complete list of endpoints that they are responsible for and how attackers are likely to target those devices and systems. Attackers must understand the multitude of options they have and how those systems are likely to be protected. In either case, cyberwar creates a challenging environment for individual systems to operate safely in.

Chapter 11 Topics

This chapter covers the following topics and concepts:

- What the role of endpoint systems and devices in cyberwar is
- What the types of endpoint systems are
- Which attacks are commonly used against endpoints in cyberwar
- What common endpoint protection concerns and strategies to protect against them are

Chapter 11 Goals

When you complete this chapter, you will be able to:

- Explain the role of endpoint devices in cyberwarfare
- Describe types of endpoint systems and devices
- Describe common attacks against endpoints
- Describe current U.S. Department of Defense endpoint protection goals
- Explain endpoint protection strategies and designs
- Explain the concept behind zero trust designs, and their role in secure infrastructure

Cyberwarfare Endpoints

Modern computer networks are often composed of devices that connect either locally or from a remote location to a network and services infrastructure that allows them to perform a variety of functions while accessing and using organizational data. When those computer networks are owned or managed by an organization, central management systems often control the devices on the network, but bring your own device (BYOD), cloud services, and remote work expectations mean that many organizations also have to account for unmanaged devices. While organizationally owned devices are held to some form of configuration and management standards, applying those to personally owned devices or those that are casually used to access organizational services is much harder. In all of these cases, large numbers of devices provide a huge number of potential targets that attackers can probe to find a flaw.

The sheer number of systems and devices found on a typical large network means that attackers are likely to find a neglected system, a system with an easily fooled user who will click on a malware-infected file in an email, a device that cannot be patched or that has been forgotten, or a system that has not been updated with critical patches and updates. The same techniques that non-cyberwar attackers use to compromise systems provide a means for cyberattacks as part of cyberwarfare. Even systems that are not on a network can be attacked, either by directly accessing the system in person or tricking a person who has access into admitting your attack tools.

Success in cyberwar frequently requires that attackers first identify vulnerabilities in protected systems and networks. Once they have found a gap in the layers protecting their target, they must then use that vulnerability to make their way deeper and deeper through defensive layers until they reach their final goal. That goal varies depending on the reason for the attack. It may be intended to provide information about the enemy, to provide a foothold for future attacks, or to allow the attacker to translate his or her electronic success into the physical world by disabling critical systems or causing them to act in ways they are not intended to.

NOTE

The United States has explored ever-increasing levels of network connectivity for its forces, including individual soldiers through the Land Warrior project and other network-connected soldier technology projects. These technologies and others were demonstrated via the Future Force Warrior project. The Future Force Warrior design included individual solider computer systems that monitored individual soldiers with health sensors, communications, sensor systems to detect opponents and gather other battlefield data, and helmet-mounted displays. Development of connected capabilities for individual soldiers as well as units continues, and cyberwarfare techniques can be expected to target such systems as they are deployed.

Attackers in cyberwarfare also target systems that are not part of the typical set of computer network devices that traditionally motivated attackers seek out. Cyberwar targets include military systems like targeting and command-and-control systems, drones, communication systems, and even the control systems of significant military assets like naval vessels and satellites. They can also include the infrastructure that supports military operations, such as weapons production, utilities, and the infrastructure support system. As the control systems for military and other assets increasingly rely on computer systems, they also become targets for cyberattacks. The potential to take them over or to make them inoperable with a cyberattack becomes more likely. A third, and increasingly important target for cyberwar is IoT (Internet of Things) devices, which include sensors, controls, and myriad other instruments embedded in, monitoring, and controlling buildings, systems, and industrial processes. These ubiquitous devices are a key target because they can provide attackers with broad access to organizations, their facilities, and their processes.

The broad array of network-connected devices and computer-based systems used for targets that nation-states may attack—as part of a computer network attack or an intelligence-gathering operation—is incredibly diverse. Fortunately, most of these targets can be broken down into a few major categories of endpoints, whose critical features can then be studied.

Types of Endpoints

Although networks have huge number of possible endpoint systems, most of them can be broken down into a handful of common types. The most common target from a cybersecurity perspective is the traditional personal computer. Whether it is a server, a laptop, or a desktop, a Windows, Linux, or Mac computer, the same basic cybersecurity concepts for attack and defense apply. Mobile devices, such as tablets, mobile phones, e-readers, and related devices, are also a critical target for attackers. Network devices like routers, switches, and wireless access points can be targeted, allowing attackers to control the flow of traffic or even to intercept it. Finally, the fastest growing set of new targets on network are the IoT devices that make up building and utility control systems, camera and security systems, and a wide variety of other components of organizational infrastructure. Computer network attack and defense also target military devices and systems, such as drones, weapons systems, communication, and command-and-control systems, adding to the potential endpoints defenders must consider in their design.

Computers

Personal computers and servers are the most frequently attacked endpoint systems and thus are the focus of many endpoint defense plans. The sheer number of possible variations of software, hardware, drivers, firmware, and other components that make up computers result in a challenge for defenders. Computers are also the way that most employees interact with data and networks, meaning that defenders must ensure that employees and computer users are part of the defensive plan.

Cyberwarfare attacks like Stuxnet, Shamoon, NotPetya, and Triton have all demonstrated the important role that endpoint computers can have in a large-scale, carefully planned attack.

During the Stuxnet infection, thousands of computers around the world were infected with the Stuxnet malware, which used a variety of attacks against vulnerable software. Stuxnet used those thousands of compromised computers to finally get its attack tools into specific systems in Iran. If the attacks had failed at any point before they got to the laptops used by engineers in Iranian nuclear production facilities, or if the laptops themselves had been able to detect and stop the malware from transferring to the USB drives those engineers used, the attack might have failed.

Shamoon targeted Saudi Arabia's Saudi Aramco and Qatar's RasGas petrochemical companies, causing significant disruption. Shamoon was a destructive tool with a built in component that wiped drives in a destructive way to ensure that data were nonrecoverable.

NotPetya targeted Ukrainian Windows systems like those used by shipping giant Maersk's global network, crippling their operations and demanding a ransom. It wasn't the start of cyber hostilities in Ukraine, however. Instead, it was an escalation of years of cyberwar conducted by the advanced persistent threat (APT) group known as Sandworm. NotPetya spread rapidly using two exploits aimed at Windows systems, and even if victims attempted to pay the ransom, it wasn't reversible. In fact, NotPetya did not retain the encryption key for the files it had encrypted, making it a purely destructive cyber weapon. Due to its quick spread and destructive capabilities, NotPetya is estimated to have caused more than $10 billion in damage, making it potentially the most costly cyberattack conducted to date. Like other cyberattacks, NotPetya's impact was felt outside of the target country as well, with worldwide impacts resulting from infections.

Triton (or Trisis) has been used to attack petrochemical plants, disabling safety systems and allowing attackers remote access, but more importantly, it was also the first time where malware had been discovered that was purpose-built to put lives at risk. This escalates what has largely been cyberwar waged on infrastructure, information, and computational assets to intentionally endangering human lives.

Personal computers provide attackers with a multitude of options when they look for a way to attack. Because the software they run is usually commercially available, attackers have the option of buying the same hardware and software, then practicing their attacks or developing new tools without actually attacking their potential victims. This means that a skilled and well-funded attacker, such as a nation-state's computer network attack team, can develop advanced tools without opponents being aware of those tools before they are used.

Mobile Devices

Mobile devices might appear to simply be another type of small computer at first glance. In fact, mobile devices like tablets and smartphones add a new set of challenges for organizations that are designing a defense strategy for their networked devices. Not only are many mobile devices personally owned, and thus difficult to centrally manage, they also come in a huge number of varieties. Most mobile devices might use Apple's iOS or Google's Android, but other mobile operating systems are in use in specialized environments and devices. They can also run any of a multitude of various versions and releases of those operating systems. Much like the challenges organizations face with desktop computers, the massive number of possible configurations and software versions can make securing mobile devices a challenge.

Organizations that choose to centrally purchase and manage only a limited selection of mobile devices still must face the limitations of the devices themselves. Operating system patches, vendor configurations, and even the basic security capabilities of the mobile devices' operating systems can limit what defenders can do with them. Many high-security organizations limit or even prohibit the use of mobile devices in their networks due to these challenges. More-open organizations must face the risk that mobile devices can create a path into their network they cannot easily control.

Industrial Control Systems

An **industrial control system (ICS)** includes the devices and systems that control industrial production and operation. ICSs include systems that monitor electrical, gas, water, and other utility infrastructure and production operations as well as the systems that control sewage processing and control, irrigation, and other processes. Commonly recognized types of ICS systems include **supervisory control and data acquisition (SCADA)** systems, **distributed control systems (DCSs)**, and **programmable logic controllers (PLCs)**.

ICS systems are a particularly attractive target for cyberwarfare attackers who want to disable a nation's power grid or damage or destroy parts of its utility infrastructure. ICS systems are often not as well secured as traditional computing infrastructure. Their high requirements for stability and continuous operations mean that they are less likely to be consistently patched and updated. In fact, some ICS system manufacturers advise their customers to not update the control systems and sensor devices. This makes protecting SCADA and DCS systems an even greater challenge requiring additional planning to overcome.

Supervisory Control and Data Acquisition Systems

SCADA systems are used to monitor and control remote equipment. SCADA systems like those shown in Figure 11-1 are very common in industries that require remote monitoring of their infrastructure and production systems, such as natural gas pipelines, power production and distribution infrastructure, and water supply control systems. SCADA systems typically include individual remote sensors known as remote telemetry units, which provide reports back to the central data collection system and provide some level of local control. The central system then uses the information

Control workstations

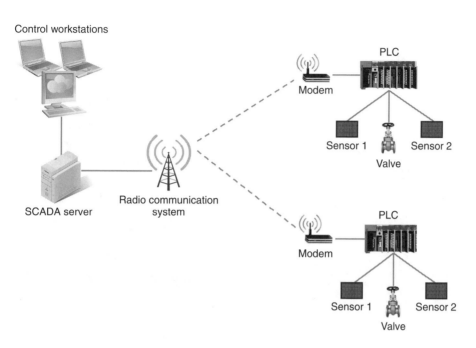

FIGURE 11-1

A sample SCADA implementation focuses on data gathering.

provided by the remote units to control the entire grid or pipeline of production and control systems. Attacks against SCADA systems can target the feedback provided to the central control system. Or they can cause the local sensor and control unit to perform an incorrect action.

Distributed Control Systems

DCSs, like those shown in Figure 11-2, are frequently used to control water and wastewater treatment and distribution systems, power generation plants, refineries, and production facilities like those that make cars, electronics, and food products throughout the world. DCSs use a combination of sensors and feedback systems to control and adjust processes as they receive feedback. Much like SCADA systems, an attack against a DCS could be as simple as providing incorrect feedback, resulting in a shutdown, overproduction, or delay in the system at a critical time.

Programmable Logic Controllers

PLCs are special-purpose computers designed to handle specialized input and output systems. They are typically designed to handle difficult environments with special temperature, vibration, or other requirements while still functioning. PLCs are designed to handle and respond to their specialized input and output requirements reliably to ensure that the processes they support occur without interruption or delay. PLCs connect to a Human Machine Interface (HMI) to provide interfaces that can interact with human operators. Typical PLCs don't have a monitor or other interface beyond buttons or lights built in to them.

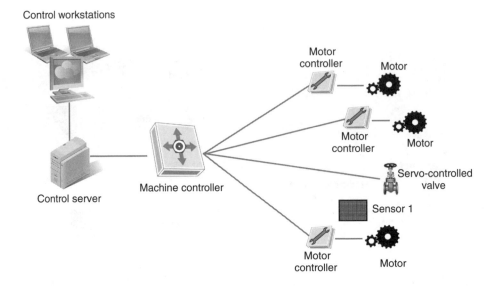

FIGURE 11-2

A sample DCS
implementation
focuses on
process control.

Control workstations

Motor
controller

Motor

Motor
controller

Motor

Servo-controlled
valve

Sensor 1

Control server

Machine controller

Motor
controller

Motor

FYI

PLCs normally use a specialized operating system known as a *real-time operating system*, which prioritizes inputs to ensure that they are handled appropriately. Like desktop operating systems, exploits exist for common real-time operating systems, but most real-time operating systems do not receive the same level of security scrutiny that desktop operating systems do.

Targeting SCADA and ICS Endpoints

Attacks against SCADA systems aren't new. In early 2000, a former employee in Queensland, Australia, used his knowledge about the water treatment software and systems to release 800,000 gallons of raw sewage into local parks, rivers, and the grounds of a local hotel. The spill killed marine life and polluted the local creek and surrounding area. An investigation showed that neither cybersecurity defenses nor security policies or procedures were in place.

Attacks against SCADA systems are a popular topic at security conventions and conferences. A 2013 demonstration at the Black Hat security conference showed how the underlying Modbus communication protocol between PLCs and their controllers could be attacked, allowing the attackers to disable protective controls and then make the system work in ways it was not intended to. Operational technologies like Modbus, BACnet, and OPC Classic continue to be significant targets for attackers, showing that the long lives of

ICS and SCADA systems makes them an ongoing target, and the number of vulnerabilities that have been discovered each year has continued to be high, with hundreds of vulnerabilities found per year.

As organizations have added more devices to their networks and systems, the number of devices that can be targeted has continued to grow. In 2018, the Mirai botnet targeted BACNet systems and leveraged devices to conduct distributed denial of service attacks against services in countries around the world. Not only are insecure systems threats to their home organizations, they can also be weaponized to attack others.

Military Systems

The term *military systems* describes a range of devices and platforms. Some use common civilian operating systems and software; others are built using custom-designed software and hardware. Due to the diversity of military systems, defenders must carefully evaluate the defensive capabilities of the endpoint devices they have to protect and then design appropriate layers of defense based on those capabilities.

> **NOTE**
> Later in this chapter you will learn about the U.S. Department of Defense (DOD) cybersecurity strategy and the stance that the DOD takes on protecting military endpoint devices and systems.

Drones and Remote Platforms

Electronic attacks against drones have demonstrated that drone platforms and their command against drones' use of the Global Positioning System (GPS) and those aimed at capturing video feeds from drones deployed in active combat. According to leaked National Security Agency documents, al-Qaeda-sponsored research has primarily been aimed at jamming GPS and infrared marker systems as well as the use of lasers to dazzle drone sensors. Because drone command-and-control system links are encrypted, drones' software and the systems used to control them are a more likely target.

FYI

While attacks against military drones haven't resulted in a publicized incident where attackers have gained full control of a drone, jamming and falsification of GPS signals may have been at least partially at fault for some unmanned aerial vehicle crashes. As drones are used more frequently as weapons platforms, they are becoming more and more attractive targets, and it is only a matter of time until one is compromised and control is seized. In addition to attempts to take control of drones, drone jamming technologies for both military and nonmilitary use are commercially available.

Weapons Systems

Military weapons systems have integrated an ever-increasing amount of computer hardware and software into their design. Table 11-1 shows the published increase in the use of software in the avionics and flight control systems of major U.S. aircraft since the 1960s. Each successive generation has seen a significant amount of additional control by computer-based systems. As additional capabilities for communications and command and control have been added, warplanes and other systems have become more likely targets for cyberattacks.

The amount of automation isn't the only concern. The F-35's software has more than 24 million lines of code, with 9.5 million lines of that code actually running onboard the fighter itself. The F-22A used about 8 million lines of code, and the F/A-18 used 4 million lines of code. That means the potential attack surface for modern planes is much larger and that software flaws and bugs could potentially exist in more places as well.

Modern military aircraft provide an excellent example of the increasing use of software-controlled systems and the increasing complexity of their underlying software. They use flight control, navigation, intelligence, and communication systems that not only have software-based controls—they can also frequently receive in-flight updates and data, providing a potential means of attack or deception.

FYI

In 2007, reports surfaced that F-22 Raptors attempting to cross the International Date Line on their way to Asia had to turn back due to a software bug that crashed their onboard computers. Although this wasn't a cyberattack, it demonstrates the potential that a carefully designed and implemented attack could have on aircraft or other weapons systems heavily controlled by software. Fortunately, the F-22 fighters were able to follow their in-air refueling tankers back to base for a safe landing. A different software issue in early 2014 caused air traffic controllers to ground flights in Los Angeles, California, to avoid a collision with a U-2 reconnaissance aircraft. Analysis later showed that multiple waypoints and altitude changes between control zones had resulted in a software error.

Command and Control

Military command-and-control systems are a particularly attractive target in cyberwar. As the DOD continues to integrate computing capabilities into a broader and broader array of military systems, it calls this **Command, Control, Communications, Computers, and Intelligence (C4I)**. C4I systems provide situational awareness—the ability to know where both friendly and enemy forces are—and to react and respond. C4I systems include capabilities like weapons targeting, including the data that are passed to weapons as they are en route to their target. The links and control systems of active weapons, or the real-time data that soldiers and commanders use to decide what they should do and where they should be in a combat situation, are obvious targets for cyberattack.

Table 11-1 Percentage of aircraft function handled by software.

AIRCRAFT	YEAR INTRODUCED	PERCENTAGE OF FUNCTION HANDLED BY SOFTWARE
F-4	1960	8
A-7	1964	10
F-111	1970	20
F-15	1975	35
F-16	1982	45
B-2	1990	65
F-22	2000	80

Due to their critical role in battlefield control, C4I systems are an important part of current and future combat strategies. Defending C4I systems while they are in use is very important. The endpoint devices that individual soldiers carry into combat, the sensor systems that provide information, and the control systems for weapons platforms and smart weapons themselves all must be protected, in addition to the more traditional computer networks and devices used by command staff.

Embedded Systems

The growth of devices built in to many of the technologies that surround people's daily lives creates a new type of target for cyberattack. Embedded systems like telephone switches, traffic light controls, vehicle-based computers, consumer electronics, computers built in to buildings, and even appliances with network access all provide a target for attackers. Such a target can provide a useful launching point for further attacks or access to disrupt lives and productivity.

Some embedded systems use traditional operating systems, but many use specialized versions that are designed to meet the specialized needs of the devices or systems they are built into. Much like the SCADA and ICS systems examined earlier in this chapter, embedded systems are far less likely to be patched or to have strong security models in place, because they are focused on performing specialized tasks. Designers didn't expect them to be exposed to network attacks or other threats.

Embedded systems create a number of unique challenges for defenders, as security standards and other common tools are rare or nonexistent for most embedded systems. Military and defense-specific embedded systems often receive special attention to help them be more resilient than civilian and commercial systems. But the resulting trade-off is evident in the DOD cybersecurity strategy: Faster, more nimble acquisition and update strategies are necessary to help ensure that outdated and devices known to be vulnerable do not remain in service.

Attacks like the Stuxnet attack demonstrate that motivated nation-states can research and exploit embedded systems as well as ICS and SCADA technologies. Embedded systems in critical locations require both awareness and dedicated security designs to ensure that they don't create a path for an attacker or an exploitable weakness in an organization's capabilities.

Medical Devices

A type of endpoint that most defenders would never have considered until recently became highly visible in 2013, when former U.S. Vice President Dick Cheney revealed that his doctors had disabled the wireless control capabilities of his pacemaker. Like many other computerized devices, modern pacemakers support wireless updates to their settings and can provide information about their current status to a nearby receiver.

Researchers reverse engineered pacemaker radio controls as early as 2008, with findings published in an IEEE symposium paper titled, "Pacemakers and Implantable Cardiac Defibrillators: Software Radio Attacks and Zero-Power Defenses." Similar research in 2011 demonstrated vulnerabilities in insulin pumps that could inject users with fatal doses of insulin, revealing yet another potential attack method.

In 2019, the U.S. Food and Drug Administration (FDA) issued an advisory for patients with Medtronic implantable defibrillators or cardiac resynchronization therapy defibrillators, noting that vulnerabilities had been found in the wireless telemetry functions. The broad range of impacted devices were revealed to not use encryption, authentication, or authorization. In essence, with minimal effort, attackers could access and manipulate impacted devices—potentially at a distance due to the wireless nature of the device. Despite the vulnerability, the FDA recommended that patients and caregivers continue to use the devices as instructed.

Although so far no one has documented malicious real-world attacks against implantable or wearable medical devices, the possibility of attacks against heads of state or others who use wirelessly accessible devices expands the boundaries of cyberwarfare to include direct physical harm. It is possible that an attacker could take even subtler action to speed up or slow down a diplomat's heart rate or modify his or her blood sugar levels.

The FDA provided guidance in 2016 on postmarket management of cybersecurity in medical devices and released draft guidance 2018, which is intended to guide how the medical device industry designs, labels, and documents devices before they are reviewed if they have cybersecurity risks. Despite this, the field remains immature, with poor design and practices in many cases.

In addition to the FDA guidance, MITRE released an incident response and preparedness playbook in 2018, with topics like medical device procurement, vulnerability analysis, cybersecurity support, detection, analysis, and incident handling processes. Medical device manufacturers, like the creators of many other specialized systems, continue to provide unique solutions and thus can create unexpected challenges for endpoint protection in an unexpected place.

Attacking Endpoints

In both computer network defense and traditional information security practice, one of the best ways to understand endpoint defense requirements is to learn to think like an attacker. Attackers analyze endpoints, their management practices, their design, and their administrators and users all as potential targets, looking for gaps in defenses. Once they find a way in, either through vulnerabilities, misconfigurations, mistakes, or simply the helpful nature of the staff who use the systems, they can conduct further attacks as a springboard into a network. They can use the device for its intended purpose, but to their own ends. Or they can just gather data from the endpoint they have under control.

The huge number of endpoint types leads to an even larger number of possible types of attacks against them. It helps to categorize the attacks into a few useful groups to match defensive techniques against the attacks themselves. Attacks can be categorized a number of ways:

- Attacks that require physical access, such as inserting a hardware keyboard logger into a system, or making a physical copy of a drive by capturing the encryption key from live memory without remote access to the machine

- Attacks that require network access, such as denial of service attacks, or attacks against services that the system provides, such as a web server or file server

- Attacks that require the user or an administrator to take action, such as phishing attacks or other forms of social engineering that lead a user to compromise the machine by inadvertently helping an attacker

- Attacks that target the hardware or firmware of the device, which often require access to the design or manufacturing process of the device to compromise the design or to insert a vulnerability or Trojan into the firmware before the system is sold

- Attacks against application software, such as web browsers; most application software attacks require the attacker to discover a vulnerability, either through vulnerability testing or by software vulnerability analysis using the actual code that makes up the program

- Attacks that use the endpoint's normal function against it, such as those that provide fake feedback to intrusion detection systems, causing them to block legitimate traffic, or the GPS attacks against U.S. drones that allegedly have caused the drones to try landing in the wrong place

Other categories can be created to match specific types of endpoints, with the primary goal of grouping attacks in a way that allows an organization to determine which defenses are appropriate and which can reasonably be put in place. Some defenses, such as awareness, can help against attacks by both detecting and helping to prevent attacks, whereas other defensive techniques and designs will only be effective against a specific type of attack or scenario.

FYI

Firmware is the software that has been stored in read-only memory, which is part of the hardware system. Firmware typically contains the basic operating processes of a device, such as the basic input/output system (BIOS) of a computer or the software that operates a pacemaker. The Stuxnet malware was specifically designed to attack the Siemens Simatic S7-300 PLC's firmware, which had one of two very specific variable frequency drive units attached to it, allowing it to target specific uses of that PLC.

Protecting Endpoints

Computer network defense strategies for endpoint protection require assessment of the threats that the endpoint systems and devices will face. Once defenders have assessed the likely threats and which methods would be effective against them, they can balance the costs of the defenses and their ability to implement them against the need of their organization to provide that level of defense.

Typical defense-in-depth strategies for endpoints include a number of common layers:

- Physical security
- Policy and procedures
- Configuration standards
- Central management
- Awareness and information sharing
- Anti-malware and antivirus software
- Configuration management, patching, and updates
- Allow listing and deny listing (sometimes called whitelisting and blacklisting)
- Testing, including penetration testing and the use of red teams

Figure 11-3 shows many of the typical layers of defense for a computer workstation. Defenses start at the firmware and hardware level, which provides the underlying environment that the rest of the software and defenses reside on. The computer's operating system and the application software that it runs provide the next layers of defense; the operating system must be properly secured, and the application software must prevent attackers from using it to compromise the system. Anti-malware software works to detect infections, whereas logging and reporting capabilities provide status information about the system. Finally, most network-attached systems now use some form of system-level firewall or other network protection that filters and sometimes monitors the network's connection to other systems and networks for attacks.

Endpoint protection also relies on non-system-specific control capabilities, such as configuration management, central monitoring, and the human side of controls via policy, procedures, and awareness. These defenses reside outside individual systems and are typically common to a larger group of systems or devices, rather than providing unique controls for each computer.

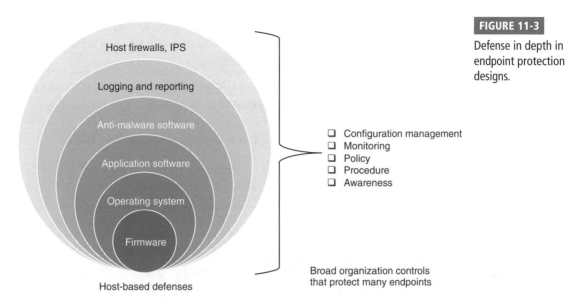

FIGURE 11-3

Defense in depth in
endpoint protection
designs.

The following sections explore each of the major defensive concepts. There are many specific techniques, and each system or device may require unique designs or special capabilities to properly protect it. Defensive techniques for common operating systems and frequently used devices are broadly available, but relatively uncommon systems or those that are not part of a typical organizational cyber defense strategy will need special attention to understand their defensive needs.

U.S. Department of Defense Strategy

The DOD's Cybersecurity Culture and Compliance Initiative (DC3I) provides some insight into the DOD's views on the importance of endpoint security. It notes that the vulnerability of DOD networks has grown in parallel to the increasing reliance on them and that threat actors have successfully penetrated networks through multiple vectors. It also notes that despite improvements in security, successful intrusions will still occur. The DOD's Cybersecurity Discipline Implementation Program expands on this, with a focus on device hardening and reducing the attack surface as well as the use of strong authentication.

The DOD has spent years working on department-wide **cyberhygiene** initiatives with three main components: the DC3I, the Cybersecurity Discipline Implementation Plan (CDIP), and the Cyber Awareness Challenge training.

- The DC3I effort focuses on four major areas: strong authentication, device hardening, reducing attack surface, and aligning to cybersecurity and computer network defense service providers.

- CDIP identifies 17 tasks that are intended to be put in place across the DOD. The CDIP's tasks include common security practices like removing unsupported software and operating systems, using secure configuration guidelines, and vulnerability management.

- The Cyber Awareness Challenge gamifies security awareness and includes video, audio, and other content presented in the form of a series of scenarios with decisions for trainees to make as part of the learning process.

The DOD is using a multipart approach to endpoint defense that takes into account technical capabilities, staffing, and personnel skills as well as awareness.

FYI

You can try the unclassified version of the Cyber Awareness Challenge yourself at https://dl.dod
.cyber.mil/wp-content/uploads/trn/online/cyber-awareness-challenge/launchPage.htm.

Defense in Depth in Endpoint Security

The concept of defense in depth remains important when deploying endpoint defenses. Each endpoint can become its own virtual castle, with protective layers providing defenses against a variety of attacks. Designing standards around how devices should be protected in various situations and configurations can provide a good baseline to give organizations with new types of endpoints a starting point to work from.

The process of designing endpoint layers for protection in a variety of situations can also be important, particularly for systems that travel. Laptops, mobile devices, and other systems that will leave their normally secure locations must be protected against threats beyond the normal attacks a workstation in an office or a server in a data center might face. Devices that attackers could access need even more protection to ensure that they haven't been modified, accessed, or had their hardware replaced while they were out of their owner's hands.

Much like a medieval castle under siege, endpoints also shouldn't have to stand alone against attacks for a long time. A strong management system, reporting and logging capabilities, testing processes, and capable administrators and defenders are all necessary to keep even a well-defended system secure in the long term.

FYI

In 2008, attackers in Afghanistan planted a USB thumb drive loaded with malware in the parking lot of a U.S. DOD facility. When the thumb drive was plugged into a laptop, it infected the laptop with a worm that spread through military networks. While the DOD never reported what data may have been exposed, the breach was described at the time as "the worst breach of U.S. military computers in history" and led to the creation of the U.S. Cyber Command. The 14-month-long defense against the attack is known as Operation Buckshot Yankee.

Zero Trust and Endpoint Security

The concept of zero trust can be applied to endpoint systems as well as to networks. Zero trust in the context of endpoint security focuses on endpoints as self-defending systems that presume that users, other systems, and the network they are connected to cannot be trusted. In an environment designed with a zero trust concept in mind, endpoints will limit user permissions to only those required, may require authentication and authorization before tasks are allowed to be performed, and will use a broad range of defensive and monitoring techniques to help prevent attacks.

Endpoints also have a role to play as part of the monitoring and detection infrastructure for a broader zero trust environment. They report on their own security status, including patch levels, applications, events like user logins and software installations, and many other datapoints.

Physical Security

Although it might seem obvious, the need to protect endpoints from falling into the hands of attackers is one of the most basic requirements of endpoint security. If attackers have physical control of a system and enough time, they can almost always compromise the system or the data it contains. If they have less time with the device, they might copy its data or insert software or hardware bugs to capture keyboard input—or even record conversations that happen near the device. Some attacks even rely on replacing the device or system with an apparently identical system that is sent on to the original owner or new purchaser of the gear.

Physical security also plays a role in the devices brought into secure environments. The civilian sector concern about staff bringing their own devices and connecting them to internal networks is even more critical for military and governmental organizations. Insecure, untrusted devices connected by staff to a sensitive network can provide a bridge into it for attackers. Even bringing in a simple USB thumb drive can cause problems if that device contains malware.

Policy

The root of most defensive strategies is a cyber defense policy. Policies assign responsibility and set the overall tone for computer network defense activities. Well-written policies don't specify specific technologies or processes. Instead they focus on the organization's strategic direction. Once a policy has been created, organizations will then create procedures and standards to ensure that their staff can implement the policy effectively. Policies are typically authorized by the highest levels in an organization to ensure that they are both appropriate to the organization and that they have support when policy issues come up.

FYI

The DOD includes a deputy assistant secretary of defense for cyber policy. The position is tasked with providing support to the DOD by "formulating, recommending, integrating, and implementing policies and strategies to improve the DOD's ability to operate in cyberspace."

Figure 11-4 shows the relationship among policies, procedures, and standards. A broad security policy provides the core guidance to develop procedures. Those procedures are used to guide the development of standards for specific configurations and actions. Together, policies, procedures, and standards help ensure that people involved in endpoint protection act appropriately.

Procedures

An organization's procedures for endpoint defense in depth are based on the policy or policies the organization has in place. Procedures typically describe things like how the organization acquires systems or develops new systems, how the organization assesses risks, or how it responds to attacks. Formalized processes that define what, when, who, and how the organization takes action ensure that responses are predictable and reliable and that the staff who are responsible take appropriate action.

The DOD has defined a number of requirements for its own procedures in response to the increasing threat of cyberattack. The DOD's acquisition strategy says that the program managers must "recognize that cybersecurity is a critical aspect of program planning. It must be addressed early and continuously during the program life cycle to ensure cybersecurity operational and technical risks are identified and reduced and that fielded systems are capable, effective, and resilient."

The strategy also points to the use of modern development and security practices: "This pathway integrates modern software development practice such as Agile Software Development, Development, Security, and Operations (DevSecOps), and Lean Practices. Small cross-functional teams that include operational users, developmental and

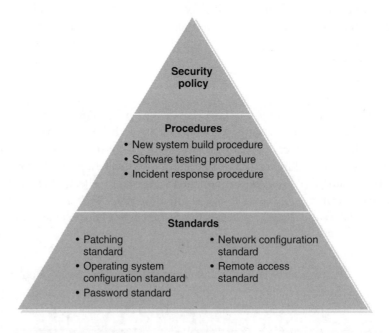

FIGURE 11-4

Relationship of policy, procedures, and standards.

operational testers, software developers, and cybersecurity experts leverage enterprise services to deliver software rapidly and iteratively to meet the highest priority user needs."

The DOD is taking advantage of procedural changes and modern development techniques to help ensure that each stage of its acquisition process is appropriate to modern cyberwarfare requirements. It demonstrates an awareness of historical issues with slow acquisition and updates, which previously resulted in out-of-date and often vulnerable systems as well as issues with slow, long-term development cycles that made software hard to fix and difficult to update when problems were found. Similarly, the DOD is showing a willingness to simplify procedures to improve responsiveness.

Configuration Standards

Configuration standards define the settings and options a system, application, or other part of an endpoint system has in place. Common operating systems and devices often have publicly available configuration standards that have been published by the manufacturer, groups of experts, or specialized organizations. Standards help to set expectations for what a given system can be expected to do, how it will respond to attacks, and how it is secured.

Organizational requirements drive configuration standards. Even when industry standards exist, organizations will typically modify the standards before adopting them to their specific needs. Systems that comply with specific configurations can also be *certified* in those standards, proving that they have reached a specific level of security compliance through a combination of their configuration and capabilities.

> **NOTE**
>
> The National Security Agency provides configuration guides and other security management information for applications, operating systems, and wireless devices at https://apps .nsa.gov/iaarchive/library /ia-guidance/security -configuration/?page=1.

Central Management

Central management systems provide the ability to enforce configuration standards, report on the status of systems, and make changes and updates to those systems from a single, central location. This allows management of large numbers of endpoints rather than requiring each individual endpoint to receive individual maintenance. Central management is a critical part of large-scale endpoint defense strategies because it helps maintain compliance with standards. A system that no longer meets those standards may be compromised or have problems that must be addressed.

Central management can also create a threat to endpoint systems due to the level of control it has over endpoint systems. A compromised central management system can allow attackers to modify configurations or to ensure that defenders are not notified of issues with their systems. Thus, central management systems must have extra layers of defense included in their own endpoint defense design.

The SolarWinds breach that was announced in early 2021 points to what can happen when a central management and monitoring solution is compromised. SolarWinds Orion platform is a power monitoring and management platform deployed at thousands of customer sites, with deep insights into organizations. Attackers leveraged SolarWinds patching process to deploy malware to customers as part of their normal updates and then were able to pick and choose which targets they wanted to expend time and resources on.

Balancing defensive strength against organizational needs for management tools like this is a challenge for defenders. At some point, the underlying infrastructure and systems will require some level of trust to operate, and very few organizations have the resources to assess every technical component of their infrastructure for possible malicious action. Attacks like the SolarWinds breaches demonstrate how powerful attacks targeted at major infrastructure and systems tools and providers by advanced adversaries can be.

Configuration Management

The basis for most central management systems is configuration management. Although configuration management systems for individual endpoints also exist, most networked systems rely on a central management system to provide configuration management.

Configuration management systems typically apply one or more prebuilt configuration templates to a device, which provide control over a variety of settings and capabilities:

- Security settings
- Software versions
- Network settings
- User rights management
- User interface settings

Patches and Updates

Many central management systems provide another critical service for endpoint protection: patch and update management. A typical computer in use today will have an operating system; numerous hardware drivers that make video, audio, and other devices work; application software such as word processing and spreadsheet software; web browsers; and a variety of other software packages. It is critical to the security of computers and other devices to have current software versions and patches that provide security fixes.

Patch and update management is often a challenge in complex environments. New software patches and versions might not be compatible with older software or might have flaws. Individual endpoints might not properly install the software update, leaving them vulnerable. Despite potential flaws, without central management capabilities for configuration and patch management, it is nearly impossible to manage and update large numbers of systems in a reasonable amount of time.

One danger of patch management systems is the reliance on the system to ensure patch installation. Some patches can fail without the patch management and configuration management
systems being aware of it, and patch management also relies on all systems being in the patch
management system and properly managed. This means that systems that are forgotten, are
not supported by the patch management system, or are added without central administration's
knowledge can all result in significant gaps. In some cases, patches may not fix the issue or can
leave parts of a flaw in place. Some patches and updates will even show the correct version
when asked, but the patch itself will not have installed properly, leaving the system vulnerable.
Although patch management and configuration management helps, it isn't always a guarantee
of correct configuration and updates.

Awareness

The DOD noted in its cybersecurity strategy that "it is just as
important for individuals to be focused on protecting themselves
as it is to keep security software and operating systems up to date."
It's critical for endpoint protection that system users—along with
the administrators and security personnel who build, support,
maintain, and defend those systems—be aware of both the status
and typical behavior of their systems. They must also be alert to
the threats and attacks they are likely to experience.

Awareness as a defensive measure requires that staff know
about threats and attacks and be inclined to use that knowledge.
Most awareness programs use training in various forms as their
basis, but training isn't the only option. More advanced awareness
programs often combine experience or testing-based experiences
for participants, including simulated attacks that allow feedback
on which issues are detected and how participants respond.

Awareness also means that those who are responsible for
administration of endpoints, including their hardware, software,
and configuration, must monitor information sources for recently
released updates and active exploits that target vulnerable parts of
their systems. An organization that has a great level of user security awareness, but does not remain up to date on software patches,
may detect a breach but will likely not have acted to stop it.

 NOTE

In March 2011, attackers used
an email phishing attack to
persuade staff at RSA, a major
security solutions provider,
to run malware. The final
estimate of data exposure was
24,000 files, including sensitive
data that impacted their
SecureID multifactor tokens.
This has a broad impact on
companies and governments
that relied on RSA tokens
as part of their security
infrastructure. Phishing is
one of the most common
types of social engineering
attacks, which focus on human
weaknesses, and is a frequent
focus of awareness efforts.

Information Sharing

Awareness of attacks and new techniques for defending and attacking systems often relies
on sharing knowledge between organizations. Information-sharing organizations include
organizations such as the U.S. government's U.S. Computer Emergency Readiness Team

(US-CERT) and Information Sharing and Analysis Centers (ISACs). Information sharing as an awareness strategy is a key part of the DOD and U.S. government's information security strategy. Information-sharing controls are in place to ensure that information about actual attacks and vulnerabilities is distributed only to appropriate individuals and organizations.

Anti-Malware and Antivirus

Many of the most successful attacks against endpoints in cyberwar have been malware-based. The Stuxnet and Flame attacks both focused on malware-based infections, and the Aurora attacks and other APTs use malware as a key element of their ongoing control of systems. Anti-malware technologies are therefore an important layer in the defense-in-depth strategy for cyberwar.

Anti-malware software uses three primary techniques to detect malware:

- **Signature-based detection** focuses on knowing the digital fingerprint of a malware package. These signatures are packaged into frequent detection package updates, which anti-malware vendors release to their customers. Unfortunately, modern malware uses a variety of techniques to help ensure that their tools don't have the same fingerprint across different installations, making signature-based detection less likely to work. Anti-malware vendors continue to develop additional capabilities in an attempt to overcome this, but attackers have an advantage when signatures are the only line of defense.

- **Heuristics** involves behavior-based detection techniques. This provides defenders with the ability to observe what a program is doing and then to determine whether that activity looks like malware activity. This offers a better chance to detect otherwise harmless-looking malware, but it also runs the risk of detecting a legitimate program that takes actions that malware might also take.

- **Anomaly based detection** relies on a baseline understanding of what a system, network, or device does and monitors for differences from that baseline that may be malicious or unwanted behavior. Of course, a good baseline is needed, and new behaviors may appear malicious despite being innocuous.

Unfortunately, traditional techniques for detecting malware have become less effective over the past decade as malware creators have used increasingly advanced techniques to hide their tools from defenders. Now defenders must use increasingly advanced, multi-layered defense and detection methods to protect systems, instead of the traditional signature-based defenses that used to provide strong protection.

Network Protection

Much of the protection that endpoint devices require against network attacks is handled as part of the design of the networks they connect to. But endpoint devices still must be able to protect themselves against the traffic that reaches them past those network filters and protective systems. Therefore, many systems provide their own host-based protections. For most devices, those take the form of two types of network protection:

- Firewalls block communication both to and from the system based on a set of rules.
- Intrusion prevention systems are software-based filters that use behavior- and signature-based detection capabilities. These capabilities are like those used for malware defense to recognize and stop attacks both against the system itself and those that might come from within the system on which they are installed.

Unified Threat Management (UTM) devices are another option, combining multiple protective technologies into a single network appliance or software device. UTM systems typically include firewalls and intrusion detect or prevention systems (IDS/IPS) capabilities at a minimum but frequently add a blend of additional security capabilities like antimalware, proxying and filtering, data loss-prevention, virtual private network, and other inspection and logging capabilities. As security appliances continue to combine functions, the distinctions between devices have blurred, and many security devices continue to add functionality.

The Failure of Anti-Malware Software

Traditional antivirus applications are a familiar tool to computer users and are typically considered a necessary part of a well-designed security infrastructure. Unfortunately, the effectiveness of traditional antivirus methods that use signature-based detection techniques has been steadily decreasing. Next-generation malware packages, particularly those used by APT organizations, can be nearly invisible to antivirus and anti-malware tools.

Researchers studying the malware applications used to compromise machines in APT attacks have discovered evidence that APT tool writers use common antivirus packages to test their applications. Once they're sure that the attack won't be caught by those protective applications, they can deploy the malware with the expectation they won't be caught. They can even continue to use publicly available updates from the vendor to continue to test their malware and can change it if and when it is detected.

This doesn't mean that anti-malware software doesn't have a place in endpoint defense strategy. Malware that isn't as advanced as that employed by highly advanced actors remains a threat that network defenders must take seriously. Less-advanced threats often recycle existing malware or make relatively minor modifications, allowing traditional detection methods to work. Thus, much like a flu shot, traditional antivirus tools can help protect against the known threats and can have some success against unknown threats—even if they're not a complete solution.

Encryption

Encryption for endpoint defense typically involves full disk encryption intended to prevent theft of the data that reside on the system or device if it is stolen or is outside its owner's possession. This doesn't prevent theft of system data when the system is in use, so encryption isn't as useful against attackers who compromise the system using a vulnerability in the operating system or software applications.

Full disk encryption for many secure systems is assisted with a **Trusted Platform Module (TPM)** chip. This is a cryptographic processor built in to the motherboard of the device. TPM modules are unique to each system. They allow the system to prove that it is the system that should be unlocking a given hard drive or other storage device. Systems without TPM chips must rely on other ways to verify their identity like a passphrase or a digital certificate stored on a USB flash drive and matched with a password that only the system's owner should know.

Unfortunately for those tasked with defending systems, most disk encryption systems that aren't built into the disk itself are vulnerable to a variety of attacks. These attacks use techniques to recover keys stored in memory, or **side-channel attacks.** These target other ways to recover the keys to the drive, such as capturing keyboard input or even recording the sounds of typing on the keyboard to determine which keys were pressed. Of course, even built-in in encryption can be problematic or flawed, and once it is built in, it can be difficult or impossible to fix. This was demonstrated in October 2020 when jailbreakers demonstrated that the checkm8 exploit used to jailbreak iOS devices could be used on newer Macs as well, potentially allowing for disk encryption, firmware passwords, and the cryptographic verification chain to be bypassed.

FYI

In 2012, NASA lost a laptop with the algorithms required to control the International Space Station. The laptop was not encrypted, meaning that whoever ended up in control of the system could access the algorithms directly.

In addition to the key role that full disk plays in defending data at rest, encryption also plays a critical role in communication between endpoints. It's also key in the processes used to provide updates and patches that are verifiably the originals from the manufacturer or another trusted organization.

Due to the extensive use of encryption techniques to secure and protect both data at rest and in transit between systems, attacks on cryptography are increasingly a focus for nation-states. The additional resources, technological tools for drive analysis, and advanced intelligence acquisition techniques that nation-states can bring to bear on individual devices mean that encryption is useful but cannot be considered impervious to attack.

Allow Listing and Deny Listing

The ability to choose which software, websites, services, or access a system provides can be very useful when designing endpoint security. Allow listing and deny listing provide this capability, allowing administrators and security staff to ensure that the decisions that they make and the policies they must enforce can be successfully put into place.

Allow Listing

Allow listing (sometimes called whitelisting) relies on the ability to build a list of trusted software, systems, networks, or other resources. Once an allow list is built and implemented, it's checked each time a request is made to use that type of resource. If the resource is on the allow list and matches it, access to the resource is allowed. If the resource is not on the list, access isn't allowed, and further action must be taken to access it. A number of types of commonly used allow list technologies are available, including software or application allow lists, website allow lists, and even user allow lists that allows specific users access to systems.

> **NOTE**
>
> Allow listing is trust-centric, whereas deny listing is threat-centric. This means that allow listing is often used when it is reasonable to know every item that should be on the allow list, without updates and maintenance becoming unmanageable. Deny listing is used when a list of threats can be built and maintained more effectively or when access should be allowed in most cases unless it is intentionally blocked by the deny list.

Allow listing has a number of advantages in a high-security environment:

* It allows administrators to prevent running of unauthorized software.

* It can provide a very strict set of requirements for websites or other resources.

* It does not require administrators to think of every possible alternative or workaround.

Figure 11-5 shows a typical application allow listing scenario. When a user attempts to run a program, the allow listing application intercepts the request and checks the program against its list. The allow list program verifies that the application being run is actually the allow-listed application by checking the file, its location on the workstation, and other details about the program. If it matches properly, which it does in this example, it is then allowed to run. If the application did not match, or had been modified, the allow-listing program would prevent the user from running it and would notify an administrator if it were configured to do so.

Deny or Block Listing

Deny listing or **Block listing** (sometimes called blacklisting) reverses the technique used in allow listing and builds a list of prohibited applications, files, sites, or other data or access. Figure 11-6 shows a deny list check for a sample endpoint system using a deny list application that tracks known malicious websites. When the system attempts to access a deny-listed site, the access is stopped, or as the example shows, it can be redirected to an awareness website.

FYI

Allow listing is typically considered a more secure solution than deny listing because it works on a foundation of trust. Unfortunately, the difficulty of keeping a allow list up to date means that blacklisting is more commonly used in most environments.

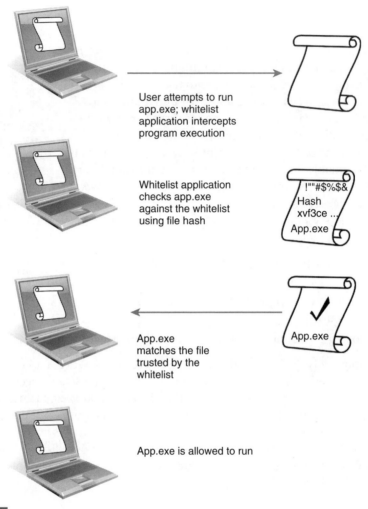

User attempts to run
app.exe; whitelist
application intercepts
program execution

Whitelist application
checks app.exe
against the whitelist
using file hash

App.exe
matches the file
trusted by the
whitelist

App.exe is allowed to run

FIGURE 11-5

Allow list check process for a software application.

Deny lists offer advantages when specific accesses must be blocked, or when the list of known files or data is well known. Deny lists can also be useful for quick responses, as adding a known problem website or file to a deny list distributed to endpoint systems can stop those systems from suffering the same issues other systems already have.

Deny lists can also create a number of problems:

- Deny lists can result in legitimate resources being blocked.
- Some deny list can be overly broad, resulting in useful resources being blocked at the same time as threats.
- Deny lists require updates and maintenance to ensure that new threats are added.
- Deny lists becomes increasingly difficult as the size of the deny list grows.

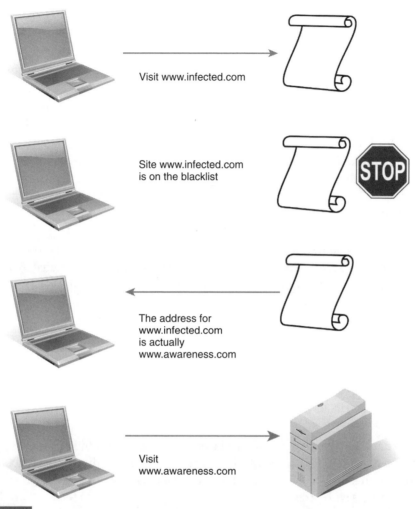

Visit www.infected.com

Site www.infected.com
is on the blacklist

The address for
www.infected.com
is actually
www.awareness.com

Visit
www.awareness.com

FIGURE 11-6

Deny list validation for a blocked website.

Testing

Once endpoint defenses are in place, they must be tested to ensure they are performing as expected. Computer network defense plans typically include a variety of testing methods that verify settings and technical configurations as well as procedures, practices, and the awareness of staff who use and support the endpoint systems. Military and government agencies use a variety of testing methods and often have access to methods and resources beyond those used for typical civilian information security operations. A few of the most common testing techniques include the following:

- **Software testing** is commonly used both during the development of software and when software is deployed, using automated and manual testing methods. Automated, software-based testing applications allow defenders to validate new software's resistance to common attacks. The actual underlying code that makes up software can also be tested for common problems using source code validation and testing tools and techniques.

- **Certification** to standards such as the Common Criteria provides extensive testing guidelines, against which certification organizations can test computer systems.

- **Port scanning** tests systems and network devices to see what network ports they have accessible. Port scanning is often used to quickly verify if a system is providing the services it is expected to provide or if it responds in unexpected ways.

- **Vulnerability scanning** is used to test the services that a system or device provides and to check those services for vulnerabilities either by checking the version information it responds with or by attempting to exploit vulnerabilities known to exist in the service.

- **Configuration management systems** can provide useful testing and reporting capabilities that allow administrators to verify that the systems under their control have the settings, software, and policies applied that they were intended to have.

- **Penetration testing** is a broad discipline that uses attacks and exploits against an organization to verify whether and how its security controls, systems, and staff respond to attacks. Two major categories of penetration tests are typically conducted. The first type is *white box penetration testing*, which provides the testers full information about the targets. White box testing is also sometimes known as *crystal box penetration testing* because it provides a complete view of the target. The second common type is *black box penetration testing*, which provides the testers no prior knowledge, requiring them to act like an attacker who doesn't know the internal workings of the organization and its systems.

- **Simulations** and **war games** are used by organizations to test their processes by creating a scenario for their staff to practice what and how they would react in the event an issue occurred. Simulations are often used to test procedures and to identify issues with processes.

Red Teams and Tiger Teams

Testing that goes beyond simulation into active tests of defenses as part of planned exercises often uses specialized teams of trained and skilled attackers. These teams, sometimes known as **tiger teams**, use the same tactics and tools as the actual attackers an organization is likely to face. This provides a better test of the defenses put in place and can reveal much more than a simulation or internal testing might.

Organizations also use the term blue team to describe defenders and *purple team* to describe a combination of aggressor **red teams** and defender blue teams. Their intention is to use the combination of teams to better understand how defense and attacks can work together and to allow shared knowledge in both scenarios.

CHAPTER SUMMARY

This chapter examined the concept of endpoint defense. There are many types of endpoints, including computers, mobile devices, ICS and SCADA systems, military endpoints such as weapons systems and command-and-control devices, and embedded systems. The existence of many types of endpoints makes defending them a challenge—configurations and capabilities vary widely between them. Due to the wide variety of endpoints and the ways they can be targeted, attacks on endpoints may consist of anything from social engineering that targets users to complex attacks that require access to systems' supply chain and design process.

This chapter also explained that endpoints are common cyberwarfare targets because they contain data and can provide access to other parts of a computer network. Defenses against attacks on endpoints often focus on a defense-in-depth strategy that combines physical security, policy, procedures, standards, configuration management, central management systems, awareness, and technical controls. All of these protections must be tested, and many organizations do so by conducting penetration tests and using red teams—groups of experts who help validate the protections their defenders put in place.

KEY CONCEPTS AND TERMS

Allow lists
Anomaly based detection
Block lists
Deny lists
Certification
Command, Control, Communications, Computers, and Intelligence (C4I)
Configuration management systems
Cyberhygiene

Distributed control systems (DCSs)
Heuristics
Industrial control system (ICS)
Penetration testing
Port scanning
Programmable logic controllers (PLCs)
Red Cell
Red team
Side-channel attacks

Signature-based detection
Simulations
Software testing
Supervisory control and data acquisition (SCADA)
Tiger teams
Trusted Platform Module (TPM)
Vulnerability scanning
War games

CHAPTER 11 ASSESSMENT

1. A tiger team is a group of external attackers who tear through defenses.

   A. True
   B. False

2. Malware detection that relies on a fingerprint of a virus is known as _____ detection.

   A. heuristic
   B. Trojan
   C. signature-based
   D. real-time

3. _____ provide a listing of settings that are set for an operating system to be compliant with an organization's desired goals.

   A. Patch management servers
   B. Procedures
   C. Policies
   D. Configuration standards

4. Organizations that provide information to each other are engaging in _____. (Select two.)

   A. Espionage
   B. Information sharing
   C. Awareness
   D. Email forwarding
   E. Network situational awareness

5. A policy drives the creation of _____. (Select two.)

   A. procedures
   B. central management
   C. encryption
   D. standards

6. A control system for a major gas pipeline is probably a _____ system.

   A. SCADA
   B. DCS
   C. Windows
   D. PLC

7. The control system for a water plant that controls the process of purifying water is likely a _____ system.

   A. PLC
   B. Windows
   C. SCADA
   D. DCS

8. A war game is only fought using physical weapons between two countries.

   A. True
   B. False

9. Patches and updates guarantee that a system cannot be attacked.

   A. True
   B. False

10. Full disk encryption is vulnerable to attacks that target the live memory where the key is stored.

    A. True
    B. False

11. The U.S. DOD's strategy for integrating technology command and control with intelligence is known as _____.

    A. ComCon
    B. C^3
    C. C4I
    D. CCI2

12. A _____ conducts attacks during practice exercises.

    A. black team
    B. guardian team
    C. red team
    D. hacker team

13. A list of allowed software is a type of _____.

    A. allow list
    B. graylist
    C. deny list
    D. rule-based list

Defending Networks

COMPUTER NETWORKS ARE A CRITICAL PART of national infrastructure. They provide communications among computers and other devices around the world, allowing command and control, infrastructure monitoring and management, and all of the other parts of modern business, government, and society to transfer data effectively. Of course, this makes computer networks one of the most important targets in cyberwarfare, as attackers can use networks to access data, to disable systems and infrastructure, and to attack vulnerable endpoints and other networks. Modern networks are often home to Internet of Things devices, building and system controllers, wearable devices, vehicles, drones, and a host of other emerging technologies. This added complexity means that network designs that were already more porous due to wireless technology and the pervasive use of personal technology are even more challenging and that the complexity of network defenses increases. Complexity also increases as networked devices layer additional ways to connect to multiple networks, adding additional pathways to networks that defenders may believe are well defended through their expected entry and exit points.

For many years, network designs were built with strong external protective layers to prevent attacks from the outside world. These designs were very much like a castle where successive walls and other defenses protected the most critical elements of the network. As networks have become more complex, defensive designs have changed, and more granular designs have emerged like the concept of Zero Trust. These new design concepts focus on a world in which no device or network should be fully trusted.

Since many network attacks focus on gaining access to the systems and devices that reside on that network, defenders employ a broad range of controls and techniques to help endpoints defend themselves, ranging from traditional firewalls to advanced behavior and anomaly detection systems that can detect and stop attacks as they occur. Network administrators and security specialists must carefully design defenses to balance the ability to defend a network and the systems it hosts against the operational needs and requirements of the network's users.

Attacks against systems are not the only challenge that networks can face. Denial of service attacks, kinetic attacks against critical network infrastructure, sabotage, or even specialized attacks against network devices are all possibilities in cyberwar. The U.S. Department of Defense (DOD) pays particular attention to the concept of mission assurance—the ability to provide continuous operations despite attacks, system failures, or other disruptions. The DOD's Mission Assurance directive (3020.45) provides a framework using four processes, identification, assessment, risk management, and monitoring, to help protect against attacks on the nation's defense assets as well as civilian capabilities and assets.

The U.S. National Institute of Standards and Technology (NIST) and the National Security Agency (NSA) also use five cybersecurity functions as part of mitigation efforts: Identify, Protect, Detect, Respond, and Recover. These functions are part of a complete cybersecurity strategy used by the U.S. government. In addition to the five functions, there are other government techniques and strategies. An example is the NSA's cybersecurity mitigation strategies, which include network defense strategies like hunting for network intruders, segregating networks, using threat reputation services, and other modern techniques. The emphasis that both the U.S. DOD and the NSA put on network defense demonstrates the importance of defending networks to national security and the defense of national assets in cyberwar.

Network defenders must consider both existing frameworks, like those used by DOD and the NSA, and the actual attacks that occur both in cyberwar activities and in traditional computer and network attacks and exploits. The same technologies and designs used for network security are frequently found in nation-state network implementations. The defenses deployed to protect military platforms and weapons systems can combine network technologies found in the civilian world with customized or unique platforms or systems. They may also be entirely distinct from the software and hardware available to the open market. This creates unique problems for network defenders in cyberwar, who cannot always rely on the existing body of knowledge and technology created to protect commercially available networked devices. At the same time, cyberwar defenders who deploy commercial products may find that the devices, systems, or software they have procured have built-in back doors for other countries—or that those back doors have been compromised and are available to an even broader group of threat actors.

All of this means that the network border remains one of the most important lines of defense in cyberwar, but it is just that: one line of defense. An effective network defense strategy that applies defense in depth, applies concepts like Zero Trust, and leverages an understanding of the technologies and concepts typically used in a well-designed network is critical to a successful defense. A complete security design that leverages the capabilities of network devices, procedures, policies, personnel, and standards is key to computer network defense strategies.

Chapter 12 Topics

This chapter covers the following topics and concepts:

- What designing defense in depth into computer network defense entails
- What mission assurance and information assurance in network defense entails
- What critical concepts in network security design are
- What technologies commonly used for network defense are
- What active defense in network security is

Chapter 12 Goals

When you complete this chapter, you will be able to:

- Describe network defense in depth
- Explain the concept of Zero Trust network security design
- Describe the Identify, Protect, Detect, Respond, and Recover functions
- Explain mission assurance and information assurance as used by the U.S. DOD in network defense design and implementation
- Describe commonly used network defense technologies like firewalls, intrusion prevention systems, and security information and event management devices

Network Defense in Depth

Effective defenses for networks typically require multiple layers of technologies, processes, procedures, awareness, and training, which operate in layers to provide defense in depth. Network defense in depth must focus on attacks from outside a network, attacks from trusted systems allowed to connect to the network or that may be compromised and then used to attack the network from the inside, and attacks from coworkers or trusted network or system administrators.

Layered defenses in a network design allow defenders to deal with a breach in one or more layers effectively by using other defenses to prevent the attack from making further progress. It allows them to detect the attack and react appropriately. In computer network defense activities, layered defenses are an expected part of almost any network design. They ensure that the network can survive attacks from nation-state-level adversaries who can bring significant resources to bear when they want to bypass network defenses.

FYI

The Stuxnet attack offers an excellent example of how a determined and skilled group of attackers can bypass strong network defenses. The Iranian nuclear facilities used an air gap, or physical separation between their computer networks, to help protect their production facilities. Physical separation of networks is usually considered one of the strongest protections defenders can put in place. In the case of the Stuxnet attack, the attack succeeded because of Iranian reliance on contracted engineers who used USB flash drives to transfer data between unprotected systems and the separated nuclear-facilities production network. The attack's success helps demonstrate the need for both technical and human factors in defense in depth. It also illustrates the need for a complete vision and ongoing testing when defending networks.

Cyberwarfare defenders may need to consider more than just the typical computer networks that most people envision when they hear the word **network**. In fact, cyberwarfare defenders must consider many types of networks when they design defenses. Among those networks, four major types of networks are most commonly found in many organizations. Their distinct needs can shape the choices defenders make:

- **Local area networks (LANs)** are used to connect systems and devices together. Defenders typically can control LANs relatively effectively, allowing them to choose technologies, implementation, practices, and policies to best suit the needs of their organizations.
- **Wide area networks (WANs)** connect local networks. WANs often use third-party service providers to interconnect local networks, resulting in risks based on the abilities of those service providers to ensure reliable, secure service.
- The Internet is a decentralized, open network that is not designed to be inherently secure. The Internet is a network of networks and provides both an immense variety of options for organizations that want to communicate and just as many possibilities for attack.
- Proprietary networks are based on specialized technologies or have some other distinguishing factor that makes them separate from other, more traditional networks.

This chapter explores both technologies and strategies for defending networks and uses examples of cyberattacks that can provide insight into defensive strategies that have worked and those that have failed.

Network defense of potential cyberwar targets creates significant challenges for military and government defenders due to the open nature of the Internet. Utility companies, Internet service providers, the supply chain for military, and civilian operations all rely on the Internet for their day-to-day business and may also be targets or used as part of attacks. This means that the security of more than just military systems and networks is a concern for defenders.

FYI

One question that is frequently asked when trying to build a defensible network is "Who owns the Internet?" In the context of cyberwarfare activities like attack and defense, the question often becomes "Who is responsible for the traffic on the Internet?" Because the Internet is intentionally designed to be decentralized and difficult for any country or corporation to shut down, the Internet is not the sole responsibility of any single organization. For many years, the United States held authority over the root zone file, which contained all of the names and addresses of the top-level domains for the Internet, providing it with greater power than other countries. That changed in 2016 when the Internet Corporation for Assigned Names and Numbers (ICANN) and the U.S. Department of Commerce agreed to remove ICANN from U.S. government oversight. Since then, control has rested with ICANN, which relies on advisory committees, including a 112-member governmental advisory committee (GAC) representing UN member countries and other states and a wide range of GAC observers from organizations like the World Health Organization, the International Red Cross, and others.

Unlike some countries that can easily identify a limited number of primary Internet connections to their national electronic infrastructure many, countries often have hundreds, if not thousands, of major connections to the Internet, including hardwired, wireless, and satellite connectivity. Tactics like those used by countries with limited connectivity or direct control of their network connections when they shut down or strictly limited their Internet connections are more difficult for heavily connected countries that have significant national business and security.

The United States and other major players in cyberwar are generally heavily connected with significant infrastructure that requires Internet access to operate successfully and to leverage their own cyberwar capabilities. The major Internet powers are therefore investing in both military and government programs to foster stronger, more secure networks through cooperation with corporate and other organizations through information sharing, standards, and training. At the same time, their intelligence agencies and computer network attack teams are actively looking for vulnerabilities in computer networks, creating situations where national interests may create a desire for both secure networks, and keeping known issues secret so they can be used by the nation that discovered them for intelligence or cyberwarfare activities.

The March 2014 discovery of the Heartbleed bug in the popular OpenSSL encryption package, which provides secure connections for many websites on the Internet, provided an excellent example of a bug that may have been known and exploited by a nation-state prior to its discovery by the rest of the world. News reports indicate that the NSA may have been aware of and using the Heartbleed bug to view unencrypted web traffic or to capture server security certificates for up to 2 years before the bug was patched.

Identify, Protect, Detect, Respond, and Recover in the Context of Network Defense

NIST provides a "Framework for Improving Critical Infrastructure Cybersecurity." The documented, updated in 2018, describes five core functions:

- Identify: This part of the framework focuses on understanding organizational context through activities like asset management, governance, and both risk assessment and risk management activities.

- Protect: The protect function includes activities like identity and access management, access control, awareness, training, data security, and protective technology.

- Detect: Detection focuses on identifying cybersecurity events. Anomaly and event detection, continuous monitoring, and detection processes and procedures are all part of the detect function.

- Respond: The respond function covers the activities to take when a cybersecurity incident is detected. Communication, analysis, mitigation, and improvements are all part of this function.

- Recover: The final function is the recover function, which combines ongoing efforts to ensure resilience for organizations and environments and the restoration of capabilities and services that were impacted by an event or incident. Communication and improvement efforts once again appear here, along with recovery planning.

Each of these functions are then broken down into categories like Identity Management or Detection Processes, and then each category has subcategories that describe the outcomes of specific technical and management activities. NIST describes these subcategories as including concepts like "Data-at rest is protected" and "Notifications from detection systems are investigated."

Framework implementation is also assessed using tiers as shown in Figure 12-1. The tiers themselves are not maturity levels, something that NIST is careful to note. Instead the tiers are intended to be matched to the organization's security goals. In a network defense implementation, this means that organizations must select the network defense capabilities that meet their risk and capability profile and that will address the threats that they expect to face.

> **NOTE**
>
> You can read the full Cybersecurity Framework at https://nvlpubs .nist.gov/nistpubs/CSWP/NIST .CSWP.04162018.pdf.

In a network defense scenario, U.S. government security practitioners can apply this framework to the design and operation of network defenses. First, defenders must understand the environment in which they are operating. They will build protective capabilities and processes, and then operations will detect and respond to events that are detected. Finally, practitioners will recover from incidents and events and will apply

Tier 1: **Partial**	• Risk management is not formalized, ad-hoc, and/or reactive. • Risk awareness is limited, and information sharing may not occur effectively in the organization. • The organization is not working effectively with external organization.
Tier 2: **Risk informed**	• Risk management practices are approved by management but may not be broadly implemented or supported by policy. • The organization is aware of cybersecurity risk but does not have an effective organizational strategy to address it. • The organization understands its role in a broader ecosystem but is not effectively operating in the ecosystem.
Tier 3: **Repeatable**	• The organization has stable, repeatable, and formally approved risk management practices supported by policy. • There is an organization-wide approach to cybersecurity, and policies, practices, and procedures are risk-informed. There is broad awareness as well as appropriate skills and communication. • The organization is effectively operating in the broader ecosystem and is both generating and sharing information in the ecosystem.
Tier 4: **Adaptive**	• The organization actively adapts to new cybersecurity threats and risks using lessons learned and other data sources. • The organization effectively manages cybersecurity risk and organizational objectives to maximize efficiency and effectiveness. Cybersecurity is part of the organizational culture, and the organization quickly adapts to new needs. • The organization uses real-time or near real-time information to shape its cybersecurity practices and responses. And communicates proactively using both formal and informal relationships.

FIGURE 12-1

The NIST Cybersecurity Framework—Tiers definitions.
Data from National Institute of Standards and Technology.

the knowledge that they have obtained to improve their environment, technology, and processes, continuing the cycle. Mapping cybersecurity defense practices to a standard framework helps to ensure that components are not neglected and that organizations are using similar practices.

That consistency, the way that people are trained, as well as the policies, standards, and processes they use, has an important role in the U.S. government's strategy—as do the operational elements such as log monitoring, certification, and day-to-day management.

Mission Assurance

The United States DOD uses a concept known as **mission assurance** as a key part of its defense-in-depth strategy. Mission assurance is defined as "a process to ensure that assigned tasks or duties can be performed in accordance with the intended purpose or plan ... to sustain ... operations throughout the continuum of operations." In short, mission assurance in cyberwarfare operations is the process by which systems and networks are kept online and functioning. DOD helps to ensure this by using a risk management process that assesses the functional requirements of networks and systems, then matches appropriate methods of control to them to ensure that the risks they face will not disrupt them.

DOD has steadily developed its mission assurance strategy over time, and in March 2014, it adopted a new standard as part of its overall **information assurance** strategy. The Risk Management Framework for DOD Information Technology is designed to provide mission assurance and is aligned with NIST's Risk Management Framework (RMF). The policy under which the RMF was adopted states that DOD will use enterprise-wide decision-making processes for cybersecurity risk management based on NIST standards and guidance.

DOD uses three mission assurance levels, known as **Mission Assurance Categories (MAC)**:

- MAC I are systems that are critical to the completion of DOD's mission and which, therefore, must have the highest possible level of integrity and availability as well as extremely strong security controls.

- MAC II includes systems that support deployed and backup forces. Although not as critical as MAC I, MAC II systems could have a significant impact on DOD's ability to complete its mission. Loss of availability is only allowable for very short periods, whereas loss of integrity through the modification of data is not acceptable for MAC II systems. DOD notes that MAC II–level systems must exceed common industry practices for security to provide high levels of integrity and availability.

- MAC III includes systems that are used for normal operations but are not used for operational support of deployed units or for mission-critical operations. Thus, although MAC III systems are important, the exposure or modification of data on MAC III systems, or issues with their availability, can be dealt with in a reasonable way. MAC III systems are expected to use common industry standards for security to provide integrity and availability assurance.

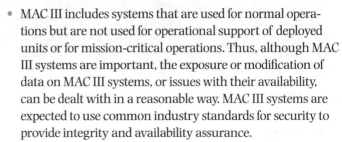

NOTE

NIST's guidance on how to apply its RMF can be found at **https://csrc .nist.gov/publications/detail/sp /800-37/rev-2/final.**

The NIST RMF provides a process for assessing risk to systems and data, then selecting, implementing, using, and monitoring appropriate controls. Figure 12-2 shows the NIST RMF process flow for risk assessment. In computer network defense operations, the categorization process typically includes assessments of the classification of data and systems as well as the mission assurance requirements of the systems and networks involved. Categorization also typically looks at relevant laws, standards, and strategic requirements, as well as what resources are available. NIST provides guidance on classification of systems for U.S. government agencies as part of FIPS 199, a standard written in 2004 for security categorization of federal information and information systems.

The DOD's continued quick pace of change in risk management and information assurance strategies demonstrates the pace at which defensive strategies and technologies have changed since DOD's Information Assurance Certification and Accreditation Process (DIACAP) was first adopted under the NSA's guidance. The level of reliance on computer networks, as well as their role in cyberwarfare and traditional kinetic warfare, has greatly increased since DIACAP was fully approved, creating an ongoing need to update and

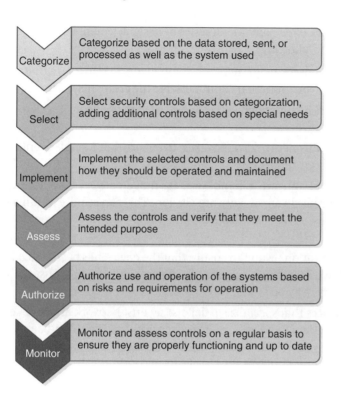

FIGURE 12-2

The NIST risk management process.
Courtesy of from National Institute of Standards and Technology.

replace existing standards. In addition to the changes made necessary by new technologies and threats, the move to the DOD RMF for information technology also better aligned the department with other U.S. government agencies that had adopted the NIST RMF for their own use. This allows shared training, easier auditing, and other advantages based on a shared knowledge base for military personnel and civilian contractors.

FYI

Prior to DOD RMF's adoption, department systems were certified and accredited under DOD DIA-CAP, adopted in 2007 as DOD Instruction 8510.01. DIACAP controls were based on a system or network's MAC and its confidentiality level (CL). The predecessor to DIACAP was the Department of Defense Information Technology Security Certification and Accreditation Process (DITSCAP).

Surviving Attacks

The need for mission assurance and the ability to react mean that surviving network attacks is an important part of defending networks in cyberwar. Networks must be both resilient and adaptable to changing network conditions—like those found in large-scale denial of service attacks conducted from thousands of systems around the world— and yet still must be able to handle an attack that targets their component parts.

Providing a resilient network often combines a series of design elements that can include capabilities such as the following:

- Diverse geographic locations for major network components to make sure that natural disasters or direct enemy action cannot disable the entire network.

- Multiple service providers if commercial services are used to provide network connectivity.

- Careful analysis of the paths network traffic takes, to ensure that a single event can't disable multiple paths at one time.

- Selection of hardware and software to handle attacks that create excessive traffic or are aimed at a specific hardware or software version. This can require multiple brands or types of devices, as well as different testing than a typical commodity device might undergo.

- Certification and configuration testing to ensure that the device has been tested for security and design flaws. This doesn't prevent the flaws, but it can help ensure that the devices are more secure by default.

- Use of commercial services that specialize in defense against network attacks like distributed denial of service (DDoS) attacks. These services use very high-bandwidth systems distributed closer to end users to ensure that an attack against a single location or data center cannot prevent the service from continuing to be available to most users.

Surviving network attacks requires both careful design and the ability to respond quickly and intelligently to attacks that do succeed. Restoring services must be carefully balanced with understanding what attackers did and what the best route to recovering may be.

Network Operational Procedures

The procedures and processes used to operate a secure network are part of the human layers of protection that go into a well-designed defense-in-depth network defense strategy. The following procedures must be in place:

- Network device management, updates, and configuration must ensure that the devices making up the network are properly patched, do not use default passwords or vulnerable configurations, and that the configurations are backed up and consistent. In addition, the way that devices are managed can help to ensure that a trusted attacker cannot modify the device to breach the organization's security and that attackers can't use a trusted system to make changes without those changes being noticed. This is done via automated configuration and validation tools or systems.

- Response processes for both breaches and vulnerabilities are needed to ensure that issues are dealt with promptly and fully. Internal disclosure and communication can help to minimize an issue's effect, whereas incorrect or incomplete information or lack of a process can lead to the problem becoming even more of an issue.

- Network entry and access management can help ensure that devices are trusted and secure. The ways in which devices are added to a network and the methods used to verify that systems or devices are allowed to connect are particularly important in a high-security environment. In DOD's compartmentalization systems, a system that is not rated for Top Secret data should not be allowed into a Top Secret network. Ensuring that a system cannot just plug into a network jack or connect to a wireless network that isn't compatible with its sensitivity rating helps to protect both the data and the network.

- Device acquisition policies and procedures are increasingly important in computer network defense. The manufacturer of devices, the code and hardware that are part of the devices, and even the shipping process they go through can be important to the security of a device.

The fear that nation-state-level organizations have influenced or modified the hardware and software of network devices has been a repeated news topic in recent years:

- In 2007, the NSA launched an investigation of Huawei, a major Chinese electronic manufacturer. The NSA's goal was to determine if the company had ties to the Chinese military and had provided, or intended to provide, a back door into devices it built. Reports in 2012 noted that 9,539 banned businesses had sold technology to the U.S. government, and that about 10 percent of those incidents involved counterfeit parts or gear.

- Information released in 2014 by Edward Snowden included pictures purporting to show routers that were diverted to facilities used by the NSA, where additional software was allegedly loaded onto the routers. The code reportedly included "phone home" capabilities, which would activate once the routers were added to a live network. Such activities can lead to devices being compromised even before they are delivered, resulting in difficult-to-detect security issues in apparently fresh-from-the-factory hardware.

- In 2016, Juniper announced that its Netscreen firewalls contained unauthorized code leveraging the NSA-designed Dual EC pseudorandom number generator (PRNG) that could allow VPN traffic to be decrypted, with the flaw possibly dating back as far as 2008.

- In May 2019, the United States banned foreign telecommunications equipment, citing national security risks. The ban saw multiple delays in full implementation as well as a licensing system, limiting its impact in many cases.

Many operational processes and policies are needed for a strong, secure network. A complete defense-in-depth design must first identify what may be needed and then must prioritize which processes are most important without leaving gaps attackers can exploit.

Network Security Design

The initial design of a network forms the basis of its security model. A strong design needs to follow organizational policies and strategies and must incorporate elements that will help handle the risks the organization faces. In cyberwar, those risks involve nation-state-level attackers, cybercriminals, organized crime, activists, insurrectionists, and even internal threats. Defenders must carefully design networks and plan additions and updates. Fortunately, many resources and design guides exist to help security designers.

FYI

The concepts and commonly used design techniques behind network defense in depth could fill an entire book. Organizations like SANS and vendors like Cisco provide frameworks for secure design, which can be useful references:

- Cisco's SAFE network design framework can be found at **https://www.cisco.com/c/dam /en/us/solutions/collateral/enterprise/design-zone-security/safe-overview-guide.pdf.**
- The SANS reading room has numerous white papers on secure network design at **http:// www.sans.org/reading-room/whitepapers/bestprac/.**

Network security designs in cyberwar typically combine the need to meet classification standards and the operational requirements of supporting government or military infrastructure. They then use traditional information security models to provide a strong, secure foundation for the network.

Classification

Nation-state-level defense networks typically add another layer of complexity to design and implementation that traditional organizational networks don't. The need for networks that can handle data classified under governmental or defense classification schemes leads to additional network segmentation. That segmentation often results in physically distinct networks. These are managed to a level of security that matches the data classifications requirements for the information traveling across them.

An excellent example of a publicly announced network scheme for handling classified data can be found in the networks of the U.S. Defense Information Systems Agency (DISA). DISA provides multiple levels of IP network services based on the data classification the network transmits. They include the following:

- SBU IP Data, formerly known as NIPRNet, is an unclassified network that provides Internet connectivity and is used for email, file transfers, and web traffic. SBU IP Data is designed to funnel traffic through security gateways, allowing monitoring, web traffic filtering, and traffic protection.

- Secret IP Data, or SIPRNet, is designed to provide a transport network for data falling under the U.S. government's Secret classification. The network is specifically designed to allow compartmentalization based on classification and supports traditional applications like email, web traffic, and other similar uses for that data. The network extends to DOD, intelligence, the civilian U.S. government, and U.S. allies. It provides centralized, monitored access to other networks.

- TC/SCI IP Data, formerly known as JWICS, provides secure communications between Top Secret and sensitive compartmented information (SCI) users who are part of the U.S. intelligence community.

The Defense Information Systems Network

The DISA operates a global telecommunications network for the U.S. military. The Defense Information Systems Network (DISN) is designed to meet specialized requirements, including security, transmission capabilities, and specialized needs based on geographic locations.

The DISN's physical network design is divided into three major areas. The first is the sustaining base, which includes the parts of the network that exist in long-term bases, facilities, and locations that host military units and other organizations. The Command, Control, Communications, Computers, and Intelligence (C4I) infrastructure on individual bases is typically maintained by each branch of the military rather than being centrally maintained by the DISA.

The second area is the long-haul infrastructure, which provides communications and network services between deployed units and task forces and the sustaining bases they come from. The DISA operates the long-haul infrastructure, including capabilities like IP telephony, teleconferencing, and various levels of secure network capabilities designed around classification levels.

The third element of the DISN is deployed units and individuals as well as their commanders. Deployed units are typically supported by their individual units. This design requires coordination among bases, the DISA, and deployed units and their support elements to ensure network security and a strong defense. Monitoring and sharing information about attacks and issues are critical, but they are increasingly complex as units use more and more networked services and platforms.

For obvious reasons, the DISA provides far less information about the more sensitive networks it operates than it does for the unclassified SBU IP Data network.

Network Defense Technologies

The technologies used to defend networks have evolved from basic network firewalls and logging systems into advanced systems that can detect attacks and take action to protect against them as they occur. Modern network defense technologies can be integrated into the underlying fabric of a network and can provide protection at any point along the network from the network edge to the port or wireless access point that individual devices connect to.

Network defense technology deployments require an understanding of the types of devices and technologies that exist and how they interact. It's easy to deploy expensive and complex network defenses without significantly strengthening a network. Attackers will exploit any gap in a network's defenses that they can find. Defenders must deploy a well-designed, monitored, and responsive network to defend against attacks that they are aware of and to help to prevent the attacks they haven't seen before.

> **FYI**
>
> Network protocols are used to determine how traffic flows between and inside networks, how traffic errors are handled, and a huge variety of network functions. This section covers two examples out of many possible choices.

Zero Trust

Once an organization has determined its security threats and risks, an appropriate design can be selected. One of the major changes in recent years has been a move toward **zero trust** designs. The zero trust concept assumes that breaches will occur, and that while traditional network defenses have a role, each system, device, or component of the network must also be protected from other components. In other words, the network is designed without a presumption of trust and with a presumption that other devices may be malicious or compromised.

In a zero trust environment, actions are validated explicitly. That means that each action must be authorized by an authenticated user, service, or server. It also means that just because that authentication has happened before, does not mean it should be trusted or allowed to occur without reauthentication. This ensures that a compromise does not allow ongoing access to resources or data. In addition to validation, least privilege access models that only provide the access that is needed to perform an action or role is are a default for zero trust. Adding on just-in-time access that only provides rights when they are needed can also add an addition layer of protection.

In a zero trust environment, these practices are combined with security technology, monitoring, and analysis tools to ensure that the likelihood of a breach is greatly diminished, and that if one occurs, the blast radius, or maximum level of impact, is limited as much as possible.

Protocols

Networks rely on protocols to control how they send and receive information, how that information is formatted, and a whole host of other elements that enable them to work. Protocols play a significant role in the security of networks by allowing data encryption, ensuring that data arrive intact and unchanged, and ensuring that attackers who have access to network traffic cannot view the data, even if they possess data decryption capabilities. The following three examples are just a small subset of the protocols used to make the Internet work, but they offer a view of the types of concerns that must be accounted for in implementations.

Border Gateway Protocol

The Border Gateway Protocol (BGP) is a protocol used between network routers to communicate information about how and where to send traffic (routing). It provides the ability to verify that other routers on a network are the neighbors they are supposed to be. Unfortunately, BGP wasn't designed from the ground up as a secure protocol. Thus, the information that it sent by BGP is not secure, and routing using BGP relies on trust between internet service providers. BGP hijacking attacks occur with reasonable frequency—sometimes inadvertently, but sometimes on purposes as well.

The issues with BGP point to one of the underlying issues with many protocols used to power the modern Internet: they weren't designed with security in mind. As more modern protocols are designed the tend to have better built-in security. At the same time, efforts are underway to add features or additional controls to existing protocols to help them be more secure as well. Unfortunately, broadly adopted protocols can be very difficult to replace in practice, meaning that insecure protocols that are in use tend to stay in use.

Choosing a routing protocol like BGP means that administrators must fully understand the implications of their choice and how to properly secure a BGP router. Each network protocol requires careful implementation and a full understanding of its specialized needs.

Transport Layer Security

Some protocols are designed from the ground-up to be secure. One of the most common protocols used to secure network traffic is **Transport Layer Security (TLS)**. It uses asymmetric cryptography to allow endpoints that have no prior relationship to securely share symmetric session encryption keys. This allows web traffic to be secured between applications and servers, providing both confidentiality and integrity of messages on a network.

> **FYI**
>
> **Secure Sockets Layer (SSL)** was the predecessor to TLS, and TLS is often still referred to as SSL. SSL itself is now outmoded, and TLS is the standard of choice of secure web traffic.

IPsec

Another example of a security-oriented protocol is IPsec, or Internet Protocol Security. IPsec is a protocol suite that provides both authentication and encryption of data packets, ensuring that systems can communicate securely. IPsec is often used for VPNs, but it also plays a role in application and routing security.

Network Access Control

An important part of network defense is controlling which devices and users can connect to the network. In addition to physical controls, networks can also use technologies that allow them to identify users and systems before allowing them to fully connect to the network. Equally useful is the ability to use the information that is provided to place a system into a network segment that is appropriate to the system's rights or status.

Network access control (NAC) allows networks to require users to authenticate to the network. It can also use software on systems to assess those systems' security and compliance status. If the user is validated properly, and his or her system meets the requirements that the NAC system is configured to require, the user can then be placed into the proper network segment based on his or her user rights. If the user fails, the network can reject the connection or can put him or her into a virtual network penalty box that limits access to only the tools needed to make the system compliant.

NAC systems can be very thorough in their testing, with some NAC software systems integrating antivirus scanning to check for malware and configuration scanning and validation tools to ensure that systems have every setting right. NAC systems can even allow users to change the network they connect to based on their needed access.

Network Firewalls

Firewalls are one of the most commonly deployed network security technologies and are widely used to provide network separation and defense. Firewalls use a set of rules about which traffic may travel through them based on the network addresses, ports, protocols, or details of the traffic that is sent. These lists of rules are known as **rulesets**, and firewall administrators manage their firewalls' rulesets carefully to ensure that only allowed, trusted traffic is passed through.

Three common types of firewalls are deployed, ranging from very basic packet filters to stateful packet inspection firewalls to advanced, next-generation, application-aware firewall devices. Each firewall type has different capabilities. The basic features of each are as follows:

- **Packet filter firewalls** are the most basic type of firewall. They inspect the IP address that is sending traffic, the receiver's IP address, the port that the traffic is sent or received from, and the protocol by which the traffic is being sent. Packet filters look at each packet that sent, repeating their check to ensure that the traffic matches what the firewall allows. They have no understanding of the content of the traffic—that means that traffic sent from a valid sender, on a valid port, and via a valid protocol to a system that will accept the sender's traffic will go through the firewall, even if it is actually an attack.

- **Stateful packet inspection firewalls**, or stateful firewalls, apply additional intelligence beyond the simple individual packet inspection applied by packet filters. Stateful firewalls maintain a mapping in memory of systems that have participated in allowed communications and use that information to allow conversations (sessions) to continue between systems. This means that instead of each packet being inspected, the firewall only needs to verify that the following packets are part of the same network conversation before allowing them to pass. It also allows the firewall to use simpler rules, as responses to messages sent can be allowed back in as part of the conversation, rather than requiring specific inbound rules to permit the traffic in. This simplifies ruleset construction and makes it less likely that firewall administrators will accidentally forget to allow traffic back through a firewall.

- **Application-aware firewalls** are even more advanced and provide the ability to inspect the actual protocols and application traffic that flow through the firewall. This means that the firewall must spend significantly more resources looking at traffic that passes through it, but it also means that it can detect attacks against the application servers that it protects. Application-aware firewalls can also verify that attacks against Internet protocols are not occurring—preventing attackers from exploiting the network software on servers and devices.

- **Unified Threat Management devices** and other advanced security devices add additional features beyond firewall capabilities, including intrusion detection and prevention, anti-malware capabilities, and other features that each device manufacturer decides are important. Many modern firewall devices are far more than just a firewall.

FYI

A well-designed firewall's final rule will always be a rule that denies all traffic from passing through it. This rule, known as an implicit deny rule, ensures that any traffic that doesn't match a rule allowing it will be stopped. The opposite rule, which allows all traffic, sometimes known as an **ANY/ANY rule**, will make the firewall useless because it will match all traffic and allow it through.

Figure 12-3 shows a sample attack from a remote system through each of the common types of firewall devices. The packet filter firewall checks to see if the traffic is allowed, and because that system is allowed to send responses to a trusted system, it passes the traffic. The stateful packet inspection firewall makes a similar check and notices that the system started

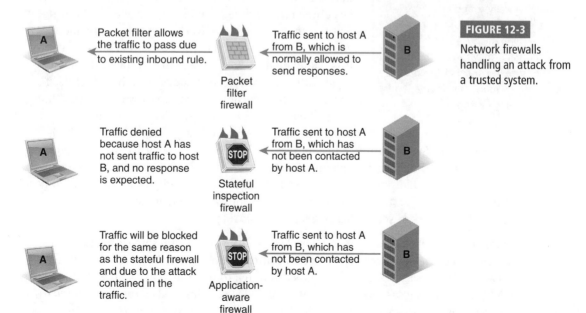

FIGURE 12-3

Network firewalls handling an attack from a trusted system.

Packet filter allows the traffic to pass due to existing inbound rule.

Packet filter firewall

Traffic sent to host A from B, which is normally allowed to send responses.

Traffic denied because host A has not sent traffic to host B, and no response is expected.

Stateful inspection firewall

Traffic sent to host A from B, which has not been contacted by host A.

Traffic will be blocked for the same reason as the stateful firewall and due to the attack contained in the traffic.

Application-aware firewall

Traffic sent to host A from B, which has not been contacted by host A.

the conversation, rather than responding to an ongoing conversation, and stops the traffic. The application-aware firewall detects that the communication does not match the requirements of the application protocol and stops the traffic because it is a known attack. It would also stop the traffic if the traffic were not part of an existing conversation started by host A.

Firewalls are used in most secure network designs to ensure that only permitted traffic is sent through network control points. In most designs, they aren't intended to react rapidly to new threats. Instead, firewalls are typically changed through a well-managed change process. This change process is designed to ensure that new rules don't prevent legitimate traffic. It also ensures that changes that are made are well understood and only allow the intended traffic to pass.

FYI

Host-based software firewalls also exist and are built in to many modern operating systems. They provide protection only for the system they are installed on rather than an entire network or groups of systems. This allows fine-grained protection for each device, but it also requires ruleset maintenance for every machine that uses one. In most cases, host-based firewalls are either controlled centrally, or users are allowed control over the firewall so that they can allow applications to send or receive traffic.

> **NOTE**
>
> The term **enterprise** here refers to a professional-grade device of a type typically found in a large organizational network. Enterprise devices typically have additional functionality, may have redundant power supplies or other features, and provide more management and configuration options.

Routers and Switches

Networks rely on devices known as **routers** to send data between computer networks. Routers are typically found at network boundaries or inside large networks to control traffic between network segments. Routers offer a wide range of features that can allow network defenders to provide access control and security for a network, including access control lists. Many modern enterprise routers provide the ability to add other network security devices to the router itself, providing advanced firewall features as well as capabilities like intrusion detection or network monitoring.

Networks also use switches to distribute data to individual systems. Switches, much like routers, are available with a variety of capabilities, including some that have routing functionality built in to them, blurring the boundaries between a traditional switch and a router. Enterprise switches often can protect their ports from various attacks; they also have options and software that can help them support additional network-based authentication and control policies.

Access Control Lists. Most routers, more advanced switches, and some other network devices can use a tool known as an **access control list (ACL)**. An ACL lists what systems or network IP addresses are allowed to pass through a router, much like a firewall ruleset can limit traffic between systems. Typical ACLs provide less-advanced features compared to the more-advanced types of firewalls and are therefore used for simple restrictions.

Most of the Internet uses a version of the IP protocol known as IPv4. The number of IPv4 network addresses in the world are limited (4,294,967,296, or 2^{32} addresses), which led to worries about an impending disaster when the last IP address was allocated. A technique known as **network address translation (NAT)** was created to allow use of a range of IPs behind a router, with one or more addresses on the outside that those addresses are translated to. The router then tracks which traffic belongs to which internal system. NAT is sometimes used to help protect systems; attackers normally can't directly target a device that is using an internal IP address behind a NAT device because the actual address of the system isn't accessible.

Network Security Boundaries

Network defense inside networks often relies on the network security boundaries that network and security administrators can build between parts of the network or systems that have different security or access requirements. Boundaries in networks are typically created at network control points, such as the following:

- Firewalls or routers, which are used to divide networks
- Endpoint switches, which can limit access to wall jacks and individual systems, as well as tag the devices connected to a port into a given network
- Wireless access points, which can authenticate connections and can allow users and devices to connect to multiple distinct networks
- Remote access systems, which can authenticate users and place those users into separate network groups or address ranges

Administrators and defenders can both use these control points to separate systems based on the rights their users have, the configuration or state of the system, or other information they have about the systems based on software running on the system or an access control system providing information about the user's rights.

NIST recognizes the complexity of network boundaries and those found in more complicated network-based information systems. In NIST Special Publication 800-37, a complex information system is broken down into multiple subsystems, as shown in Figure 12-4. In the NIST example, breaking a complex system into subsystems allows network segmentation and security domain separation. The NIST standard notes that organizations can then develop security at the subsystem level, with privacy plans and controls set at each subsystem, for groups of subsystems, or for the system as a whole as appropriate.

Virtual LANs

Network security relies on two major types of separation: logical and physical. One of the most common ways to logically separate networks is using a virtual LAN (VLAN). VLANs are network segments that are logically separated using information known as a VLAN tag, which is applied to network traffic. VLANs are configured on network switches and routers, allowing networks to place systems into appropriate groups based on the tag provided.

Physical separation, on the other hand, is done by using completely separate hardware and wiring. Obviously, physical separation is typically both more complex and more

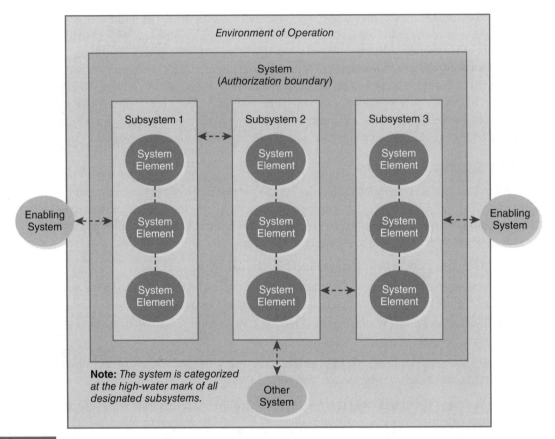

FIGURE 12-4

NIST's use of network security boundaries.
Data from National Institute of Standards and Technology.

expensive, but when networks have a very strong set of security requirements, physical separation can provide greater protection—unless your opponent can access the network in person or through social engineering.

VLANs are used when the security requirements for separation of networked systems and devices do not make physical separation necessary. They can provide a moderate level of separation, although they cannot guarantee separation because misconfiguration, access to the switches or routers, or VLAN hopping attacks could provide attackers with the ability to see traffic on other VLANs.

FYI

Attacks against VLANs known as **VLAN hopping attacks** can allow attackers to send traffic to other VLANs. Fortunately, modern network devices provide protection against VLAN hopping, although it is not always enabled.

Intrusion Detection and Prevention Systems

The ability to detect attacks as they occur is very useful when defending a network. **Intrusion detection systems (IDSs)** are devices created to detect attacks based on their distinct network signatures or the behaviors they exhibit. After creating IDSs, the next step was to put the IDS between the attackers and their potential victims and allow the system to both detect and then prevent the attacks. This resulted in the creation of **intrusion prevention systems (IPSs)**.

Both IDS and IPS software and devices rely on two major types of detection:

- Signature-based detection looks for specific attack signatures that malware or attacks use. Signature-based detection requires that the defender know what the attack looks like before detecting or stopping it. This means that signature-based systems tend to be reactive instead of proactive. Properly built signatures will catch all examples of a known attack that they can see but can't stop unknown attacks.

- Heuristic or behavior-based detection systems look for patterns or types of action that match attacks. They can stop unknown attacks that fit those patterns, but they can also stop legitimate traffic that happens to match the same pattern.

Most IDS and IPS systems can combine signatures and heuristics, and administrators have to choose which rules and methods they want to use to balance the security requirements of their network against the detection schemes' strengths and weaknesses.

> ### The Dangers of Detection
>
> DOD's mission assurance strategy creates a challenge for IPS operators. One of the frequent challenges for IPS operators is balancing the need to block attacks against the need to allow legitimate traffic. Unfortunately, the balance between identifying attacks and not stopping legitimate traffic can be difficult. Often, new attacks will appear to be very similar, if not identical, to normal traffic. As administrators and network defenders write new rules to identify attacks, it is easy to make mistakes or to create an overly broad rule.
>
> Once a rule is in place and active, it can result in significant network outages. A single IPS placed in a critical location that blocks important traffic, or that identifies a common network protocol as a source of attack, can bring network operations to a halt. Much like a hardware device failure, a misconfigured IPS or one with a bad rule in place can knock an organization offline.
>
> This doesn't mean that an IPS isn't worth the risk. But it does mean that defenders whose primary goal is to provide mission assurance must carefully consider the balance between the risk of attack and the risk of network outages. Some defenders may opt for an IDS that will make them rely on manual responses to attacks. Others may design a strong testing process that will slow down deployment of potentially important new protection rules.

Security Information and Event Management Systems

Network devices can provide information about their status using automated logging capabilities that send details of what they are doing, whether errors have occurred, and what their configuration is. This data, as well as traffic data from the network, firewall, IDS, IPS, and other device status, and user actions are often gathered centrally. Once that huge amount of data is securely copied to a central location, defenders must then use it. Traditionally, that required huge amounts of time and manual searches through data. In many cases the data were simply ignored or used only after a compromise or attack occurred.

Fortunately, software and devices that can automatically manage the data and identify events and issues have become more very common and capable. **Security information and event management (SIEM)** devices typically include the ability to monitor network traffic and traffic flows, logs, security information provided by machines, vulnerability scanning data, and other security information. They match the information to known attacks, as well as patterns that may indicate attack, and allow administrators to create rules to set off alarms or to take preprogrammed actions when those rules are matched.

Managing the overall security posture and status of network devices can allow defenders to determine if a compromise has occurred or whether an attack is in progress. It can also allow them to see abnormal traffic or behaviors that might help show that a trusted user is acting outside his or her role. SIEM systems help to sort through the massive amounts of data that a well-configured and logged network provides.

A relatively new category of security solution that may sometimes be confused with SIEM solutions is a **SOAR**, or **Security Orchestration, Automation, and Response** tool. Where SIEM focuses on information and events, SOAR focuses on case management, incident response workflows, and playbook-based responses using alarm and event data.

Physical Network Protection

Physical access to networks provides an ideal situation for attackers. Once they have physical access to the network or its component devices, they can do anything from monitor the traffic on the network to insert their own devices into it. Monitoring miles of cables and interconnected devices spread across huge distances can be a real challenge, which means that most physical network defense is done at the building or installation level, with traditional physical security boundaries like walls, fences, and security guards employed to ensure that attackers cannot gain access.

Access to the physical cables that data travels over is harder to prevent. The United States deployed listening devices into undersea cables using submarines in the early 1970s as part of Operation Ivy Bells. It tapped undersea fiber-optic cables in the 1990s, demonstrating the lengths that nation-states will go to in acquiring intelligence. Detecting devices that can simply read traffic passing through a cable can be nearly impossible. This makes it a necessity to strongly encrypt traffic that passes through unsecured network infrastructure, wireless connections, or where traffic has to be carried through commercial providers.

Network defenders are not always responsible for the physical-security design of their network. But they must be aware of the threats physical access to the network and its component devices can create. Nation-state-level resources can make otherwise unlikely scenarios a reality when the traffic that passes through a network is sufficiently valuable.

Wireless Network Security

Wireless networks create additional security issues for defenders due to the inability to restrict their signals. Wireless signals can be accessed from far-greater ranges than users may expect by attackers using directional antennas and other tools. This makes data sent over wireless connections much more easily accessible than physical network traffic in a well-secured building.

In addition to the threats that defenders must take into account with the spread of wireless signals, a number of other wireless threats exist:

- Fake wireless networks, which lure unsuspecting users into sending their data through untrusted access points

- Tools that exploit unencrypted wireless connections to capture data or to modify data sent by users

- Fake cellular towers, which capture data seemingly sent through to legitimate carriers

With a broad range of attacks possible against wireless networks, physical wires remain a popular solution for the highest-security environments in use today. Where wireless is used, defenders often layer encryption and additional tools that help to prove the network identity of both the wireless infrastructure and the systems that connect to it using tools like digital certificates.

> **FYI**
>
> The U.S. NSA and other high-security organizations sometimes design or modify their buildings to limit, or entirely prevent, the spread of wireless signals through the walls and floors of the buildings themselves. This is an expensive solution, but one that can be necessary to prevent wireless signals from being monitored by those outside the building or even between parts of the same structure. The NSA uses the code name Tempest for technologies related to both capturing and protecting information from capture through the use of leaked emanations like vibrations, radio frequency emissions, and sound waves.

Remote Access and Administration

Attackers can cause major damage if they can gain remote access to a network or take over the network's administrative systems. Network management and remote access capabilities are both important to the network's defensive strategy.

Network Management. The tools used to manage network devices play a major role in the maintenance, configuration management, and monitoring of a network. Network management tools help ensure that routers, switches, wireless access points and controllers, and other network devices have the correct configurations, that the devices are functioning properly, and that administrators can properly control them.

TOR: The Onion Router

A common tool used for anonymity is TOR, the Onion Router. TOR combines elements of a virtual private network with routing and encryption technologies to create an anonymizing network that permits the origin, destination, and content of traffic to be concealed from observers.

TOR passes encrypted traffic through TOR entry nodes, which then pass that traffic through other TOR nodes, each stripping off a layer of encryption to learn where to send the traffic next. The final node doesn't know where the traffic came from, only where it is going, and where along the chain to send responses. This conceals the full path from any participating node and provides anonymity to TOR users. For this reason, TOR is used for both legitimate and illegitimate uses, including cyberattacks and cyberwarfare activities.

TOR also uses its own internal Domain Name System, allowing the creation of systems that are only accessible via the TOR network. TOR addresses have been used for a variety of purposes, including online drug and weapons sales, secret message boards, and other criminal enterprises. Due to the combination of criminal activity and anonymity, the TOR network has become a prominent target for intelligence and law enforcement agencies.

Various nation-state organizations, including the NSA and UK Government Communications Headquarters, have created tools designed to defeat TOR. The toolkits they have employed include compromising the originating systems via browser-based tools, using fingerprinting to uniquely identify browsers, and even adding government-owned and monitored TOR nodes. TOR remains resistant to purely network-based attacks; attackers would have to own a majority of the TOR nodes in the world to have a reasonable chance of identifying both the sender and receiver in any given communication. Attacks that focus on the originating system or the server that it communicates with have generally been more successful because they allow unique identification of the traffic. Attacks against the server don't reveal the source of the traffic without additional data, but attacks against the sender have the potential to reveal everything sent and where it is sent.

Network management tools can also provide a useful means for managing standards. Many manufacturers provide standard configuration guidance that can ensure that devices meet security baselines, and their management tools can then use those prebuilt profiles. Network defenders who use administration tools can often use the time they save by using those tools to improve the security of their network in other ways.

Virtual Private Networks. One of the most common ways to remotely access a network is through the use of a virtual private network (VPN). A VPN uses software and network protocols to build a private channel through other networks. VPNs are often, but not always, encrypted, allowing a separate, secure network path through otherwise public or untrusted networks. VPNs are also used to connect remote networks together through third-party infrastructure such as the Internet. A VPN set up between two sites can allow those sites to appear to be directly connected, despite the fact they are not actually both connected to the same physical network.

VPNs are an attractive target for attackers; VPN-connected users appear to be on the network they are connecting just like a local physically connected user would be. Defenders often choose to place VPN users in a distinct zone with different trust or monitoring rules. They may also make VPN users employ additional methods to verify their identity, such as two-factor tokens or additional logons.

There are many examples of VPNs being targeted, including the November 2020 posting of a list of systems vulnerable to an issue discovered in 2018 that impacted Fortinet security devices. The security flaw resulted in attackers potentially being able to gain VPN credentials from almost 50,000 unpatched VPN devices. With those credentials in hand, attackers could log in via the VPNs and access protected networks. While exploits using those credentials have not been reported at the time of the writing of this book, the scale of exposure and the broad visibility of the issue mean that the opportunity exists for a significant number of compromises based on the flaw.

Active Defense

The U.S. Department of Defense has increasingly required its networks to use an **active defense** paradigm, as shown in its 2011 strategy document. The basic design requirement is that a network must adapt to attacks. If an attacker is using a specific attack, the network should respond by blocking that attack. If a system is compromised, the network must detect it, remove the system from its trusted network zone, notify administrators that the system needs investigating, and then allow normal operations for the systems that were not part of the incident.

Active defenses require a carefully constructed network and provide many challenges for defenders. The more active defense capabilities a network has, the more likely it is that the network could inadvertently take actions that interfere with normal network operations. The risks are matched by advantages that drive DOD's push for networks that actively defend themselves. A well-built network with active defense capabilities can in theory respond to most attacks faster than human monitoring and management teams can. It can adapt to attacks that might overwhelm the network before it is taken down. In a cyberwar scenario, a **self-defending network** that responds to attacks as they occur and changes itself to prevent, defeat, or at least minimize the assault would be a major advantage against an attacker who did not have the same capabilities on his or her own network.

FYI

Self-defending networks have been proposed for years; however, a network with true end-to-end, self-defense capabilities doesn't yet exist. Still, networks often have some self-defense capabilities, and the advent of SIEM and SOAR systems that integrate well into network management tools has brought the idea closer to reality. As automation has continued to advance and ephemeral systems and artifacts like virtual machines, containerized applications, and cloud hosting have all become the norm, automated responses are becoming more common.

DOD's active defense strategy envisions the following:

- Real-time detection of threats so that they can be handled as they occur
- Attribution of threat actors, providing the ability to deal with the actual adversaries as they attack and expanding the potential for effective, real-time responses
- Flexibility of response actions with choices available to defenders about how to respond and on what scale
- Intelligence dissemination

Honeypots, Honeynets, and Darknets

In addition to self-defending networks, defenders can benefit from understanding the attacks their networks face. Nation-state-level network security organizations must remain constantly aware of new threats. One way to do this as part of an active defense design is to use a **honeypot**. A honeypot is a vulnerable system that is monitored and equipped with logging capabilities. When an attacker compromises the honeypot system, defenders are provided with copies of any files or software the attacker used as well as complete information about how the system responded to the attack.

Honeypots can provide defenders with up-to-date information about attacks by showing the complete process that an attacker uses (see Figure 12-5). **Honeynets** take the

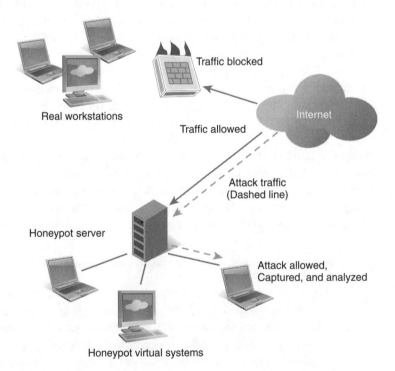

FIGURE 12-5

A honeypot captures and allows analysis of attack traffic.

Real workstations

Traffic blocked

Internet

Traffic allowed

Attack traffic
(Dashed line)

Honeypot server

Attack allowed,
Captured, and analyzed

Honeypot virtual systems

concept of a honeypot to the network layer and provide a network honeypot for attack-ers. Honeynets can combine monitored network devices with honeypot systems, often in a virtual environment. Defenders can easily restore compromised systems to their original state while retaining all of the information about how the system or network device was attacked.

Organizations that have unused network space sometimes also set up darknets. A **darknet** relies on the fact that unused space should have no legitimate traffic sent to it. Monitoring unused networks allows defenders to see what scans and attacks occur against systems and can help identify attackers who are targeting broad ranges of network space.

> **NOTE**
> The term *darknet* is sometimes applied to TOR networks because they operate in an otherwise difficult-to-detect part of the Internet.

Defenders who want to take a more active role in their defense can use honeypots, honeynets, and darknets to provide insight into attacks and who the attackers might be. Defenders can use the information they gain from honeypots, honeynets, and darknets to build firewall rules, to improve the defenses of the systems they deploy, and to understand what new attacks may be used on their real networks and devices.

The Honeynet project provides prebuilt and configured honeypots and honeynet sys-tems. You can read more about this at https://www.honeynet.org.

Active Response

In traditional information security scenarios, an **active response**, or "hacking the hacker" is normally considered to be a bad idea. Civilians and corporations are typically prohibited from attacking others under local, national, or international law. Cyberwarfare opens a new possibility for defenders who may be authorized or even expected to use active responses as part of their computer network defense responsibilities.

Active responses are simply computer network attack capabilities linked to computer network defense. When an attack occurs, or when one is known or suspected, defenders may strike at the systems or networks that are attacking or will attack. This can allow the defenders to weather an attack relatively safely if their own systems are not attacked or if their actions blunt the attack.

Active responses require a careful balance by defenders, however, as many attacks come from systems or networks shared by civilians. Although international law has not solidified around cyberwarfare activities, attacks against civilian infrastructure can have unexpected and undesired effects.

CHAPTER SUMMARY

This chapter discussed computer network defense in cyberwar. U.S. civilian agencies and the military use the concept of mission assurance, or the ability to provide available and usable networks to the organization, as a key element in network defense. Critical to mission assurance are, first, a strong network security design that supports the classification of data and, second, systems based on assigning roles and rights to users and objects, such as programs and computers.

Network defenders must also be familiar with a variety of network security devices commonly used in secure networks. These include border defenses like firewalls, intrusion detection and prevention systems, and security configurations deployed on network routers and switches. They also include systems designed to gather network logs and security events, bait systems designed to gather information about attacks, and a wide range of other tools. The U.S. military envisions a secure network that uses these tools and detection capabilities to react to threats and provide an active response to attacks by reconfiguring itself to stop or limit them as they occur.

KEY CONCEPTS AND TERMS

Access control list (ACL)

Active defense

Active response

ANY/ANY rule

Application-aware firewalls

Darknet

Enterprise

Firewalls

Honeynets

Honeypot

Information assurance

Intrusion detection systems (IDSs)

Intrusion prevention systems (IPSs)

Local area networks (LANs)

Mission assurance

Mission Assurance Categories (MAC)

Network

Network address translation (NAT)

Packet filter firewalls

Routers

Rulesets

Secure Sockets Layer (SSL)

Security information and event management (SIEM)

Security Orchestration, Automation, and Response (SOAR)

Self-defending network

Stateful packet inspection firewalls

Transport Layer Security (TLS)

Unified Threat Management devices

VLAN hopping attacks

Wide area networks (WANs)

Zero trust

CHAPTER 12 ASSESSMENT

1. Which of the following is not a key concept for zero trust networks?

A. Actions are validated explicitly.
B. Breaches are assumed to occur.
C. Access should be granted just in time and with least privilege.
D. All operations must be done as an administrator.

2. What are the three key parts of the NSA's defense-in-depth strategy?

A. Protect
B. Prevent
C. React
D. Detect

3. What are the NIST framework for improving critical infrastructure capability core functions?

A. Assess, Review, Design, Implement, Monitor, Report
B. Identify, Protect, Detect, Respond, Recover
C. Assess, Design, Implement, Review, Report
D. Categorize, Analyze, Select, Implement, Monitor

4. Which of the following networks implements the highest level of network security for DOD use?

A. SBU IP Data
B. TC/SCI IP Data
C. TopNet
D. SIPRNet
E. ARPANet

5. Technology that determines whether a user should be allowed to connect to a network and the computer's current security profile is known as _____.

A. AAA
B. a firewall
C. an IP gateway
D. network access control

6. A firewall that knows whether a system was previously contacted by a protected computer and allows responses from it is a(n) _____.

A. stateful packet inspection firewall
B. packet filtering firewall
C. application-aware firewall
D. access control list

7. DOD uses separated network segments known as _____ to contain different levels of data or users.

A. nets
B. blocks
C. suburbs
D. enclaves

8. A system that can block attack traffic by reviewing the content of the traffic and identifying the attack based on behavior or a fingerprint is known as a(n) _____.

A. intrusion prevention system
B. firewall
C. intrusion detection system
D. active response system

9. Wireless networks are as secure as wired networks.

A. True
B. False

10. A device that collects both network logs and monitors for security events is known as a(n) _____.

A. SEM
B. SIM
C. IDS
D. SIEM

11. Technology that allows a network to defend itself as attacks occur by reconfiguring systems and devices or taking other actions is known as _____.

A. SIEM
B. active defense
C. active response
D. an IPS

12. False systems known as _____ are used to observe the methods attackers employ to break into them.

A. darknets
B. syslogs
C. honeypots
D. bear traps

Defending Data

C OMPUTER NETWORK DEFENSE, OR CND, is the defensive component of cyberwarfare. In a world filled with complex environments with multiple forms of connectivity and a multitude of attack surfaces for malicious actors to target, it is also incredibly complex. Defenders have the seemingly impossible job requirement of defending systems, networks, and data in many forms in a massive number of locations without making it unusable.

Defensive operations in cyberwar often have a strong focus on protecting data that are stored on workstations, devices, and in network file storage. Network defenders must protect data from both external attackers and unauthorized internal access. This involves data classification, roles and rights, and need-to-know standards as well as encryption and other data defense techniques. If employed by nation-states, network defenders protect data such as military and civilian government secrets and strategies, as well as everyday communications and business information. Although some of that information might sound unimportant, leaks of U.S. diplomatic communications, emails belonging to political campaigns, and other lower security data have shown that leaked day-to-day data can result in diplomatic embarrassment or even larger issues for a country.

Defenders in cyberwar also must consider the larger context in which information warfare takes place. For example, other nation-states will target nongovernmental organizations, including targeting defense contractors with the goal of acquiring military technology, designs, and specifications. Universities and other research organizations are often targeted due to their work with government contracts and research, and uninvolved organizations may be attacked as a stepping stone or incidental target of the attack. Even apparently unrelated data thefts can result in attackers gaining greater knowledge about supposedly secret information or access to systems they can then use to more effectively reach their final goals.

The need to protect data has resulted in the creation of a number of techniques and technologies that attempt to make data easier to manage and protect. Government agencies use data classification to label data based on its need for secrecy and the handling requirements that it requires. This allows defenders to focus on data that should receive more time and effort, while providing less protection to data that has a lower classification level. It also provides defenders with an opportunity to focus on the human side of computer network defense by ensuring that individuals given higher levels of clearance understand the importance of the data and networks they have access to. Finally, the vetting and review processes involved in clearance for access to data can help ensure that human threats to data are decreased—although these processes cannot prevent individuals with access from abusing their privileges.

Technology to protect data can detect changes in data, identify data are at risk of being sent or transferred, and prevent data that are lost from being recovered. Technologies for data protection include data integrity-checking tools, which verify that data have not changed, and a means of restoring data after they are changed. They also include data loss prevention tools, which monitor network traffic and transfers to removable media and then stop the transfer or send an alarm to monitoring systems and staff. Encryption also has a major role to play by ensuring that data and devices that are lost or stolen cannot be easily recovered.

Defenders use tools, techniques, and processes such as classification and data loss prevention systems to protect data throughout its life cycle. Defenders must handle data at rest, data in transit, and data that have reached the end of their life-span in ways that help to protect confidentiality and integrity, while ensuring the data are available for users who need them. Attackers will attempt to counter each of the techniques that defenders use. In many cases, despite knowing the protections are necessary, even the users of the data will work around the protections that are in place if they make work difficult to accomplish.

Chapter 13 Topics

This chapter covers the following topics and concepts:

- What data classification is and why it is used to provide data security
- What data loss prevention technologies and strategies are
- What data integrity and availability are, and how they are preserved
- What data retention and disposal policies are
- What response strategies for data loss in cyberwar are

Data Classification

Data **classification** is the process of labeling data based on its sensitivity and handling requirements. The U.S. government uses three primary classifications, with additional compartmentalization and security labels added based on specific organizational or handling needs (see Figure 13-1). Data classification and compartmentalization help the government ensure that only authorized individuals can access data and that the data are appropriately protected.

The U.S. government uses classification authorities—individuals trained to identify and assign the proper level of classification to data. Classification authorities also ensure that the data are classified for a valid and proper reason, and then ensure that they are marked with the proper classification identifiers. This helps to ensure that documents can be declassified when appropriate.

Classification follows the data it is attached to. Thus, if data with a higher classification level are added to data with a lower classification level, the data will typically be classified at the higher level while the higher-classification data element or elements are present. In paper documents, this could make an entire document Top Secret if a single piece of information were added to an otherwise unclassified document. If done properly, each individual element of data may be classified, making it easier to understand the document's or data's classification context.

 NOTE

The National Institute of Standards and Technology (NIST) provides guidance on Security Categorization in FIPS-199. This isn't the same as data classification, but there are parallels in the analysis required to determine how data and systems should be classified based on their risks and impact. NIST Special Publication 800-60 provides more detail on how to analyze mission-based information types and how the information is delivered. You can review FIPS-199 at https://nvlpubs.nist.gov/nistpubs/FIPS /NIST.FIPS.199.pdf.

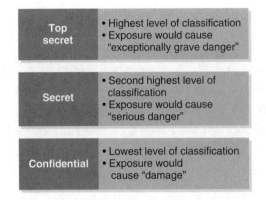

U.S. government data classification levels.

Data that have been removed from the classification scheme are **declassified**. Data that have been reclassified at a lower level but remain classified are known as **downgraded** data. The U.S. government uses a combination of sensitivity and the age of the data to make many decisions around declassification. In some cases, it declassifies much of a document or data set, while **redacting** or **sanitizing** the remaining classified information.

The government uses other data labels to provide additional compartmentalization of data and has historically had as many as 100 different designations for data. Examples include the following:

- Sensitive But Unclassified (SBU) is for data that should not be released casually but are not considered Secret and would not cause significant damage if released.

- Sensitive Security Information (SSI) is for data that are sensitive but not Secret information. It's subdivided into 16 categories. It includes information such as security programs and plans, threat information, confidential business information, and information related to the security of transportation.

- Critical Program Information (CPI) is defined by DOD to include information from a research, design, and acquisition program that "if compromised, could cause significant degradation in mission effectiveness" or could cause a variety of other problems.

FYI

Redaction of both physical and digital information can be challenging. Physical documents are often blacked out using a marker, and similar processes were once used for digital documents like PDFs. In multiple cases, government-published data in PDF form were redacted using a tool in Adobe Acrobat. The redaction tool simply placed black blocks over the original data, and it was a simple matter to remove that layer to reveal the redacted data. Adobe has since modified how the redaction tool works to actually remove the data instead of simply adding a block over it. This issue was so prevalent that the National Security Agency (NSA) released a document on how to properly sanitize reports. You can read it at *https://apps.nsa.gov/iaarchive/library/ia-guidance /security-configuration/applications/redaction-of-pdf-files-using-adobe-acrobat-professional-x.cfm*.

Two of the most commonly referenced models for data security enforcement in government and military systems are Bell-LaPadula and Biba. Bell-LaPadula uses security labels for objects and clearances for subjects to provide data confidentiality by ensuring that users with lower levels of security cannot read higher-level data, while users with higher levels can't write to lower levels. This is often written as "no write down, no read up." Biba protects integrity by preventing high-security-level users from reading lower-level data and lower levels from writing upward, or "no read down, no write up."

- For Official Use Only (FOUO) is for data that should be used only for official purposes.
- Law Enforcement Sensitive (LES) is for information that should be restricted to law enforcement officers.
- Public Safety Sensitive (PSS) is for information that is appropriate for public safety officers, including emergency medical services, fire safety, and law enforcement.

An attempt to redesign and limit the number of categories resulted in the creation of the designation **Controlled Unclassified Information (CUI)**, which was published as a final rule in September of 2016. Prior to the CUI program, federal departments, agencies, state and local, tribal, and other organizations all used different practices for handling and sharing data. The CUI effort is intended to allow sharing in a more effective way while providing standards that help to ensure it is handled appropriately. It includes categories such as critical infrastructure, defense, intelligence, export control, and other types of data.

Examples of the limited dissemination controls as well as their descriptions as provided by the U.S. government that are part of CUI include the following:

MARKING	CONTROL	DESCRIPTION
NOFORN or NF	No foreign dissemination	Information may not be disseminated in any form to foreign governments, foreign nationals, foreign or international organizations, or non-U.S. citizens.
NOCON	No dissemination to contractors	No dissemination authorized to individuals or employers who enter into a contract with the United States (any department or agency) to perform a specific job, supply labor and materials, or for the sale of products and services. Note: This dissemination control is intended for use when dissemination is not permitted to federal contractors but permits dissemination to state, local, or tribal employees.

13

Defending Data

MARKING	CONTROL	DESCRIPTION
FED ONLY	Federal employees only	Dissemination authorized only to (1) employees of U.S. government executive branch departments and agencies (as agency is defined in 5 U.S.C. 105) or (2) armed forces personnel of the United States or Active Guard and Reserve (as defined in 10 USC 101).
FEDCON	Federal employees and contractors only	Dissemination authorized only to (1) employees of U.S. government executive branch departments and agencies (as agency is defined in 5 U.S.C. 105), (2) armed forces personnel of the United States or Active Guard and Reserve (as defined in 10 USC 101), or (3) individuals or employers who enter into a contract with the United States (any department or agency) to perform a specific job, supply labor and materials, or for the sale of products and services, so long as dissemination is in furtherance of that contractual purpose.
DL ONLY	Controlled by a dissemination list	Dissemination authorized only to those individuals, organizations, or entities included on an accompanying dissemination list. Note: Use of this limited dissemination control supersedes other limited dissemination controls but cannot supersede dissemination stipulated in federal law, regulation, or government-wide policy.
DISPLAY ONLY	Only for display	Information is authorized for disclosure to a foreign recipient, but without providing the foreign recipient with a physical copy for retention, regardless of medium to the foreign country(ies)/international organization(s) indicated, through established foreign disclosure procedures and channels.

Other data labels for unclassified and classified information exist beyond those that are centrally used by the U.S. government. Each is intended to ensure that only the individuals or groups requiring access to the information receive that access and that the data are easily identifiable during their life-span. This also allows data to be declassified appropriately based on the creation date and sensitivity, which helps to simplify life cycle management for those who have to deal with massive amounts of sensitive data.

A common issue with classification and labeling systems is overuse of classification. Organizations will label more data as sensitive than actually is because it is often convenient not to have to determine if the data actually require that label. Over time, this degrades the ability of the organization to use the label effectively. Careful use of classification and labels is an important part of data security life cycle management. Awareness of this issue can be seen in CUI, which requires that information "be decontrolled as soon as possible" and that "CUI that has been publicly released via authorized agency procedures shall be considered decontrolled."

 NOTE

The CUI blog maintained by the National Archives provides useful information and commentary about CUI and can be found at https://isoo .blogs.archives.gov/. The CUI Registry, including information like the CUI category list, markings, controls for dissemination, and de-control principles, can all be found at https:// www.archives.gov/cui.

Data Loss and Prevention

Data exposure from attacks and compromises, or simply from lack of care and due diligence, is a major concern for defenders in cyberwar. Because digital data can be copied without any evidence that it was duplicated, defenders are often unaware that an illicit copy exists. Unlike physical theft, digital data loss can't be detected by the lack of an object or a device, which makes the job of defenders even more challenging. Defenders must carefully watch data to ensure that they know where it is used, where it is transferred, and how many copies exist.

The ease with which significant amounts of information can be transferred—and the ability to easily change its form into something that doesn't resemble the original—can make detecting data loss difficult even in well-instrumented networks and devices. Encryption and other tools can disguise data or make it unreadable by detection systems. Defenders are often left to monitor for inappropriate behaviors in carefully built systems that try to limit what users can do with the data they have to protect. In order to address these issues, secure facilities implement controls, including prohibiting mobile devices, air-gapped systems without network connections to the outside world, preventing the use of external media, searching staff on their way into and out of the facility, and other security measures in an attempt to keep the most sensitive data from being exposed.

Data Spills

The NSA provides recommendations for what it refers to as *data spillage events*. These events occur when classified or sensitive data are transferred to unauthorized or unaccredited systems, individuals, or applications. The NSA's guidelines cite the increase in information sharing, as well as issues with data management, which allow higher-classification data to transfer to lower levels, or which result in remnant or hidden data being sent in a legitimate transfer.

Using Data Security Tools When Stealing Data

The other side of data loss is the targeting of data. Attackers also face challenges when attempting to acquire data:

- The sheer volume of data that exists
- Locating where the data reside
- Acquiring access to the data
- Transferring the data in a way that won't be visible to defenders
- Defeating encryption and other technical defenses

Some attackers use similar tools to those that defenders use to identify data that need protection. The same tool that can help a network defender find files with sensitive data can also help an attacker find interesting data.

The NSA points to four major reasons for data spillage:

- Improper handling of compartments, thus allowing data to be transferred between security zones, exposing it to unauthorized individuals or systems
- Problems with controls for the release of data, which allow data that shouldn't be released to escape
- Issues with the privacy of data, including releases of information that provide details on individuals that should not be made available
- Improper handling of proprietary information

 NOTE

The NSA specifically calls out releases to the media as a data spill, despite the media already fitting into the list of unauthorized individuals. The NSA's security tips for handling data spillage can be found at *https://apps.nsa.gov/iaarchive /library/ia-guidance/security-tips /securing-data-and-handling -spillage-events.cfm.*

To combat data spills, the NSA recommends a few critical controls. First, the NSA suggests network management to ensure that networks are defensible and have the proper levels of security. Second, it suggests that organizations create and enforce data protection policies. It suggests use of existing standards like the Defense Information Systems Agency's (DISA's) Security Technical Implementation Guides (STIGs) to ensure that devices and systems are properly secured and managed. This suggestion includes use of encryption to handle data at rest and in motion and methods for providing data in use with appropriate controls to ensure that the data aren't exposed beyond allowed sharing and usage models. Finally, the guide recommends the use of data loss prevention tools to make sure that data are monitored as they are accessed and transferred.

Data Loss Prevention

The information security industry has responded to the need to control data leaving systems and networks by creating technical solutions called *data loss prevention (DLP) systems*. These systems include both software installed on computers and devices and network devices that are placed between networks and the network edge. Integrated hybrid systems combine protection for both networks and individual systems, along with reporting and management tools to provide insight into the status of an organization's data usage and protection.

> **FYI**
>
> Some DLP systems use specialized methods to strip encryption from web traffic by using attack techniques that represent the DLP system as the remote website so it can decrypt the secure traffic. This man-in-the-middle-style attack can allow a DLP system to successfully monitor traffic, but it also makes it harder to trust the security of traffic that flows through the DLP system. Since an increasingly large percentage of Internet traffic is encrypted, network-based DLP systems rely on this capability in many cases. The alternative is host-based DLP that can check data after they are decrypted for user or system access.

DLP systems can be categorized into three major categories as follows:

- *Network-based DLP systems* are installed at network edges. Network DLP systems scan outbound traffic, such as web, email, and file transfers for data leakage. Network DLP systems often have problems when the traffic they are exposed to is encrypted, making them less useful when used to monitor many types of modern web traffic.

- *Host-based DLP systems* are installed on individual workstations and devices. They scan data that are transferred by the device. This can include data sent via a network or data transferred to removable media, such as a USB thumb drive or a DVD. Host-based DLP systems often provide local classification and discovery capabilities in addition to monitoring and data-transfer control capabilities like those a network DLP system provides.

- *Discovery DLP systems* are used to find sensitive information in file stores, workstations, servers, and other network locations. A discovery DLP system is typically used to help administrators locate sensitive data they are unaware of rather than to protect existing data in a known location.

Figure 13-2 shows the installation of a typical data loss prevention system for a network-based DLP system monitoring traffic at a network boundary. When systems attempt to send outbound email, web traffic, or file transfers, the DLP device reads the traffic and checks it for data that meets its filtering rules. If the rules are matched, the DLP takes the programmed action. In Figure 13-2, the DLP is set to block traffic that carries a Top Secret tag in its metadata description.

Figure 13-3 shows a typical host-based DLP system. Host-based DLP often includes both discovery capabilities and data loss prevention technology. In Figure 13-3, the host-based

FIGURE 13-2

Data loss prevention at a network boundary.

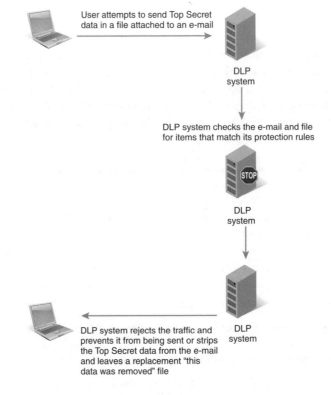

User attempts to send Top Secret data in a file attached to an e-mail

DLP system

DLP system checks the e-mail and file for items that match its protection rules

DLP system

DLP system rejects the traffic and prevents it from being sent or strips the Top Secret data from the e-mail and leaves a replacement "this data was removed" file

DLP system

FIGURE 13-3

Data loss prevention on a workstation.

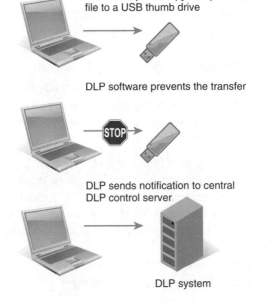

User attempts to copy a Top Secret file to a USB thumb drive

DLP software prevents the transfer

DLP sends notification to central DLP control server

DLP system

DLP system detects a user attempting to copy to a USB thumb drive a file that matches protection rules. It prevents the transfer, notifies the user, and sends a notification message to a security administrator. Host-based DLP provides a greater level of control than does network-based DLP alone. However, because of the large number of systems found on a typical network, it also increases the overall workload for security administrators.

DLP, much like intrusion detection and prevention systems, can require significant amounts of maintenance. The types of data handled by the organization and the ways in which that data are distinctive can make DLP useful. On the other hand, they can make it very hard to use a DLP system that recognizes data types based on rules that describe what it contains, how it is labeled, or where it is coming from.

Despite challenges in deploying them, DLPs are becoming increasingly common, and the technology continues to improve. DLP deployments on endpoint computers and centralized network DLP systems provide a useful tool to help prevent data loss. They also build awareness for users who have to deal with sensitive or classified data on a regular basis.

DLP Deployment Challenges

Data loss prevention systems are often a key part of security design for organizations that want to ensure that their data don't escape through their network or devices. They use network- or drive-monitoring capabilities to inspect data and to allow or prevent it from being sent—based on rules that identify protected information.

Like many complex security devices, DLP systems require significant amounts of tuning and configuration to provide an effective defense. Administrators must maintain and update the rules and tags that DLP systems use to flag data that are not allowed to exit systems and networks. Administrators also must identify false matches to make sure the DLP system doesn't prevent legitimate work from occurring—while at the same time ensuring that new data needing protection are consistently detected by the DLP system.

DLP systems can only detect data leaving a network or system that they can access. Encrypted data, or data that are in a format that the DLP system can't read, can bypass a DLP system's detection capabilities. The same issue exists when sensitive data can't be reliably detected, as distinct from data allowed to leave a network or system. This means that a DLP system can't be expected to stop all data loss, but a properly configured and maintained DLP system can significantly improve an organization's chances of detecting such loss.

13

Defending Data

FYI

DISA's Continuous Monitoring and Risk Scoring (CMRS) web system monitors DOD systems for their compliance with configuration requirements, including software. The DISA website notes that their Host Based Security System (HBSS) uses a commercial DLP product: McAfee's Data Loss Prevention tool.

Encryption and Data Loss

One of the most common methods recommended by DOD, NSA, and other organizations with significant cyberwarfare responsibilities is the use of encryption to protect data at rest and data in transit. Encryption can prevent attackers from reading the data it is protecting, although it cannot guarantee that they will never be able to access it if they have enough time and resources.

Encryption is particularly effective when data are at rest, such as when a computer or media is turned off or unused, or when data are in transit between systems or networks. These are times when the data are likely to be exposed to potential attackers and possibly outside of the system's normal protective layers of defenses. Keeping data encrypted while in use is far harder, as the programs and operating systems used to access and modify data normally require the data to be in an unencrypted state to work with it. This means that encryption can't provide a complete shield for data. But it remains an important part of the arsenal that defenders can bring to bear.

The use of encryption is very important from the perspective of defenders, but ensuring its use can be difficult. This is particularly so because data classification and handling requirements can be difficult to assess for every piece of data that users access. For this reason, organizations with high data-security requirements use automated encryption tools that can be managed centrally to enforce encryption policies.

Figure 13-4 shows an automated email encryption system. When an email containing data marked as Secret is sent, the system encrypts the content of that message

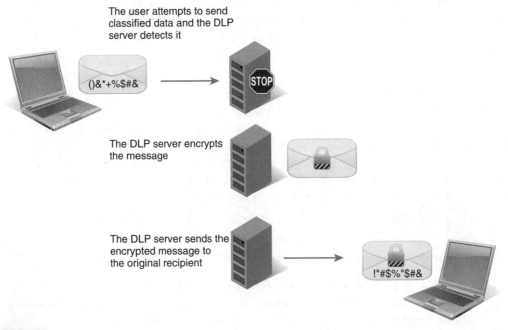

The user attempts to send classified data and the DLP server detects it

()&*+%$#& STOP

The DLP server encrypts the message

The DLP server sends the encrypted message to the original recipient

!"#$%"$#&

FIGURE 13-4

Enforcing encryption automatically on exit.

and then sends the receiver the encrypted message. In some systems, even receivers who are not previously known to be able to decrypt messages can receive encrypted messages. The secure email system simply sends them a special link to a website, which they can log on to and view the message over an encrypted connection. This helps to ensure that the email with sensitive data never travels over the network in unencrypted form.

Using encryption to protect files and drives also requires careful consideration of the data's life-span and the protection that the data need over their life-span. If the data must remain secure for more than a decade, current encryption is likely to be significantly less protective than it was when the file or drive was first encrypted. As computers become faster and access to additional computing power becomes less expensive, the ability increases to attack encryption using brute-force attacks that simply try every possible key to decrypt the file. If the encryption algorithms are also found to be flawed, or if a new technology is created that can more easily break the encryption, it could fail to protect the data more quickly than anticipated.

> **NOTE**
>
> Despite the best efforts of defenders, even strong encryption systems can be defeated if users move the protected data outside the protective system it resides in. Decrypted files left in unsecured locations, screenshots of decrypted email, or even printouts can all work around encryption systems. Even careful defenders find it hard to protect against these behaviors without preventing users from doing their jobs.

13

Defending Data

FYI

In 2019, payroll data for 29,000 Facebook employees were stolen after unencrypted hard drives were left in an employee's car. The data included names, bank account numbers, Social Security numbers, salaries, and other financial and personal data. Encrypted external drives wouldn't have prevented the theft, but they would have ensured that the data were unlikely to be exposed. Of course, other security controls would also have helped. Policies and practices that control where data are stored and how they are transported or accessed would also have helped.

This challenge means that defenders must understand the life cycle and useful life-span of the tools they deploy. They may have to decrypt and re-encrypt data with an exceptionally long life-span many times as technologies change. This means that they must carefully manage the technologies and secret keys they used to ensure that they are available throughout the life-span of both the tools and the data.

Table 13-1 shows the life-span of a number of current and past encryption algorithms, including when significant vulnerabilities were found and when they were officially retired. Although some encryption algorithms have very long life-spans without the discovery of flaws, others have had flaws discovered relatively quickly.

TABLE 13-1 Encryption algorithm life spans.

CIPHER	ADOPTION DATE	ATTACKS DISCOVERED	CIPHER CRACKED	OFFICIALLY RETIRED
Data Encryption Standard (DES)	1977	1992	1999	2002
Advanced Encryption Standard (AES)	2001	2009		
RC4	1987	1995	2005	2013 (Microsoft recommendation)
RSA	1977	1995	*	

*Current technology may be able to defeat shorter RSA keys, so longer keys are advised.

Data Integrity and Availability

Defenders tasked with preserving data must worry about more than whether attackers have accessed the data. They must also be concerned about whether attackers have been able to *change* the data. This is known as the data's *integrity*. Preserving data integrity is a key part of data protection. Attackers can also destroy the data or systems that the data resides on or otherwise prevent access to the data through a variety of attack methods. Defenders concerned with data protection must build into their plans capabilities ensuring that legitimate users can access the data and that they can provide data availability.

Integrity

> **NOTE**
>
> Tools that monitor the integrity of files allow defenders not only to monitor files for changes but also to ensure that change management requirements or policies are being met. They do this by sending alarms when files are changed. This can help detect unapproved changes made by authorized staff.

Data integrity is commonly checked using a cryptographic hash function that uses an algorithm to calculate a unique value for any given set of input. The algorithm ensures that any given file or document will only hash to that specific value, and if the file or document is modified, it will result in a change of the hash. When the monitoring software detects a change in the hash of a file, it reports the change, allowing administrators to determine why the file was changed and whether it should be restored.

Data integrity can also be maintained by making sure that files cannot be changed and that additional files cannot be added to a drive or file system. This is a less-popular approach because it makes maintenance and updates impossible without replacing the media or removing the protection. But systems that must absolutely remain the same can use it as a valid strategy. This is often the case with firmware—embedded software that is part of a device or system's underlying operation. Attackers have targeted the writable firmware of computers with malware in the past, demonstrating that if files can be changed, they are a potential target.

Availability

Meeting data availability requirements means that organizations preparing for cyberwarfare defense must assess how and where their data are accessed and stored. They then need to ensure that attacks against the data, or systems used to store or access the data, can't prevent that access. Attackers may use a variety of approaches:

- Malware attacks like ransomware that encrypt information, preventing attackers from accessing data
- Denial of service attacks, including distributed denial of service attacks that prevent access to networked resources by overwhelming networks or servers that are used to access the data
- Attacks that directly target data by deleting it, corrupting it, or otherwise making the data unusable
- Attacks against systems that take the systems offline by disabling their services
- Attacks that per organizational policy require the compromised servers or other devices to be taken offline for investigation

Knowing an organization's response policies can provide attackers with an insight into what result their attacks are having. Many organizations don't publish their detailed response plans to avoid giving attackers a game plan to use in their attacks.

Although individual organizations may use a variety of techniques and technologies to meet their availability requirements, two common strategies stand out when protecting the availability of data: ensuring proper **backups** and providing **redundancy**.

Backups

Backups provide copies of data that can be restored if data are lost. Typical backup methods require that backups not be kept in the same physical location as the live data they are copies of. Backup systems include tape, spinning disk, and even cloud-based systems and are usually selected based on the organization's requirements, such as time to restore, capacity, and how long the data must remain stored and intact.

Network defenders must take attacks against backups into account when designing their data security plans. This typically means that backup media must be encrypted and the ability to decrypt the data be preserved in the event of a disaster or attack. It also means that the way in which backups are transferred must be protected. In some instances, organizations have had to announce data breaches due to the loss of unencrypted backup tapes or other media.

> **NOTE**
>
> A common rule of thumb is that the physical location of backups and other disaster recovery systems be at least 90 miles away. This ensures that a single natural disaster or attack against a facility can't easily destroy the backups at the same time.

Redundancy

Availability assurance is often designed into servers and services through *redundancy*. Redundancy can take many forms, from simply having more than one server providing a service to complex designs that use multiple data centers in different geographic regions to ensure availability. The broad use of **virtualization** and **containerization** has allowed organizations to create redundancy using reduced amounts of hardware while retaining the ability to move virtual systems or containers between physical devices with great ease.

Modern redundancy designs often rely on or incorporate cloud services, whether they are commercially provided or built by organizations themselves. Public (or commercial) and private cloud services allow organizations to treat large amounts of hardware as virtual environments in which they can easily set up and manage systems. With proper prior design, when redundancy is required, the organization can simply create more instances of the server or service.

> **FYI**
>
> The use of cloud services specially designed to reduce or stop the impact of distributed denial of service attacks has made it possible for organizations to continue to provide services despite large-scale attacks. Previously this required massive investments in bandwidth and servers for an organization to build its own infrastructure in an attempt to counter large attacks.

Backups and redundancy can look very similar at first glance, but they play distinct roles in an organization. Backups are used to restore service, functionality, or files after a loss. Redundancy is intended to keep systems or services online and running by providing multiple copies, systems, or other components. A redundant system or service is less likely to fail, while backups can be used to recover from a disaster or event.

Data Retention and Disposal

Organizations generate data from a wide variety of sources. In addition to the data an organization creates as part of its normal course of business, the systems and infrastructure that support the organization also create data as they run. The sheer amount of data leaves defenders with several kinds of data:

- Data that they may need
- Data that the organization needs to conduct its operations
- Data whose retention may be required by policy, legal, or compliance reasons

Each of these requirements means the organization and its defenders must have a data retention policy and process in place to ensure that data have a known and enforceable life cycle. As data progress through this life cycle, they must be handled appropriately to ensure that they are not leaked, and that attackers can't access the data by simply finding a part of the life cycle that leaves the data unprotected.

Data Life Cycle Management

The use of policy to manage the life cycle and flow of data through an organization is known as **Data Life Cycle Management (DLM)**. DLM follows data from its initial creation

FIGURE 13-5

U.S. military records
life cycle.

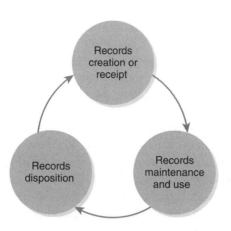

to the end of its life cycle, when it is deleted and removed from storage. DLM software and management tools exist that manage data through every part of its existence, allowing defenders and storage management staff to determine the appropriate handling methods for any given data file or element.

Figure 13-5 shows the U.S. military's basic data life cycle from creation to use and final disposal. The use of a cycle indicates that most data will be created and used, then replaced with updated data. Or new data may be used for purposes not envisioned when the original data were created.

Date reside in three major points during their life cycle:

- *Data creation* is usually from a device or worker. In a DLM environment, data are usually labeled when they are created and that label follows the data throughout the life cycle. In an environment using data classification rules, the data typically have an initial classification that matches the environment in which they were created. Data can move up or down depending on the content and usage.

- *Data at rest* are data that reside on a storage device or system. Data at rest are often encrypted to protect against the data storage media being stolen or otherwise exposed, and in fact, many organizations have handling requirements that require that sensitive data always be encrypted at rest—but despite that it may not always be in practice. Data in use may be unencrypted if on media accessed by a user, such as a workstation hard drive. Data at rest may have the data labels updated as part of a review or update process.

- *Data in motion* are typically protected by encryption as the data transits the network. Data are usually moved as part of a data transfer, including email, file transfer, or even web browsing.

Data Labeling

Even data that haven't yet been classified may need strong protections. At times, this is determined by the environment in which the data are created or gathered. This is typical of classified-data collection and storage systems, which assign their classification to any data that they contain. In other cases, data collection systems are designed to apply a data label to provide guidance on how it should be handled throughout its life cycle.

The idea of labeling data at creation or classification is known as **data labeling**. The classification labels used in the U.S. government's classification scheme can also be applied to digital data by adding tags. This is often done by adding additional data to a file, and the additional information contained in a file that describes the data is known as *metadata*. Metadata frequently includes the creator of a file, when it was created, and other information about the file. In environments where data labels are important, the metadata may also contain the data classification, allowing automated systems to scan for that label and apply appropriate controls based on what it says.

Data labeling can also be used as part of a data life cycle by using creation dates and rules that identify what type of data it is. The data labels an organization uses can indicate when the data can be disposed of or the rules for handling it. Enforcing the use of data labels can be done in an automated fashion using tools known as *digital rights management (DRM)* software. DRM software can be used to help control how and where labeled data and programs are used.

Digital Rights Management

Digital rights management (DRM) is most frequently associated with games and movies in the minds of many. But it can be a useful tool when a digital data file must be protected throughout its life cycle. DRM includes technologies intended to prevent unauthorized viewing, copying, printing, or modifying digital files or devices.

DRM software tools can flag files to run or be accessible only under specific circumstances—such as on trusted networks or workstations or when a password or other authenticator is provided. They can also prevent the data from being shifted to another format by preventing printing, screenshots, or other access.

As many users of hacked software and games will tell you, a determined attacker can almost always defeat DRM. This makes DRM a tool best suited as part of a rights management process for trusted staff rather than a tool to stop attackers from using or modifying files given sufficient time and resources.

Drives and Media Management

The devices and drives that store data are of particular concern to defenders in two phases of the data protection life cycle. First, defenders work to acquire drives and devices that can support their organization's security requirements. Some modern drives often offer built-in encryption capabilities as well as data destruction capabilities that provide a first layer of defense for the data residing on the drive. Second, defenders must consider length of the life cycle of the drives and media. Some types of media—such as writable DVDs or older media like floppy disks and burnable CDs—have a relatively short life cycle.

This means defenders must track their useful life to ensure that they don't create an availability and integrity risk to the data they contain. Defenders must also consider how they will dispose of the drives and media when the life cycle is over.

Defenders must track the life-span of many types of media and drives as well as properly dispose of them as they are replaced:

- *Magnetic media*, such as tapes, store data using magnetically charged particles. Magnetic media is particularly susceptible to damage from magnetic fields. Tapes are typically expected to have at least a 10-year life-span, but premium tape media may last as long as 50 years if stored properly.

- *Optical media*, such as DVDs, Blu-ray discs, and CD-ROMs, reads and writes data using a laser. Optical media such as CDs and DVDs are estimated to have a life-span from 2 to 5 years, although published life expectancies are sometimes as long as 10 or more years.

- *Flash memory*, such as USB thumb drives, are normally expected to last for 10 years or longer.

- *Spinning disks*, or hard drives, have a life-span that typically ranges from 3 to 5 years, but which may last up to 7 years or more.

- *Solid state drives (SSDs)* have varying life-spans but are typically expected to last 3 years or more. SSDs have a limited number of write cycles to each of their memory cells, and typically monitor their own usage to track their remaining life-span.

> **NOTE**
>
> An often-forgotten risk for many organizations is exposure of data when drives are sent back with machines as part of a lease or when systems are returned to manufacturers or off-site support facilities for repair. Reports continue to surface periodically about copiers and office multifunction devices with drives full of data from photocopies and faxes they have sent and print jobs that they handled.

> **FYI**
>
> Magnetic media can be erased using a tool known as a *degausser*, which uses very strong magnetic fields to wipe it. You can find the NSA's guide for media destruction at *https://www.nsa.gov /Resources/Media-Destruction-Guidance/*. For total destruction, a drive shredder is often used.

Manufacturers of each type of media typically provide a measure known as the **mean time between failure (MTBF)**. This number provides the average life-span of the media or device before it fails, which can provide defenders with a reasonable estimate of the storage's usable life.

Once storage devices have been removed from service, they must be disposed of. In most cases, high-security organizations resort to drive destruction at the end of a drive or storage media's life-span. But they may use a wiping utility or tool to wipe the data from the drive for reuse if it hasn't been retired yet. In most cases, this is a relatively simple process as long as the data stored are at the same security or classification level as the replacement data.

Solid state drives have created a new problem for defenders who have previously relied on drive-wiping software. Solid state drives are typically equipped with more flash memory than they actually use, allowing the drives to spread memory wear over time. This allows the drive to meet its life-span requirements. It also means that heavily used sections of the drive will be remapped to use spare memory, leaving the old data there and unseen. Attackers who can bypass the drive's usage management software could recover data that its owner thought were no longer there.

> **Drive Destruction**
>
> Total destruction is one of the few ways to entirely ensure that data cannot be recovered from drives and removable media. Other data-destruction methods include drilling holes through spinning disk drive platters, disassembly, and drive wiping using software to attempt to ensure that data cannot be obtained from the drive. Unfortunately, given nation-state-level resources and technical capabilities, methods like these may be susceptible to data recovery.
>
> Fortunately for most organizations, commercial drive-destruction companies now provide secure hard drive destruction services, which shred or crush drives and media before the pieces that remain are disposed of. Some drive shredders are even mobile, allowing them to be brought to the organization and to perform the drive destruction in front of employees. Even with the ability to shred drives on-site, organizations that have very high confidentiality require-ments may not find commercial disposal companies secure enough for their use, thus creating a need for internal disposal processes and driving a market for powerful drive shredders.
>
> In cases where a drive has not been physically destroyed, the commercial world provides some insight into the capabilities that exist for data recovery. Commercial services offer the ability to recover data from drives encrypted with popular commercial encryption software, and some even offer the ability to decrypt that data in some cases. They also specialize in recovering data from damaged and erased drives, including those that have physical dam-age to the platters inside hard drives. It is reasonable to expect that a nation-state-backed organization will be able to recover data at least this effectively, making destruction of media a necessity to ensure data security.

Data Loss Response

The NSA's data spillage event guide summarizes a typical response to data loss. A well-prepared organization should have a pre-existing set of policies and procedures to follow in the event of data loss or exposure. This helps to ensure that responses are appropriate and measured rather than hasty and implemented in panic mode.

The NSA's process is as follows:

- Assess if the spill has occurred; which data may have been exposed; which users, data, and systems were involved; and how sensitive the data were.
- Report the spill or suspected spill to the groups or individuals listed in the organiza-tion's data loss policy and procedures.

- Isolate the compromised systems to limit the chance for further damage and to allow investigation of the breach. In many cases, this requires taking parts of an operation offline and may disrupt the organization's operation. This requires the severity of the spill and the need for investigation to be balanced against the organization's operational requirements.
- Contain the breach and remediate the issue that allowed the breach to occur.

A strong incident response plan is required when dealing with data exposure. Specific information about the data—whether it was encrypted or otherwise protected when lost, and how it may have been exposed—is critical to a timely and appropriate handling process.

The NSA's guide notes that "affected media/devices take on the classification level of the compromised data until ... assessed." In some circumstances, this could mean that normally unclassified systems require special handling.

CHAPTER SUMMARY

In this chapter, you learned about data loss prevention, including data loss prevention technologies and techniques. You explored classification and data labeling, which helps governments and other organizations compartmentalize data and ensure that it receives the right level of security. You also learned that the U.S. NSA provides guidelines on preventing data loss. This includes techniques for securing networks, providing a secure data life cycle, and dealing with data spills after they occur, through incident-response and cleanup processes. As part of that life cycle, you learned how ensuring the integrity and availability of data is an important part of computer network defense in cyberwar.

In addition, you reviewed the need to provide secure backups and redundancy to ensure that attackers cannot disable an organization by preventing access to data or systems. You also learned that a complete data life cycle security plan must protect data when created, at rest, and in transit. Such a plan must also have a clear means of handling data disposal. Finally, you reviewed methods for dealing with data loss as part of a response process when all defensive strategies have failed.

KEY CONCEPTS AND TERMS

Backups	Data labeling	Redacting
Classification	Data Lifecycle Management (DLM)	Redundancy
Containerization	Declassified	Sanitizing
Controlled Unclassified Information (CUI)	Downgraded	Virtualization
	Mean time between failure (MTBF)	

13

Defending Data

CHAPTER 13 ASSESSMENT

1. Common secure drive-wiping software completely removes all data from solid state drives.

   A. True
   B. False

2. Encryption of data helps to meet what two data protection goals? (Select two.)

   A. Deniability
   B. Confidentiality
   C. Integrity
   D. Availability

3. When is encryption typically not usable?

   A. At rest
   B. In use
   C. In email
   D. In transit

4. What are the three common types of data loss prevention systems? (Select three.)

   A. Host
   B. Network
   C. Discovery
   D. Wireless

5. Encryption algorithms are always retired and removed from use when they are found to be vulnerable to attacks.

   A. True
   B. False

6. What are the three common phases in the data life cycle?

   A. Receipt, classification, declassification
   B. Creation, use, disposal
   C. Creation, classification, downgrading, disposal
   D. Creation, maintenance, declassification
   E. None of the above

7. A drive's MTBF measures _____.

   A. how hard it is to destroy the drive
   B. how hard it is to break the encryption on the drive
   C. how long it will take the average drive to fail
   D. the data density of the drive

8. The best way to ensure that data on used drives are not recovered is to _____.

   A. physically destroy the drive
   B. wipe the drive
   C. securely erase the drive
   D. use the drive's embedded SATA commands to ensure no data remains

9. A technology that allows data rights management is known as _____.

   A. DMA
   B. DRM
   C. RMA
   D. TLA

10. Availability is often ensured by providing _____.

    A. redundancy
    B. data integrity
    C. encryption
    D. DLP

11. Redundant systems and backups help to ensure _____.

    A. availability
    B. integrity
    C. confidentiality
    D. data security

12. Integrity of data can be verified using what tool?

    A. Encryption
    B. Hashing
    C. Data validation
    D. Bit checks

PART THREE

The Future of Cyberwarfare

Cyberwarfare and Military Doctrine

O VER THE CENTURIES, MILITARY PLANNERS and strategists have developed a comprehensive body of knowledge regarding the conduct of warfare. This knowledge is known as military *doctrine*. The Chinese general Sun Tzu wrote one of the earliest known treatises on military doctrine, *The Art of War*, sometime around 500 BC. In it, he outlined basic strategies that every military commander should follow. In the book, he described the importance of developing plans, the use of deception as a key component of warfare, methods for exploiting an enemy's weak points, maneuvering and marching with an army, the effects of terrain on an army, and many other essential topics.

Of course, doctrine has evolved substantially in the time since Sun Tzu created his seminal work. Although *The Art of War* is still heavily cited today, modern military planners now have millennia of experience to draw upon. After Sun Tzu's work helped planners better understand the enemy, a German general, Carl von Clausewitz, changed the way planners understand the impact of terrain on warfare. The U.S. military, through the office of the Joint Chiefs of Staff, maintains a large set of doctrine publications that provide guidance to commanders in all of the military services. These publications are supplemented by service-specific doctrine documents that provide additional detail.

In the days of Sun Tzu, the military consisted of little more than soldiers on foot and on horseback along with rudimentary naval ships. The Chinese general could never have dreamed of tanks, destroyers, or fighter jets. Modern doctrine takes these advances into account, drawing upon the ancient sources but supplementing them with more recent knowledge about battlefield advances. Every time a military engages the enemy, planners learn more and incorporate this new knowledge into revisions of the doctrine to guide future commanders.

Cyberwarfare presents yet another frontier, one that is new to military commanders—just as the exploitation of the air and space domains was new to commanders in the world wars. Cyberwarfare doctrine is a rapidly evolving field, and relatively few publications exist. The core document on the subject, *Joint Publication 3-12, Joint Cyberspace Operations*, remains classified and unavailable to nonmilitary cyberwarfare students.

Chapter 14 Topics

This chapter covers the following topics and concepts:

- What military doctrine is
- How the U.S. military is organized for cyber operations
- What the five pillars of cyberwarfare are

Chapter 14 Goals

When you complete this chapter, you will be able to:

- Describe the role and purpose of military doctrine
- Identify the major documents covering cyberwarfare doctrine
- Explain how the U.S. military is organized for cyber operations
- Explain the major concepts that must be addressed in cyberwarfare doctrine
- Describe the five pillars of cyberwarfare

Military Doctrine

In the words of General George H. Decker, who served as army chief of staff from 1960 until 1962, "Doctrine provides a military organization with a common philosophy, a common language, a common purpose, and a unity of effort." **Doctrine** provides commanders with a way to share thousands of years of accumulated military knowledge with their fellow officers in a consolidated format. Doctrine describes the manner in which a military will operate and provides commanders with guidance and advice on issues of strategic, operational, and tactical importance.

Doctrine is only advice, however. It provides a general philosophy for military operations, but it does not provide all of the answers. It remains the role of the military commander to read, understand, interpret, and apply doctrine in the unique circumstances of a particular battle, campaign, or war. In their *Compendium of Key Joint Doctrine Publications*, the Joint Chiefs of Staff sum up this important concept with a simple, catch phrase: "Doctrine Guides, Commanders Decide!"

In the Army's *Unified Land Operations* doctrine, the authors elaborate on this concept of a commander's discretion by stating:

> Doctrine is also a statement of how the Army intends to fight. In this sense, doctrine often describes an idealized situation and then contrasts the ideal with the reality Army leaders can expect. Doctrine provides a means of conceptualizing campaigns and operations, as well as a detailed understanding of conditions, frictions, and uncertainties that make achieving the ideal difficult. Doctrine also helps potential partners understand how the Army will operate. It establishes a common frame of reference and a common cultural perspective to solving military problems, including useful intellectual tools.

In the remainder of this section, you will learn some of the fundamental points of military doctrine accumulated over the centuries. These include the nine principles of war, the forms of warfare, and the levels of warfare.

Principles of War

According to U.S. military doctrine, **war** is "socially sanctioned violence to achieve a political purpose." The U.S. military believes that the effective execution of military operations relies upon nine core principles of war. These principles, as articulated in *Army Field Manual 3-0: Operations*, are as follows:

- **Objective**—Every military action must have a clearly defined purpose. The **objective** of an operation should be both decisive and attainable. Commanders must clearly understand what they are attempting to achieve in a military action and use that knowledge to guide their decisions and instructions to subordinate commanders. The existence of a clear objective ensures that operations at lower levels are consistent with the higher-level commander's intent.

- **Offensive**—Military commanders must seize, retain, and exploit the initiative. By using the principle of **offensive**, they may dictate the nature, tempo, and scope of operations. By seizing the initiative, a military may compel its opponent to react. This provides the attacking force with greater freedom of action and increases the likelihood of successfully achieving its objective.

- **Mass**—Military commanders should concentrate their combat forces at the decisive time and location. This may take place in two ways. Commanders may choose to **mass** their forces *in time* by simultaneously striking multiple significant targets. Alternatively, commanders may opt to mass their forces *in location* by bringing overwhelming force against a single location. Each of these tactics may overwhelm the adversary and lead to victory.

- **Economy of Force**—Commanders should use the minimum amount of force necessary to achieve secondary objectives. Using **economy of force** allows commanders to achieve the principle of mass by allocating all remaining forces to achieving the primary objective.

- **Maneuver**—Applying power in a flexible manner keeps the adversary off guard. When performed effectively, **maneuver** keeps enemy troops at a disadvantage by forcing them to deal with new problems and dangers faster than they can react to the changing dynamics of the battlefield.

- **Unity of Command**—A single commander should be responsible for achieving each military objective under the principle of **unity of command**. In many cases, the nature of joint and multinational operations does not permit the assignment of a single commander. In those situations, multiple commanders must cooperate, negotiate, and build consensus.

- **Security**—Commanders must never allow enemy forces to gain an advantage. This is essential to preserving and protecting a military's power. **Security** may be bolstered by deception and results from defensive efforts against surprise, sabotage, interference, and intelligence operations conducted by the enemy.

- **Surprise**—When attacking, military forces should strike the enemy at an unexpected time or place. **Surprise** brings shock value by surprising an enemy and offering attackers a temporary battlefield advantage. Surprise does not require a completely unaware enemy—it only requires that the attack take place in a time or place where the enemy can't immediately react in an effective manner.

- **Simplicity**—Military plans should be simple, clear, and concise. The more complex or ambiguous a set of orders, the more likely it is that the unit will fail to achieve its objective. **Simplicity** reduces confusion and misunderstanding.

The principles of war are not consistent from nation to nation. Every country's doctrine embodies a similar set of principles, but there are often additions, modifications, and deletions. For example, the British Defense Doctrine outlines 10 principles of warfare: Selection and Maintenance of the Aim, Maintenance of Morale, Offensive Action, Security, Surprise, Concentration of Force, Economy of Effort, Flexibility, Cooperation, and Sustainability.

In fact, the principles of war are not even consistent among military forces of the same nation. Although the U.S. Army embraces the nine principles outlined above, the Joint Chiefs of Staff use a different list, known as the *Principles of Joint Operations*. The 12 principles in that document include the 9 Army principles, along with 3 additional items:

 NOTE

After the 1984 passage of the Goldwater-Nichols Act, responsibilities for U.S. military operations were divided between the individual military services and joint commands. Services are responsible for training and equipping troops, whereas joint commands fight wars through combatant commanders.

- **Perseverance**—Military commanders must take measures to ensure that the commitment exists to achieve the desired end state. Some military operations may take years to achieve their desired objectives. When applying the principle of **perseverance**, commanders must be patient, resolute, and persistent when employing their forces in support of long-term objectives.

- **Legitimacy**—Military operations must take place within the confines of appropriate authority. The actions of military commanders must have **legitimacy**, or be consistent with national laws as well as international treaties and obligations. Commanders have a particular obligation to ensure that the actions of units and individuals are consistent with the Law of Armed Conflict.

- **Restraint**—While conducting military operations, commanders should use **restraint**, or limit collateral damage and prevent the unnecessary use of force. Military forces must be careful and disciplined. The unrestrained use of force may violate the law and antagonize both friendly and neutral parties, damaging the legitimacy of a military action.

These principles of warfare and joint operations form the core of military doctrine. Although different nations and military services may interpret and articulate them differently, they are generally respected as fundamental truths. Many of the principles date back thousands of years, appearing even in Sun Tzu's *The Art of War*. For example, Sun Tzu described the modern principles of Offensive and Security when he stated, "The clever combatant imposes his will on the enemy, but does not allow the enemy's will to be imposed on him."

> **▶ TIP**
>
> Students trying to remember the principles of warfare often rearrange them to form the acronyms MOSS MOUSE or MOOSE MUSS to help them remember the first letter of each principle.

Forms of Warfare

As military strategists think about war, they organize it into two general forms: traditional warfare and irregular warfare. These are not neat subdivisions but rather broad generalizations that allow planners to discuss the different forms of warfare using a common context and language. They define the two primary forms of warfare in these ways:

- **Traditional warfare** includes large-scale conflict between nations or groups of nations. U.S. military doctrine defines traditional warfare as "characterized as a violent struggle for domination between nation-states or coalitions and alliances of nation-states. This form is labeled as traditional because it has been the preeminent form of warfare in the West since the Peace of Westphalia (1648) that reserved for the nation-state alone a monopoly on the legitimate use of force." Nations engage in traditional warfare for the purpose of imposing their will on other nations or preventing other nations from doing the same to them.

- **Irregular warfare** can be just as violent as traditional warfare but is marked by several differences. In U.S. military doctrine, irregular warfare is defined as "characterized as a violent struggle among state and non-state actors for legitimacy and influence over the relevant population(s)." Notice two important differences in this definition. First, the participants in irregular warfare may include nonstate actors. Second, the purpose is not the imposition of wills between nations but the quest for legitimacy and purpose over a population.

14

Cyberwarfare and
Military Doctrine

FYI

Although U.S. military doctrine divides warfare into the traditional and irregular forms, not all military historians and strategists agree with this approach. Other common schemes divide warfare into different categories:

- Regular versus irregular warfare
- Traditional versus nontraditional warfare
- Traditional versus untraditional warfare
- Conventional versus unconventional warfare

The military settled on the traditional versus irregular language as its standard. The term *conventional* warfare was discarded because it was widely used during the Cold War to refer to non-nuclear warfare. Doctrine developers thought that the previous widespread use of the term for another purpose would cause confusion.

It is important to note that nation-states may engage in both traditional and irregular warfare, and in fact, the military engagements of the United States since September 11, 2001, have been primarily irregular in nature. The U.S. military presence in Iraq and Afghanistan during this century has focused extensively on influencing the populations of those countries and encouraging them to adopt democratic forms of government favorable to U.S. interests. The military has also been heavily engaged in the "war on terrorism," which has clearly been a national priority. It does not meet the definition of traditional warfare because the adversary is not a nation-state.

> **NOTE**
>
> The field of battle is complex and full of uncertainty and chance. Military scholar and general Carl von Clausewitz described this confusion as "the fog of war."

Cyberwarfare can fit within either of these forms of warfare. Nations may engage in cyberwarfare operations against each other as a component of traditional warfare. At the same time, nations may use cyberwarfare operations against nonstate actors, and nonstate actors may use cyberwarfare against nation-states when engaging in irregular warfare.

Levels of Warfare

The *forms* of warfare distinguish between warfare actions based on both the types of adversaries engaged in the war and their strategic objectives. At the same time, military commanders achieve these objectives by engaging in actions at different *levels* of warfare. These levels, the strategic level, operational level, and tactical level, are used to set loose boundaries around warfare at the appropriate echelons of command authority.

The **strategic level** of warfare addresses the national-level objectives of the war. It includes very broad goals, including both the national policy issues at stake and the strategies used to engage the enemy across an entire theater of operations. In the United States, strategic-level objectives are set by senior political and military officials, including the president, secretary of defense, and commander of the Joint Chiefs of Staff.

The lowest level of warfare is the **tactical level**. Tactical activities include the specific use of military units that engage each other in battles and small-unit actions. Commanders at this level must be concerned with the "nuts and bolts" of warfare. They direct the movements of individual units, personnel, and weapons systems. Military engagements at the tactical level are typically short term in duration and have extremely specific objectives. "Take that hill!" is an example of a tactical military objective.

In between the strategic and tactical levels lies the **operational level** of warfare. At this level, commanders set the objectives for military campaigns and major operations. These objectives link together the national policies set by senior leaders with the movements of military units at the tactical level. When referring to this level of warfare, U.S. military doctrine refers to the term *operational art* as follows:

> The focus at this level is on the planning and execution of operations using operational art: the cognitive approach by commanders and staffs—supported by their skill, knowledge, experience, creativity, and judgment—to develop strategies, campaigns, and operations to organize and employ military forces by integrating ends, ways, and means.

Cyberattack Strikes the Air Force

In the Air Force's *Cyberspace Operations* doctrine document, the authors cite an example of a real-world cyberattack against the Air Force's assignment management system. This excerpt from the Air Force Office of Special Investigations briefing on the incident describes the risk of cyberattacks against military information systems:

> In May [20]05, an unknown subject obtained unauthorized user level access to the Assignments Management System (AMS). Using that access, the subject was able to view information contained within the AMS, but was unable to alter information or gain access to any other Air Force computer systems. Computer records indicate that the subject gained access to AMS via a senior Air Force official's account. The compromised AMS account was set with privileges which allow the user to review any active duty Air Force member's single unit retrieval format (SURF) data from anywhere in the world with an Internet connection. SURF records contain sensitive data, such as assignment history, security clearance, personal identification information, rank, position, and duty status. The subject gained access to the web based account using the "forgot password" function to answer the challenge questions required to change the account password. The challenge questions asked for biographical information on the senior official, which was readily available on the Internet.
>
> Upon review, it was determined that the senior USAF official's account had been used to view the SURF records of 37,069 Air Force members. Log analysis indicates the intrusion initially originated from forty-one different source IP addresses throughout the duration that the compromised account was used by the subject.
>
> Throughout this duration, the subject's activity originated from approximately twelve additional U.S. based Internet Protocol (IP) addresses, which were later determined to be open proxies that the subject used to mask their true place of origin. There were no foreign-based IP addresses used after the incident was reported. Court order subpoenas were served on all U.S.-based source IP addresses from which the compromised AMS account was accessed; fifty in total. Information obtained via court order subpoenas identified the last known point of the origin. However, local law enforcement indicated that the information required to further identify the subject was no longer available.

The scope and duration of this attack against a sensitive military computer system reflect the sophistication and selective targeting used by adversaries engaged in cyberwarfare.

Figure 14-1 illustrates the relationship between the various levels of warfare. Note that the boundary lines drawn between them overlap. The lines between actions at the strategic and operational levels are blurred, as are the lines between operational and tactical activities.

As with the forms of warfare, cyberwarfare operations may take place at any of these levels of warfare. At the strategic level, cyberwarfare may include intelligence activities that provide critical information to senior government officials as they set national policies. At the operational level, a military campaign might include a significant cyberwarfare component that disables critical enemy defense systems just before friendly forces launch a massive offensive strike. Finally, at the tactical level, commanders might engage in cyberwarfare activities designed to support individual units on the ground as they engage the enemy in a battle.

FIGURE 14-1

The levels of warfare: strategic,
operational, and tactical, as
described in U.S. military doctrine.
Courtesy of Joint Chiefs of Staff.

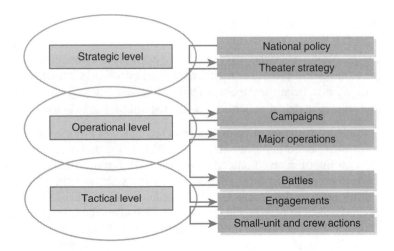

Organizing for Cyber Operations

Military forces around the world typically organize themselves based upon the types of
operations in which they engage. For example, consider the military organization of the
United States, which consists of four primary components:

- The U.S. Army is primarily engaged in operations covering the land domain
  of warfare.
- The U.S. Navy is primarily engaged in operations covering the sea domain of warfare.
- The U.S. Marine Corps is engaged in operations that cross between the land and sea
  domains of warfare.
- The U.S. Air Force is concerned with both the air and space domains of warfare.

There is, however, no "U.S. Cyber Force" that bears primary responsibility for the cyber
domain of warfare. Cyberwarfare responsibility is widely distributed among military units,
with sometimes confusing and overlapping lines of authority. In addition, military and
nonmilitary agencies that fall outside the armed services also participate in cyberwarfare
operations. For example, the U.S. National Security Agency (NSA) is a component of the
Department of Defense (DOD) and is led by a military flag officer but is also somewhat
independent of the traditional military command structure. Figure 14-2 illustrates the
overlapping relationship between cyberwarfare and the traditional domains of warfare.

FYI

The DOD released an updated version of the joint publication *JP 3-12: Joint Cyberspace
Operations* in June 2018. This document is intended to address the uniqueness of military
operations in cyberspace, clarify cyberspace operations-related command and operational
interrelationships, and incorporate operational lessons learned.

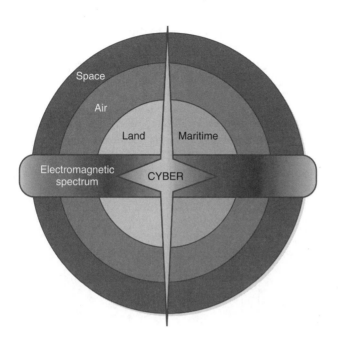

FIGURE 14-2

The cyber domain of warfare functions across all of the traditional domains of warfare.

Courtesy of Air War College, Air University.

How does the U.S. military organize for cyberwarfare? Each military unit has its own cyberwarfare component and the actions of these components are supposed to be coordinated by the U.S. Cyber Command (USCYBERCOM). In its mission statement, USCYBERCOM "plans, coordinates, integrates, synchronizes and conducts activities to direct the operations and defense of specified Department of Defense information networks and prepare to, and when directed, conduct full-spectrum military cyberspace operations in order to enable actions in all domains, ensure U.S./allied freedom of action in cyberspace and deny the same to our adversaries."

That is a very confusing and unclear mission statement, and it is somewhat reflective of the confusion surrounding cyberwarfare activities in a military that is organized to fight in the traditional domains of warfare. In July 2011, the U.S. Government Accountability Office (GAO) performed an assessment of the DOD's cyberwarfare efforts and found that they were severely deficient. In one of the findings, GAO identified a lack of consistent doctrine:

> Several joint doctrine publications address aspects of cyberspace operations, but DOD officials acknowledge that the discussions are insufficient; and no single joint publication completely addresses cyberspace operations. While at least 16 DOD joint publications discuss cyberspace-related topics and 8 mention "cyberspace operations," none contained a sufficient discussion of cyberspace operations. DOD recognizes the need to develop and update cyber-related joint doctrine and is currently debating the merits of developing a single cyberspace operations joint doctrine publication in addition to updating all existing doctrine. However, there is no timetable for completing the decision-making process or for updates to existing doctrine.

The second significant finding in GAO's assessment relates to the extremely complex and overlapping organizational responsibilities for cyberwarfare. Figure 14-3, taken from the GAO report, illustrates the complexity of the situation, with many components having multiple reporting relationships that do not meet until they reach the cabinet-level post of secretary of defense. In the words of the GAO:

> DOD has assigned authorities and responsibilities for implementing cyberspace operations among combatant commands, military services, and defense agencies; however, the supporting relationships necessary to achieve command and control of cyberspace operations remain unclear. In response to a major computer infection, U.S.

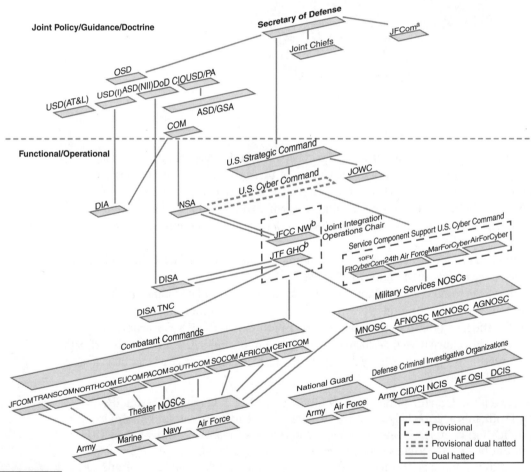

FIGURE 14-3

Organization chart of Department of Defense cyberoperations structure, as assessed by the Government Accountability Office in 2011.

Courtesy of Government Accountability Office.

Strategic Command identified confusion regarding command and control authorities and chains of command because the exploited network fell under the purview of both its own command and a geographic combatant command. Without complete and clearly articulated guidance on command and control responsibilities that is well communicated and practiced with key stakeholders, DOD will have difficulty in achieving command and control of its cyber forces globally and in building unity of effort for carrying out cyberspace operations.

U.S. Strategic Command (USSTRATCOM)

As shown in Figure 14-3, one of the nexus points for command of cyberspace operations in 2011 resided with the U.S. Strategic Command (USSTRATCOM). When it was formed during the Cold War, USSTRATCOM was focused on defending the United States against the threat posed by the Soviet Union. In particular, USSTRATCOM had responsibility for the nation's nuclear forces.

With the end of the Cold War, the mission of USSTRATCOM changed and broadened in scope. First, the command merged with the U.S. Space Command and assumed responsibility for military satellites and other space operations. Second, the command was chosen to take on the cyberwarfare mission for the U.S. military. The USSTRATCOM mission statement now reads, in part: "The missions of U.S. Strategic Command are to deter attacks on U.S. vital interests, to ensure U.S. freedom of action in space and cyberspace, to deliver integrated kinetic and non-kinetic effects to include nuclear and information operations in support of U.S. Joint Force Commander operations."

U.S. Cyber Command (USCYBERCOM)

At the time of the GAO assessment, USSTRATCOM managed cyberspace operations through the use of a "subordinate unified command," the U.S. Cyber Command (USCYBERCOM). As a subordinate unified command, USCYBERCOM had full command authority over cyberspace operations, but reported through the commander of USSTRATCOM.

In response to the confusing command structure for cyber operations, President Obama elevated USCYBERCOM to status as a full unified combatant command in December 2016, moving it out from under USSTRATCOM in an effort to streamline command and control of cyberspace operations.

14

Cyberwarfare and
Military Doctrine

FYI

A careful examination of the USCYBERCOM emblem reveals that it contains the string "9ec4c12949a4f31474f299058ce2b22a" written in the circle between the name of the command and the inner logo. This is the MD5 hash of the command's mission statement.

USCYBERCOM is located at Fort Meade, Maryland, home to the NSA, and is led by a four-star flag officer. To date, three officers have held this position on a permanent basis. General Keith Alexander held the position from the command's creation in 2010 until his retirement in 2014. Admiral Michael Rogers assumed command in April 2014 and held it through 2018, when U.S. Army General Paul Nakasone assumed command. All three officers held the command of USCYBERCOM concurrently with their service as director of the NSA.

As a unified organization, USCYBERCOM is composed of military service components representing each of the uniformed services:

- Sixteenth Air Force (Air Forces Cyber)
- Army Cyber Command (ARCYBER)
- Navy Tenth Fleet (Fleet Cyber Command)
- Marine Corps Cyberspace Command

Each of these component commands has responsibility for cyberspace operations within its respective military service.

Five Pillars of Cyberwarfare

As the nation continues to chart its course in cyberspace, the military has outlined a strategy for cyberspace operations that consists of four major pillars. These pillars are the strategic initiatives that the government considers essential to building and maintaining the nation's cyberwarfare superiority. Described in the *National Cyber Strategy*, the four pillars are as follows:

- "Protect the American people, the homeland, and the American way of life."
- "Promote American prosperity."
- "Preserve peace through strength."
- "Advance American influence."

The ability of the United States, or any nation, to effectively engage in the new domain of cyberspace depends upon the successful fulfillment of these four strategic objectives. The DOD further refined this strategy by describing five lines of effort they will engage in to achieve these objectives:

- Build a more lethal force.
- Compete and deter in cyberspace.
- Strengthen alliances and attract new partnerships.
- Reform the department.
- Cultivate talent.

Ten Things Every Airman Should Know

The Air Force continues to educate its officers and enlisted personnel about the effects that their personal actions can have on national security. General Norton A. Schwartz, former chief of staff of the Air Force, issued the following 10 cyber tips to airmen, titled "Ten Things Every Airman Should Know":

1. The United States is vulnerable to cyberspace attacks by relentless adversaries attempting to infiltrate our networks at work and at home—millions of times a day, 24/7.

2. Our adversaries plant malicious code, worms, botnets, and hooks in common websites, software, and hardware such as thumb drives, printers, etc.

3. Once implanted, this code begins to distort, destroy, and manipulate information, or "phone" it home. Certain code allows our adversaries to obtain higher levels of credentials to access highly sensitive information.

4. The adversary attacks your computers at work and at home knowing you communicate with the Air Force network by email or by transferring information from one system to another.

5. As cyber wingmen, you have a critical role in defending your networks, your information, your security, your teammates, and your country.

6. You significantly decrease our adversaries' access to our networks, critical Air Force information, and even your personal identity by taking simple action.

7. Do not open attachments or click on links unless the email is digitally signed, or you can directly verify the source—even if it appears to be from someone you know.

8. Do not connect any hardware or download any software, applications, music, or information onto our networks without approval.

9. Encrypt sensitive but unclassified and/or critical information. Ask your computer security administrator for more information.

10. Install the free Department of Defense anti-virus software on your home computer. Your computer security administrator can provide you with your free copy.

While these tips were issued by the Air Force, they apply equally to anyone engaged in cyberwarfare or who may be a target of cyberwarriors!

CHAPTER SUMMARY

Military doctrine provides commanders at all levels of the military with the guidance they need to understand the objectives and strategies of higher-level commanders. Doctrine provides a blueprint for those commanders as they seek to understand the strategic and operational environment. Commanders are then responsible for interpreting that environment and making decisions that are consistent with both battlefield realities and military doctrine.

As a new domain of warfare, cyberspace presents a challenge to military commanders, similar to the challenge they faced when adapting to the new air domain during the world wars. The current state of doctrine and organization surrounding cyberwarfare is somewhat confused. As pointed out in the GAO audit, many different doctrine documents discuss cyberspace operations, offering sometimes-conflicting guidance. The military also has many different components; each seeks to address cyberwarfare issues but does so within a confusing and overlapping command structure. As the nation becomes more sophisticated in its cyberwarfare operations and capabilities, these issues must be resolved.

KEY CONCEPTS AND TERMS

Doctrine	Offensive	Strategic level
Economy of force	Operational level	Surprise
Irregular warfare	Perseverance	Tactical level
Legitimacy	Restraint	Traditional warfare
Maneuver	Security	Unity of command
Mass	Simplicity	War
Objective		

CHAPTER 14 ASSESSMENT

1. What name is given to the collected body of knowledge that provides commanders with a shared philosophy and language for warfare and military operations?

   A. Doctrine
   B. Strategy
   C. Tactics
   D. Operations orders
   E. Presidential directives

2. Which principle of warfare states that every military action must have a clearly defined and articulated purpose?

   A. Mass
   B. Offensive
   C. Surprise
   D. Objective
   E. Economy of Force

3. A military commander decides to attack six strategic targets simultaneously, using all available offensive forces at the same time. Which principle of warfare does this embody?

   A. Mass
   B. Offensive
   C. Surprise
   D. Objective
   E. Economy of Force

4. Which principle of warfare is more difficult to achieve when conducting joint and/or multinational operations?

   A. Surprise
   B. Offensive
   C. Economy of Force
   D. Mass
   E. Unity of Command

5. Which of the following principles of warfare are found in joint doctrine but not in Army-specific doctrine? (Select three.)

   A. Perseverance
   B. Legitimacy
   C. Simplicity
   D. Restraint
   E. Maneuver

6. When nations and nonstate actors engage in the irregular form of warfare, what are their intended strategic purposes? (Select two.)

   A. Establishing legitimacy
   B. Exerting influence over a nation-state
   C. Preventing a nation-state from exerting influence over them
   D. Influencing the population
   E. Blocking access to transportation lines

7. Who may engage in traditional warfare? (Select two.)

   A. Nonstate actors
   B. Nongovernmental organizations
   C. Nations
   D. Coalitions of nations
   E. Independent individuals

8. Which level of warfare includes national-level objectives set by the president of the United States and other senior government officials?

   A. Operational
   B. Traditional
   C. Strategic
   D. Irregular
   E. Tactical

9. Which level of warfare includes the planning of military campaigns and major military operations?

   A. Operational
   B. Traditional
   C. Strategic
   D. Irregular
   E. Tactical

10. Which level of warfare is concerned with the movement of individual units and the conduct of battles and engagements?

    A. Operational
    B. Traditional
    C. Strategic
    D. Irregular
    E. Tactical

11. The position of commander, U.S. Cyber Command, is held by the director of which federal agency?

A. Defense Intelligence Agency
B. Central Intelligence Agency
C. Department of Defense
D. U.S. Air Force
E. National Security Agency

12. Which one of the following is *not* one of the component commands of the U.S. Cyber Command?

A. Sixteenth Coast Guard
B. Army Cyber Command
C. Tenth Fleet
D. Sixteenth Air Force
E. Marine Corps Cyber Command

13. Which principle of warfare requires that commanders act within the confines of national laws and international treaties?

A. Perseverance
B. Legitimacy
C. Simplicity
D. Restraint
E. Maneuver

14. Which joint doctrine publication contains information about the U.S. military's strategy for cyberspace operations?

A. JP 1
B. JP 2
C. JP 3-12
D. JP 1-10
E. JP 2-1

15. Which of the following is *not* one of the five line of cyberwarfare effort, as outlined by the Department of Defense?

A. Build a more lethal force.
B. Engage primarily in offensive operations.
C. Compete and deter in cyberspace.
D. Strengthen alliances and attract new partnerships.
E. Cultivate talent.

Pandora's Box: The Future of Cyberwarfare

S INCE 2004'S TITAN RAIN ATTACKS, nation-states have employed cyberattacks more and more as part of the accepted array of weapons. Advanced tools like Flame, Duqu, and Stuxnet are only a few of the weapons in this sometimes quiet battle between attackers and defenders. Nation-state sponsored groups like Sandworm, Hidden Cobra, Stone Panda and other advanced persistent threat (APT) actors have grown in number, capabilities, and complexity. Other attacks, like those against U.S. drones and infrastructure and services in Ukraine have helped demonstrate the risks that increasingly networked and computer-based military units, weapons systems, and critical civilian infrastructure face.

International law in the form of the Geneva Conventions and numerous other treaties has not yet been updated to accommodate the new ways in which international cyberwar affects both military targets and civilian infrastructure and populations. The lack of clear rules means that nation-states are left to use their best interpretation of the existing body of law. Obviously when nations are left to interpret laws and standards, they often choose the interpretation that best fits their own needs and desires. This makes cyberwarfare an exercise in the careful application of international law, or in many cases, the lack of fully relevant law. As nations continue to take advantage of the lack of established law and precedent, the danger of an attack resulting in large-scale deaths or major impacts to national infrastructure—either on purpose or accidentally—continues to grow.

Because the laws of cyberwarfare remain unsettled, the use of advanced malware to disable or destroy infrastructure and systems as part of covert activities between nation-states who may not be in an active state of war has made the need for strong cyber defenses a priority. While the countries with the best-developed cyberwar capabilities in the world have enjoyed a significant lead in the development of advanced tools, skills, and techniques, other countries have worked to develop their own capabilities. More and more countries continue to invest in advanced computer network defense and attack capabilities as part of their national defense spending, and the proliferation of tools that have been used in visible attacks have left components accessible to actors who might not otherwise have had access to the technology previously. In fact, the broad use of advanced cyberwarfare tools have democratized access to them in a way similar to how the Internet itself has made access to data and populations broadly available. Individual actors as well as organized groups can now acquire and use tools that were previously only the domain of nation-state sponsored groups.

As advanced tools and techniques are deployed, they leave remnant portions of their code and capabilities behind them. At the same time, breaches and exposure of toolkits from organizations like the U.S. National Security Agency (NSA) have also made technology and techniques available. This creates the possibility that any organizations or individuals with access to copies of advanced malware packages can analyze and possibly use the malware for their own purposes. Reverse engineering of cyberwarfare attack packages means that the advantages that nation-states have due to their well-funded and staffed computer network attack teams may exist only for a short time after their tools are deployed.

This analysis and reuse process has been observed over and over, with malware components being adopted by other organizations as well as re-use by their original creators. Attribution becomes even more difficult as tools are used by new threat actors, and of course deniability and working to help ensure misattribution are key goals for many threat actors. The commercialization of advanced tools also continues, as tools and capabilities that were once the purview of nation state actors are adopted and used by ransomware and other financially motivated threat actors as well.

The massive growth and ongoing prevalence of APTs indicate that computer network attack capabilities are an increasingly important element in nation-state intelligence gathering. APTs have been found throughout civilian infrastructure for businesses, higher education, and government system. These deeply embedded malware packages have been discovered with evidence pointing to long-term compromise and data gathering. Network and computer-based intelligence gathering can blur the line between nation-state interest, industrial espionage, and cybercrime—leaving defenders needing to defend against advanced adversaries without a clearly identified opponent.

Today's strategists, defenders, and attackers must all operate in a quickly changing world of cyberwar capabilities, agreements, and adversaries. Attackers enjoy many advantages in a networked world, as they often need only find a single vulnerability to make their way into a protected network. Defenders face greater challenges, as it is often difficult to determine who is attacking, from where, and for what purpose. Defenders can be faced with an almost infinite number of potential threats, making strong defensive design, capable and skilled defenders, and a strong response plan a must.

Chapter 15 Topics

This chapter covers the following topics and concepts:

- What the future of cyberwar may look like
- What the role of nonstate actors in the future of cyberwar is
- What the future of international law in cyberwar is
- What the impact of pervasive network connectivity for devices and systems is
- What the effects of cyberwar on civilian infrastructure could be
- What the role of advanced tools and training in the future of cyberwar is
- What the future of defensive cyberwar may look like

Chapter 15 Goals

When you complete this chapter, you will be able to:

- Explain the future of cyberwar as envisioned by U.S. and other nations' strategies
- Describe potential future issues caused by nonstate actors in cyberwar
- Explain the risks created by advanced persistent threats
- Describe the role of future international laws related to cyberwar
- Explain what is meant by the Internet of Things and what it means in terms of potential cyberwar targets
- Describe the future threat cyberwar poses to cloud services and the U.S. government's response
- Describe tools and methods of future computer network defense and attack
- Discuss the future of defensive cyberwar

The Future of Cyberwar

The future of cyberwar has been debated for almost 30 years. In 1993, John Arquilla and David Ronfeldt predicted that the information revolution would change how institutions are designed, how power is distributed, and how space and time effect the way war is waged. They noted that it would change large, static organizations, and the survivors would change into nimble, flexible organizations that used new, network-based models to operate.

All of this, they said, would change how war is waged and how societies enter conflicts. In fact, changes in the U.S. Department of Defense (DOD) and the creation of cyberwar-focused organizations in other countries reflect their predictions three decades ago. New shared groups have been created that distribute information and provide shared defensive and offensive services and capabilities. These are beginning to throw a spotlight on the organizations they serve.

Over the past three decades, many of Arquilla and Ronfeldt's predictions have come true. Now cyberwar tools that can attack countries far away from the originating country's borders are common, and the wide availability and use of existing tools makes them easier to create and to obtain than traditional kinetic weapons that could reach similar distances. They are also far less expensive than kinetic weapons systems, both to create and maintain, and lack of effective law and ease of attribution means they are used with less likelihood that the user would be held responsible for their deployment. This means that cyberwar weapons are more accessible to countries that want to develop an offensive cyberwar capability.

FYI

Arquilla and Ronfeldt defined a new information-related conflict they described as *netwar*: "trying to disrupt, damage, or modify what a target population 'knows' or thinks it knows about itself and the world around it." They believed that information and communications would form the core of this new type of conflict. They believed it would be fought not only between nations, but also nonstate actors like insurrectionists and corporations. Modern uses of social media, including Twitter and Facebook account exploits, have fulfilled their netwar prophecies.

The Israeli Defense Forces pointed to a number of important cyberwar issues that echo concepts that Arquilla and Ronfeldt described:

- The fact that everything is networked, making everything a possible target
- The advantages provided by human capital, such as attackers and defenders with strong computer network attack and defense skills, and the need to have a well-trained, capable group of cyber defenders
- The increasing use of computers and automation, which make cyberattacks more useful, and the increasing role of social media and online communications for propaganda
- The ease of developing tools, which allow nations to quickly become cyberwar rivals

- Unpredictability, due to attacks that are unknown and may have already succeeded and the lack of knowledge about who is conducting attacks

"Cyber attackers always advance faster than the defenders," says Major General Uzi Moshkovitz, head of the Israeli Defense Forces Telecommunications Branch. U.S. cyberwar experts agree: In 2014, General Keith Alexander, commander of the U.S. Cyber Command, described the major gaps the U.S. military faces in preparing for cyberwar and how those gaps would affect USCYBERCOM's priorities and strategies for future computer network defense, attack, and intelligence operations. In a statement to the U.S. Senate Committee on Armed Forces, General Alexander called out a number of key points:

- Cyberspace is more hospitable to attackers than defenders.
- The U.S. legacy information architecture and some weapons systems are not designed to be robust and to survive in a hostile cyber environment.
- Commanders do not always understand the cyber-risks they face.
- Commanders can't reliably gain situational awareness of their own systems or the global threat environment.
- Communications systems are vulnerable to attacks.

General Alexander noted that the future of USCYBERCOM depends on implementing a defense-in-depth strategy involving a defensible architecture staffed by trained and ready cyberforces. He also emphasized that meeting those risks will require global situational awareness and the ability to share information and knowledge. This can provide a "common operating picture," which will allow DOD to take action.

Future cyberwar activities will require nation-states and nonstate actors to meet these challenges. This will require balancing a defensive strategy with offensive capabilities, as well as proceeding cautiously to ensure that actions taken are not violations of the existing and evolving rules of war.

Blurred Boundaries: Cyberwar and Nonstate Actors

In conventional kinetic warfare, it is sometimes hard for defenders to determine an attacker's identity. In cyberwarfare, it can be almost impossible to determine the attacker—if that attacker is careful to use systems that cannot be linked back to him or her and if he or she removes identifying features from the tools and methods used. As the attack methods and advanced exploit tools and techniques used by nation-states have become available to cybercriminals, insurrectionists, and individuals, it has become increasingly difficult to determine if an aggressor is a traditional nation-state opponent or a group that doesn't abide by the traditional and accepted rules of war.

Nation-states may be forced to decide whether to respond based on the threat the attack creates rather than the aggressor. Their responses may affect civilian and nation-state resources and infrastructure. The fact that attackers often hide behind a series of compromised systems to conduct their attacks means that responses are likely to involve individuals and organizations who were not the actual aggressors but whose security failures have resulted in their networks and systems participating in attacks. This changes how international law and customs view these responses, and the issue of nation-state retribution against compromised civilian infrastructure and systems will likely become a critical issue.

Because they know that their opponents can hide the attacks' source, nation-states may be tempted to use analysis of the attacks to determine which of their potential adversaries conducted them. Analysis of attack tools is often based on pinpointing the networks and systems from which the attack appears to have originated—as well as the knowledge required to create the tools and processes used. Even the attack tools' source code can reveal clues about the native language and programming habits of the attack code's developers.

An entire industry has sprung up around APTs, identifying threat actors, analysis tools and techniques that attempt to attribute techniques, malware, and targeting practices to APT groups. Fingerprinting existing groups and identifying new groups remains challenging, as re-use of tools and techniques can make even the best identification processes difficult. Attribution and understanding the motivations and capabilities of attackers is important enough that this remains a priority for both commercial and government teams of defenders and attackers.

Identifying Attackers

Defenders who find advanced malware on their systems and networks will often attempt to reverse engineer it to identify the attackers as well as to determine what the malware can do. In the case of Stuxnet, the code contained a reference to May 9, 1979, the date that a prominent Iranian Jew was executed by the Iranian government. This led analysts to believe that Stuxnet likely had a link to Israel. Later statements by officials from both the U.S. and Israeli governments indicated that Stuxnet was a joint effort. In many cases, however, information planted in malware can mislead defenders, particularly if the malware is part of a repurposed package.

Tools like Duqu, Flame, and Stuxnet as well as many modern APT tools contain cleverly written code and advanced techniques that point to the developers being skilled professionals. Experts who analyzed Duqu and Stuxnet noted that the code stood out from other malware code by the level of skill and specialized knowledge the code demonstrated. In fact, both Duqu and Stuxnet share many commonalities, infection methods, and some source code, although the two have very different purposes. Flame was an advanced cyberintelligence gathering tool, whereas Stuxnet was a highly specific computer-based attack tool. Many analysts believe that they were either written by the same authors or that Duqu's creators had access to Stuxnet source code.

Unlike Duqu and Stuxnet, Flame appeared to use an entirely unique code base, although it did leverage two of the vulnerabilities that Stuxnet used. This reinforces the fact that attackers can repurpose attacks and software, making it even more difficult to determine who they are. Nonstate actors can repurpose tools previously used by a nation-state, and a nation-state could use the same tools as nonstate actors to deceive its opponents.

> **FYI**
>
> If Duqu and Stuxnet were actually are based on the same framework and shared libraries, it implies an additional layer of sophistication in the approach the attackers used. Information that links attacks can be a great help in determining who may have written the malware and chosen its targets.

> **FYI**
>
> Ralph Langner's detailed analysis, which includes details of the supervisory control and data acquisition (SCADA) systems used in Iran, provides history and design insight into the plants that Stuxnet attacked. You can find it at *https://www.langner.com/wp-content/uploads/2017/03/to-kill-a-centrifuge.pdf*.

Unfortunately for nations that might want to adopt this approach, Ralph Langner's analysis of the Stuxnet attack concluded that nation-state resources would not be required to duplicate a complex attack like Stuxnet. He also said that while Stuxnet's development required nation-state-level resources for intelligence gathering, infiltration, and testing, similar attacks by nonstate actors are entirely possible. In fact, Langner's analysis points out the following:

- Intelligence gathering is not difficult, and it would not take nation-state-level resources to gather information about many facilities from underpaid or disgruntled employees.
- Attacks can be reused or can be used against multiple targets. Langner notes that smaller attacks widely used can be just as effective as a single attack in one facility with a major impact.
- Exploit code is likely to become modular and user-friendly rather than remaining so complex it requires deep knowledge to implement or use.

The ability to conduct attacks using massive computer systems is also easily attainable for both nation-states and nonstate actors. The growth of large-scale cloud computing environments—which will rent massive amounts of computational power and storage connected to some of the fastest Internet connections in the world—means that attackers can simply pay for machines to conduct their attacks. Using stolen credit cards, compromised accounts, or difficult-to-track payment methods such as **cryptocurrency** makes cloud service providers an ideal and common platform for both nation-state and nonstate actors to attack from while concealing their identity. Figure 15-1 shows an example of using multiple data-center cloud computing resources. These can leverage hundreds, thousands, or potentially tens of thousands of virtual computers, which can target a single system or network from sites around the world. This type of attack can leverage massive computational resources that easily outweigh any system likely to be hosted in a traditional data center.

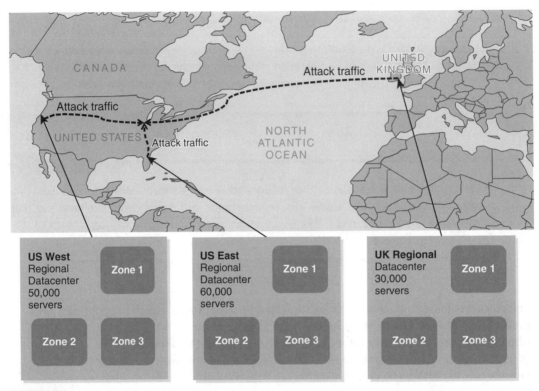

FIGURE 15-1

Leveraging commodity cloud computing for attacks.

Attackers also continue to make large-scale use of botnets, including large collections of compromised computers as well as an increasing number of compromised **IoT** (Internet of Things) devices. A single vulnerability in devices like home network routers, smart appliances, or even Internet-connected light bulbs or cameras can provide attackers with large networks of devices they can control and leverage for distributed attacks that are incredibly hard to track or stop. With many different possibilities for systems that can launch attacks at their disposal, attackers continue to have effectively free, large scale, and hard to attribute attack resources available to them.

Nonstate actors will also continue to make cyberwar a challenge for nation-states, as they are harder to target and can be nearly impossible to hunt down. The rise of the **hacktivist** group, Anonymous, as a threat to both national and civilian infrastructure, as well as businesses and commercial networks, demonstrated the threat that distributed associations organized over the Internet can create. While Anonymous has faded, new groups will continue to appear, applying the lessons that Anonymous

> **NOTE**
>
> The Commission on the Theft of American Intellectual Property estimates that losses due to stolen intellectual property are between $225 billion and $600 billion a year, in addition to lost business and jobs.

taught its members. The challenges found in guerrilla warfare against an enemy that can blend into the population with ease are even more evident when the guerrillas can hide among millions of systems on the Internet—any of which could be used as a launch pad for a cyberattack.

Advanced Persistent Threats

Attackers who seek to retain long-term access to gain intelligence and to control infrastructure and systems, along with the tools they use, are known as APTs. Future cyberwar intelligence activities will include ever-more capable APTs.

Intelligence gathering and long-term exploit of systems will likely continue to heavily involve the techniques APTs now commonly employ. As the malware packages APTs use evolve, they will continue to add new capabilities and more advanced techniques to help them evade detection; to maintain control of systems; and to detect, capture, and transfer sensitive data and intelligence.

Information security researchers have shown that many APTs already have successive layers of upgrades and modular package designs. Features like these allow them to use only the capabilities needed to evade defenders, as shown in the preservation phase of a typical APT life cycle (see Figure 15-2). When defenders become aware of the threats, the APT's controllers or the software itself can upgrade the packages used to remain on compromised systems. It is reasonable to expect that APTs will target new areas of the computing and network technologies market, such as mobile devices and cloud computing. An APT that

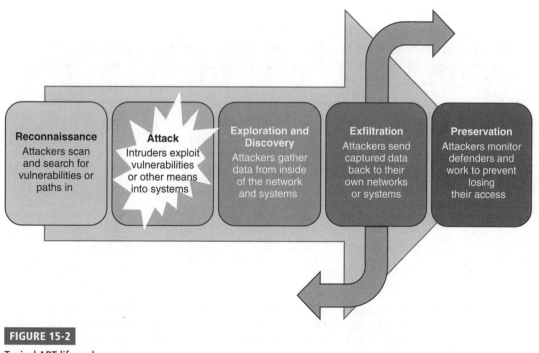

Reconnaissance	Attack	Exploration and Discovery	Exfiltration	Preservation
Attackers scan and search for vulnerabilities or paths in	Intruders exploit vulnerabilities or other means into systems	Attackers gather data from inside of the network and systems	Attackers send captured data back to their own networks or systems	Attackers monitor defenders and work to prevent losing their access

FIGURE 15-2

Typical APT life cycle.

can leverage an organization's own capabilities to make itself stronger can remain hidden for months or years. That APT could then use the organization's entire infrastructure against it in addition to providing long-term intelligence about the organization.

U.S. Charges Against APT Actors

In May 2014, the United States indicted five Chinese military officers from Unit 61398 of the Third Department of the People's Liberation Army on a variety of charges, including computer hacking and economic espionage. This is the same group that Mandiant identified as APT1. The charges cited the use of compromised systems to steal information, including trade secrets that were useful to state-owned enterprises and other businesses in China. In addition to trade secrets, prosecutors alleged that the attackers had stolen internal communications that could give Chinese companies insight into corporate strategies and vulnerabilities.

The FBI listed all five military officers on its Cyber Most Wanted list, a major step since the officers were active duty military personnel.

Victims identified in the charges included Westinghouse Electric Co., SolarWorld AG, United States Steel Corp., and Alcoa, Inc., a major aluminum producer. Prosecutors alleged the attacks occurred across an 8-year period, from 2006 to 2014.

The charges are the first ever brought against a state actor for this type of hacking, according to U.S. Attorney General Eric Holder. "The range of trade secrets and other sensitive business information stolen in this case is significant and demands an aggressive response," he said in a Federal Bureau of Investigation (FBI) press release. This form of long-term cyberespionage blurs the lines between intelligence gathering and corporate espionage, and highlights the difficulty of separating nation-state cyberwar attacks and nonstate corporate espionage.

Shortly after the U.S. charges were announced, Chinese state media released a report calling U.S. surveillance "unscrupulous" and claiming that U.S. intelligence-gathering activities violated international laws. *China Youth Daily* claimed in an editorial that Cisco, a major network technology company, had aided in the intelligence-gathering activities and had willingly helped the NSA in those activities.

The United States has continued to press charges against APT actors since this time, including charges field in October 2020 against Russian military officers who were charged due to their actions aimed at a French presidential election, the South Korean Winter Olympics, and other targets.

APT groups frequently target firms like SolarWorld for industrial espionage. In fact, SolarWorld found competing in the U.S. market difficult after successful hacks provided their Chinese competitors with information about pricing and marketing strategies, in addition to other private corporate information. In 2012, the U.S. FBI contacted SolarWorld to alert it to a persistent threat. At the time this was a relatively uncommon even, but the increased impact of nation-state sponsored APT actors has resulted in the U.S. government and commercial security teams increasing their efforts to notify businesses of APT attacks. In multiple cases since the 2014 charges against China, the United States has leveled charges at state-sponsored APT hackers, noting that U.S. intelligence gathering does not directly share information with U.S. companies that would benefit from the information, unlike the Chinese program that targeted SolarWorld and other companies.

Future cyberwarriors will continue to use long-term attacks to acquire and maintain access to government and civilian systems, both to provide intelligence information and to gain a competitive advantage for business. Nations other than the United States are responding to the threats that APTs create. In 2012, Japan's Ministry of Economics, Trade, and Industry, along with eight electronics companies, created a Control System Security Center (CSSC) to improve the security of control systems for infrastructure. The CSSC's "major business activities" are very similar to those expressed by both military and government agencies tasked with providing cybersecurity against APTs and other similar threats:

1. System security verification
2. Establishment of structures and technology to enhance security
3. Security international standard
4. International standard compliances certification
5. Incident support
6. Human resource development
7. Promotion (awareness)

The common themes found here will remain a constant for organizations faced with a combination of advanced threats and increased targeting of their previously non-network-connected devices and systems. APTs like Russia's attacks against Ukraine have proven that both industrial control systems (ISCs) and SCADA platforms can be an important target for attackers.

Continuous Warfare

Defenders in cyberwar have to be constantly aware that their networks and systems are under attack. Further, the future of cyberwar will continue to rely on defense and attack using the same methods commonly used today. **Network attacks**, **malware**, **vulnerabilities**, social engineering, and even internal threats from trusted employees provide a constant threat to the organizational resources that they are employed to protect. Figure 15-3 shows these threats, which must be balanced and defended against in future computer network defense scenarios. The sheer variety of possible threats that an organization may face will make cooperation between nations, industry, and security experts a necessity.

The ease with which basic cyberwarfare tools can be acquired and employed means that nation-states and civilians must ensure that their defenses are always active and monitored. In essence, continuous low-level warfare is always occurring. A typical

FIGURE 15-3

Continuous threats to systems and networks.

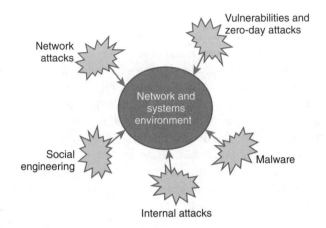

network can see thousands or even tens of thousands of network scans a day. Network-edge firewalls connected to the public Internet often block more than half of the traffic that is sent to systems behind them. Continuous attacks aren't limited to network attacks and scans, however. Defenders must keep pace with an ever-growing rate of malware attacks and viruses. AV Test reported that the first three months of 2020 saw more than 43 million new malware applications—a rate of almost a half a million new malware programs per day or 4.3 samples per second. The continued ability of malware writers to use prepackaged tools allows them to create new malware quickly and then package it in a way that makes it distinct from other malware packages that were created using the same tool, ensuring a continuing high pace of malware growth. The same malware creation packages are also likely to include tools and techniques used by APTs to provide malware with even greater capabilities. This means the research into vulnerabilities and exploits conducted under the guise of nation-state intelligence and cyberwar activities will place advanced exploit tools into the hands of cybercriminals, hacktivists, and insurrectionists. The blurred boundaries of future cyberwar will become even more difficult to determine as the attacks employed by last week's APT or nation-state attacker show up in commodity malware creation packages for relatively unsophisticated attackers to use.

Software, hardware, and firmware all provide attackers with possible attack vectors when they seek to compromise an organization's network and systems. Balancing the development of new software and devices with providing updates and fixes for current systems is a challenge for defenders. They are often left to rely on the vendors and manufacturers of the software and hardware they provide to their users and organizations. Attackers need only identify one vulnerability that will permit them to get past existing defenses to succeed in their mission of obtaining a foothold in a well-defended network.

Defenders in cyberwar must balance updates and patch management against the needs of their operations and the risks patches can create when installed. Even if they remain fully patched and updated, attackers can identify vulnerabilities manufacturers are unaware of and successfully attack devices and software that is as secure as defenders can make it. This means that defenders must constantly be on guard and must use monitoring and detection systems in addition to handling vulnerabilities they are aware of.

A sometimes-neglected aspect of the continuous attacks computer network defenders have to face is the attempts to exploit human vulnerabilities through social engineering. The strongest network defenses can be defeated by a trusted internal employee or administrator who falls for a social engineering attack. Ruses like phishing attacks, or even more-directed threats using specific knowledge about an organization, can provide attackers with an easy route into a protected environment. Continuing education and awareness efforts are often a significant part of computer network defense activities in many organizations.

Integrating Cyberwar and Kinetic Warfare

One possible result of the constant use of cyberwar capabilities will be the integration of cyberwar capabilities into kinetic warfare. Over time, cyberwar is becoming an expected part of normal warfare and intelligence activities, much like air combat and submarine warfare before it. Those once-revolutionary game changers have become an expected part of combined arms combat. Strategy that doesn't take existing technologies and capabilities into account is likely to fail.

Cyberwarfare provides a new way to attack civilian and military infrastructure as well as the weapons and information systems themselves. This means that it will likely be used at every opportunity as a relatively low-cost means to military and intelligence ends. A technology that becomes that closely integrated into every act is not as remarkable as one that nations fear to use, such as nuclear, chemical, or biological weapons.

FYI

Russian cyberattacks in early 2014 against Ukraine did not create severe disruptions. Later attacks did. In fact, Ukraine has become a proving ground for cyberwarfare techniques and tools with massive impacts felt across the country as both military and civilian targets are impacted. Election software, the power grid, and the commercial sector have all fallen victim to advanced cyberattacks. Groups like Fancy Bear, Cozy Bear, and others are constantly in action. In response, the United States has provided funding and U.S. companies have provided technical know-how to help protect systems and infrastructure. You can read more about the ongoing efforts at https://www.politico.eu/article/ukraine-cyber-war-frontline -russia-malware-attacks/.

Alliances and Partnerships

The need for a coordinated effort to provide greater security has received increasing attention from the U.S. government. The U.S. Cyberspace Policy Review points to efforts to develop an international cybersecurity policy and strengthened international partnerships. Other efforts like U.S. Cyber Command's CYBER FLAG exercise allow allied nations to participate in cyber defense training with U.S. military units. The U.S. Department of Defense, Department of Homeland Security, and NSA all engage in partnerships with universities and colleges across the United States to develop talent and increase cybersecurity skillsets. The NSA's Centers of Academic Excellence in Cybersecurity are a particularly important component in this cybersecurity talent development effort.

> **NOTE**
>
> The CCD COE is responsible for the creation of the Tallinn Manual, and hosts CyCon (https://cycon.org/), the International Conference on Cyber Conflict.

Cybersecurity concerns are driving the creation of cooperative cyber defense agreements between NATO members as well as the improvement and growth of organizations like the Cooperative Cyber Defence Centre of Excellence (CCD COE). Such concerns have also resulted in increasing cooperation between the United States and the European Union. This includes focus on cybersecurity capability development for third-party countries and focused work by specialized cybersecurity working groups on critical infrastructure cybersecurity and awareness.

The need for cooperation extends into industry and has resulted in the creation of groups like the Control System Security Center (CSSC), the Health Information Trust Alliance (HITRUST), and other vertical industry-centric organizations for cooperation and sharing of security information. In some cases, information-sharing groups are created after a breach. For example, the U.S. banking and retail industry created the Financial Services Information Sharing and Analysis Center (FS-ISAC) after large-scale breaches were announced in early 2014. Groups such as these work to share threat and event information and alert their members. They provide expertise and awareness and serve to coordinate the efforts of industries where organizations and companies have traditionally competed with, rather than assisted, each other. The constant state of cyberwar and the broad range of targeted organizations mean that cooperative response and shared knowledge are far more appealing than the lost competitive advantage that sharing information may cost.

FYI

The FBI has operated InfraGard, a partnership between the FBI and the private sector since 1996. InfraGard works with 16 critical sectors, including the chemical, communications, manufacturing, defense/industrial, and transportation systems sectors. Critics of InfraGard have questioned the information-sharing restrictions that the group uses to help preserve interorganizational trust, and have asked whether U.S. government information about threats that is shared with members should be made public.

Gaining access to an information-sharing partnership is an obvious target for future attackers. The inside knowledge and response information shared between industry partners could provide attackers with useful details about which attacks are successful and which have been detected. Although cooperative organizations are careful to review their members, those members may not always be able to protect themselves, their systems, or their email from talented and capable attackers. This means future attacks will likely occur through the very trusted groups that used to share information about the attackers.

International Law and Cyberwarfare

The need for updated and clarified international laws on cyberwarfare has been reinforced by the analyses of groups of experts such as those who wrote the Tallinn Manual. The ways cyberwarfare is conducted create new issues in many areas. Therefore, the means of conducting cyberwar will face legal challenges in a number of areas, including, but not limited to, the following:

- The requirement to respect other nations' territorial sovereignty and jurisdictional rights
- How and when cyberattacks count as a use of force
- When and what type of self-defense is reasonable, proportionate, and appropriate when a cyberattack is occurring, has occurred, or may occur
- What constitutes cyberweapons, and how they are regulated under international law
- How existing international laws and agreements like the Geneva Conventions apply to cyberwar
- How the roles and laws are applicable to nonstate actors in cyberwar
- Which services and systems, such as those used for health care and journalism, are protected and how they should be treated in cyberwar

NOTE

A second version of the Tallinn Manual is already envisioned, and updates based on cyberwar activities since its original publication should point to additional clarifications needed in international law.

The need for greater attention to international law, as well as the significant differences found between the laws and how many countries interpret them, has been expressed in a variety of ways in recent years. This indicates agreement on a need for action. In fact, the U.S. Cyberspace Policy Review states, "Law applicable to information and communications networks is a complex patchwork of Constitutional, domestic, foreign, and international laws that shapes viable policy options." It also notes that three critical areas are related to the laws and accepted rules of cyberwarfare that need international cooperation:

- Acceptable legal norms for territorial jurisdiction
- Sovereign responsibility
- The use of force

The Cyberspace Policy Review also points to a need to work to make the differences among the laws covering investigation, prosecution, data preservation and protection, and privacy less of an issue at both the international and interjurisdictional levels. The Cyberspace Policy Review doesn't call for legal action until its mid-term action plan is put in place. At that time, it sets a goal to "[i]mprove the process for resolution of interagency disagreements regarding interpretations of law and application of policy and authorities for cyber operations." The plan doesn't include a timeline for efforts to create international legal consensus.

NOTE

You can find the U.S. Cyberspace Policy Review in its entirety at https://www.nitrd.gov/cybersecurity/documents/Cyberspace_Policy_Review_final.pdf.

Networks Everywhere: Cyberwar in a Highly Connected World

The increase in network connectivity for almost every possible system and device will continue to expand the potential targets in cyberwar. Past attacks have targeted industrial control systems as well as those controlling power grids and other infrastructure, in effect duplicating surgical strikes against those parts of infrastructure. As vehicles, households, and individuals continue to add more connectivity, they will become easier targets. The idea that more and more devices and systems will be Internet-connected is referred to as the Internet of Things (IoT). Future attacks enabled by pervasive networking might include the following:

- Targeted or untargeted compromises of network-connected vehicles. Attacks against vehicles are already being demonstrated to the media, and network-connected and self-driving cars are being tested. The same concepts apply to airplanes and naval vessels, which provide internal networks for control as well as networks for passengers. Now designers must implement network security controls in their vehicle designs.

- Complete compromise of drones and other weapons systems through multistep attacks focused on the drone's control systems. As drones enter wider use in both the civilian and military world, this results in the potential for attackers to cause armed drones to attack unintended targets. It could also result in the use of unarmed civilian drones for intelligence gathering—or even attacks using the drone itself as a weapon.

- Increased focus on mobile devices, which allow tracking of an individual's location. This potentially allows attackers to listen to nearby conversations using built-in microphones and take pictures or capture video. Mobile devices such as cell phones and tablets also often contain significant amounts of information about the individual, from daily schedules to contacts. For convenience, they are often set to automatically log on to remote tools and websites, also permitting an attacker to gather useful information.

- Compromises of home automation and control systems. Whereas changing the temperature of an individual's house might seem relatively unimportant, changing the environment around a key negotiator or public figure can make him or her less able to respond in an emergency.

- A focus on remote access to medical devices. News articles about the potential issues with former U.S. Vice President Richard Cheney's pacemaker pointed to the potential to use cyberattacks against politicians and other key personnel who use medical devices. Attacks could range from simple modification of the device's settings, like speeding or slowing down a pacemaker, to far-greater impacts—such as disabling a pacemaker or using its built-in defibrillator to cause a heart attack.

The increased connectivity of devices not previously connected to networks means computer network defenders must consider the capabilities and design of each device added to a network. As the numbers and types of devices increase, this will make defenders' jobs harder. Managing the software updates, vulnerabilities, and configuration of more devices not designed for central management and protection will present significant challenges.

Network design and centralized protection capabilities will become increasingly important as devices without additional protective software and management capabilities enter networks. This will require network defenders to design their networks to isolate devices that cannot be secured effectively from trusted or protected devices—without making the difficult-to-protect devices less secure. Although some civilian, military, and government network defenders may be able to restrict the devices that appear on their networks, many other networks will use them.

Hacking a Car

In 2013, *Forbes* magazine reported that Charlie Miller and Chris Valasek had figured out how to hack the operating software of a Prius sedan. Miller and Valasek were working to discover security vulnerabilities as part of research into vehicle security systems funded by the U.S. Defense Advanced Research Projects Agency (DARPA).

The video clip that *Forbes* released showed that the car's brakes could be disabled, the horn could be sounded, and the steering wheel and seat belts could be manipulated remotely. Although the attacks were successful, they did require extended physical access to the Prius. Other hacks that were not DARPA funded have attacked other vehicular systems, such as locks that use encrypted codes to allow remote opening.

Since then, attacks against cars have increased, with Israeli firm Upstream reporting 150 incidents in 2019, a 99 percent increase in the field in a single year. As vehicles become increasingly Internet connected and as driverless and automated features improve, vehicles will continue to be a major target.

Tesla and other manufacturers have begun to use online updates to provide software fixes and patches as well as feature upgrades, and constantly connected vehicles are increasingly becoming a norm. At the same time, manufacturers that haven't learned security lessons well are finding that their vehicles are vulnerable to attacks just like other network connected devices.

Integration of network defense between civilian and government agencies to ensure that networks are not easily compromised or used as launching points for further attacks will continue to be a difficult task. Individual home networks and network-connected devices will never receive the same level of attention as a centrally managed and protected network. This means the many endpoints connected to the Internet will remain a source of infrastructure for attackers. These hostile actors can compromise network devices, using them and their collective bandwidth and resources to attack other sites and facilities while using the compromised systems to hide their own identity.

Cyberwar and Infrastructure

One of the most terrifying threats in future cyberwar will be the threat to civilian infrastructure—including power and utility grids, communications, and transportation infrastructure. Software flaws and configuration mistakes have already demonstrated the potential effect that a cyberattack could have on existing infrastructure, with results ranging from flooding to power-grid shutdowns. Intentional attacks against increasingly computerized infrastructure and control systems could create massive disruptions due to the complexity of power grids and utility infrastructure.

The advent of large-scale commodity cloud computing means that the cyber infrastructure companies and governments rely on is moving to massive data centers designed to allow virtualized infrastructure to be shared by many users. This means that infrastructure has begun to centralize in the form of hosted services and systems, making the cloud providers themselves attractive targets. The concerted attacks on Google's systems, which resulted in exposure of both back-end systems and source code, show a potential future in which the underlying infrastructure of cloud services is a preferred target for intelligence activities as well as cyberattacks.

The U.S. government has attempted to preempt this problem for cloud services with its Federal Risk and Authorization Management Program (FedRAMP). FedRAMP provides a set of common security controls and verification processes that help ensure that providers meet known standards. This avoids common issues, but it doesn't promise guaranteed security for government use of cloud services.

The lines between kinetic war and cyberwar will likely continue to blur as infrastructure attacks can be executed at a lower cost and with less-direct exposure for nations with stronger cyberattack capabilities. At the same time, infrastructure defenders who have historically protected infrastructure not connected to the network will have to learn the hard lessons that network defenders in other disciplines have had to learn over the past two decades.

Advanced Tools and Training

The increasing need for capable computer network defenders, as well as the growth of cyberwarfare as a viable tool for attack, has meant that countries are increasing their efforts to develop people with those skills. The U.S. government has increasingly required certification using commercially available certification packages for many jobs that involve cybersecurity. These certifications range from entry-level to advanced and provide a demonstrable foundation of knowledge for cybersecurity practitioners.

At the same time as the growth in certification and skills, the U.S. government and military are increasing the number of cybersecurity staff. Significant growth has continued for the U.S. military, allowing greater support for cyberwarfare's increasing role in military operations. The same type of growth has been visible in China, with the reported growth of cybersecurity and hacking schools that support the Chinese military and the continued growth of China's cybersecurity operations. It is reasonable to expect that countries that currently have significant cyberwarfare and cybersecurity capabilities will expand them, continuing to project power into cyberspace.

Advanced Training on Cyber Ranges

Threats to commercial and governmental organizations have resulted in the creation of virtual environments designed to mirror the same systems and devices found in the real world. These virtual environments, like the Michigan Cyber Range and the SANS NetWars CyberCity, provide environments that allow practitioners to attack and defend the same types of networks they would encounter without the risk of taking production systems offline.

The SANS NetWars CyberCity environment includes SCADA controls for electrical power distribution, water, transportation, medical, and commercial environments, as well as typical residential infrastructure. These are all built using physical models with the help of a hobby shop. It even provides streaming cameras for defenders and attackers to check their progress when attacking the model city. SANS also provides Cyber Situational Straining Exercises (Cyber STX), which are live, full-scale, scenario-based attack and defense training. The U.S. Army uses SANS NetWars Cyber STX to train for attack and defense, including hybrid cyber and kinetic warfare activities.

As cyberwar training becomes increasingly important, the ability to accurately simulate entire cities down to individual houses and infrastructure control nodes allows practitioners to gain hands-on skills that otherwise require years of experience. They can review critical parts of training scenarios and even recreate actual attacks in ways that allow students to study each part of a compromise, then respond in different ways as they gain skills and knowledge.

The growth of cyberwarfare as a means of asymmetric power projection—which empowers organizations and nation-states that would normally have had a relatively small ability to conduct kinetic warfare against opponents—means that the growth of advanced tools and training will be an attractive option. The relatively low cost of purchasing an exploit on the open market that provides access to an opponent's critical infrastructure is a low barrier to entry into cyberwar. Even a simple exploit can allow attackers to gain access to less-important systems—which they can then use to expand their access and gain further intelligence. This means relatively poorly funded organizations with little support can have a disproportionate effect on their targets.

Examples like these and those deployed around the world demonstrate that military use of both computer network attack and defense techniques have become part of the normal mission preparedness regime for cyber and kinetic warfare units. Commanders and national leadership consider cyberwar techniques to be part of their arsenals and plan for cyberwar as part of operations the same they previously planned for combined intelligence, deception, and kinetic activities.

The Future of Defensive Cyberwar

Successful defense in future cyberwar will require even greater integration of capabilities between the government and civilian sectors. Integrated, zero trust, self-defending networks that can apply data-centric techniques to adapt to the state of the systems, and components that make up network parts will have to respond to threats as they appear. Defenders will have to monitor and respond based on complicated inputs and detection capabilities while maintaining their networks' and systems' ability to perform their required functions.

For now, visions of a complete, smart, self-defending network continue to be more of a desire than a reality, although small-scale implementations exist. Defenders must balance design, training, and response capabilities with current technologies. Defenders will have to be increasingly alert to suspect network devices and gear that have been modified or compromised prior to delivery by nation-state actors with access to the IT supply chain or who can work directly with vendors to deliver systems already under their control.

The need for international agreement on cyberlaw—particularly updated and, in some cases, entirely new laws—to handle cyberconflict between nations and between nations and nonstate actors will become increasingly clear. As conflicts escalate and have greater effects on countries worldwide, it's likely that efforts like those that produced the Tallinn Manual will be leveraged to create a new body of international law dealing specifically with the ways cyberwarfare is waged.

Cooperation among industry, governments, and the manufacturers and creators of networked devices is crucial to a survivable infrastructure and IoT. As more and more of the world becomes network connected, the potential for a cyberattack to have a massive impact on national infrastructure increases. The likelihood of cyberattacks crossing over into kinetic warfare in the form of attacks that directly cause fatalities—either from harm to infrastructure and controls or by causing combat platforms such as drones or weapons systems to activate—continues to increase.

Cyberwar will require trained, experienced, and talented professionals to preserve and protect networks and national infrastructure. The policies and standards you apply and the way that you design and manage your networks and devices will be crucial to their ongoing security and safety in a world of constant attacks and increasing access to cyberattack tools.

CHAPTER SUMMARY

This chapter discussed the future of cyberwarfare. It examined the blurred boundaries among cyberwarfare, cybercrime, and attacks by nonstate actors. It reviewed the growth of APTs and constant attacks against networked systems, resulting in a constant low-level form of cyberwarfare. This requires attackers to be constantly aware of the environment in which they are operating. Those same attacks demonstrate the need for clearer international law and general agreement on the accepted rules of cyberwar. The legal future of cyberwar will be guided by the responses of the United Nations and individual nation-states to cyberattacks.

Future cyberwar has the potential to target even more systems and infrastructure due to the growth of networks. Devices and systems, from vehicles to medical implants and industrial control systems, will become increasingly network connected. This means that governments and nongovernment organizations must increase their network defense and offense capabilities through the creation of tools, standards, and trained personnel. Those added capabilities could help with future defensive cyberwar, but attackers are likely to keep an upper hand in a heavily networked world.

KEY CONCEPTS AND TERMS

Cryptocurrency	Internet of Things (IoT)	Network attacks
Hacktivist	Malware	Vulnerabilities

CHAPTER 15 ASSESSMENT

1. According to the commander of USCYBERCOM, cyberspace is more hospitable to attackers than defenders.

   A. True
   B. False

2. China charged U.S. government officials with criminal hacking charges related to SolarWorld's use of intellectual property.

   A. True
   B. False

3. Long-term attacks using advanced malware are known as _____.

   A. advanced threat agents
   B. advanced persistent threats
   C. persistent network attacks
   D. network persistence threats

4. Attackers can use what technologies to help hide their identity? (Select two.)

   A. IP spoofing
   B. Network address translation
   C. Cloud computing
   D. Hacked machines

5. The expansion of the Internet to devices everywhere is known as _____.

   A. IoT
   B. network profusion
   C. IPv6
   D. IPEverywhere

6. The secure operating system used in most modern cars has a built-in firewall to keep attackers from taking over locks and other mechanisms remotely.

   A. True
   B. False

7. What makes malware-based attacks like Stuxnet and Flame dangerous after they have been used?

   A. The component parts of the malware can be reused.
   B. Other countries can determine who wrote the malware.

   C. Reverse engineering the malware can allow attackers to find new vulnerabilities.
   D. A and B only
   E. None of the above

8. Anonymous is a group of nonstate actors often described as _____.

   A. hacktivists
   B. cyber protestors
   C. LOICs
   D. cyberCons

9. U.S. policy statements have noted that the existing international laws of warfare apply to cyberwarfare.

   A. True
   B. False

10. Similarities in _____ and _____ indicate that the same team may have written them.

    A. Flame
    B. Duqu
    C. Code Red
    D. Stuxnet

11. The U.S. government is emphasizing _____ as a means of verifying computer network defense proficiency.

    A. certification
    B. proficiency testing
    C. ethical hacking
    D. proven hacking experience

12. The U.S. government's _____ program certifies cloud computing service providers for security.

    A. CYBERCOM
    B. OCTAVE
    C. FedRAMP
    D. GOVCloud

13. InfraGard is an information-sharing partnership operated by the U.S. _____.

    A. NSA
    B. FBI
    C. CIA
    D. DOD

APPENDIX

A

Answer Key

CHAPTER 1 Information as a Military Asset

1. B   2. D   3. B and C   4. B   5. B   6. A   7. D   8. D   9. A, C, and E
10. A   11. A   12. A   13. B   14. B   15. C

CHAPTER 2 Targets and Combatants

1. B   2. C   3. B and D   4. A and C   5. E   6. C   7. D   8. A   9. B   10. A

CHAPTER 3 Cyberwarfare, Law, and Ethics

1. B   2. C   3. A   4. B, C, and E   5. Kinetic   6. Treachery   7. B   8. A
9. C   10. B

CHAPTER 4 Intelligence Operations in a Connected World

1. A   2. A   3. B   4. D   5. A, C, D, and E   6. A   7. E   8. B   9. A, C, and
E   10. B   11. C   12. C   13. E   14. A   15. A, B, and E

CHAPTER 5 The Evolving Threat: From Script Kiddies to Advanced Attackers

1. A   2. Reconnaissance   3. A   4. A, B, and D   5. D   6. E   7. D   8. B
9. A   10. B   11. B   12. E   13. A   14. A   15. Opportunistic

CHAPTER 6 Social Engineering and Cyberwarfare

1. A   2. B   3. A   4. D   5. A, D, and E   6. A, B, and D   7. B and D   8. E
9. B   10. E   11. A   12. B   13. A and B   14. B   15. D

CHAPTER 7 Weaponizing Cyberspace: A History

1. A   2. C   3. E   4. A   5. B   6. E   7. A   8. A   9. A   10. C   11. A, C, and
D   12. A   13. A   14. E

CHAPTER 8 Nonstate Actors in Cyberwar

1. A   2. B   3. A   4. B   5. C   6. D   7. E   8. A   9. B   10. A   11. E
12. B   13. C   14. B   15. B

CHAPTER 9 Defense-in-Depth Strategies

1. Enclave   2. A, C, and D   3. B   4. B   5. Common   6. B   7. A
8. Human   9. B   10. A

381

CHAPTER 10 Cryptography and Cyberwar

1. B   2. C   3. B   4. A   5. D   6. B   7. B   8. D   9. B   10. A   11. A and D   12. A   13. C

CHAPTER 11 Defending Endpoints

1. B   2. C   3. D   4. B and C   5. A and D   6. A   7. D   8. B   9. B   10. A   11. C   12. C   13. A

CHAPTER 12 Defending Networks

1. B   2. A, C, and D   3. B   4. B   5. D   6. A   7. D   8. A   9. B   10. D   11. B   12. C

CHAPTER 13 Defending Data

1. B   2. B and C   3. B   4. A, B, and C   5. B   6. B   7. C   8. A   9. B   10. A   11. A   12. B

CHAPTER 14 Cyberwarfare and Military Doctrine

1. A   2. D   3. A   4. E   5. A, B, and D   6. A and D   7. C and D   8. C   9. A   10. E   11. E   12. A   13. B   14. C   15. B

CHAPTER 15 Pandora's Box: The Future of Cyberwarfare

1. A   2. B   3. B   4. C and D   5. A   6. B   7. A   8. A   9. A   10. B and D   11. A   12. C   13. B

Standard Acronyms

ACD	automatic call distributor
AES	Advanced Encryption Standard
ALE	annual loss expectancy
ANSI	American National Standards Institute
AO	authorizing official
AP	access point
API	application programming interface
APT	advanced persistent threat
ARO	annual rate of occurrence
ATM	asynchronous transfer mode
AUP	acceptable use policy
AV	antivirus
B2B	business to business
B2C	business to consumer
BBB	Better Business Bureau
BC	business continuity
BCP	business continuity plan
BGP4	Border Gateway Protocol 4 for IPv4
BIA	business impact analysis
BYOD	Bring Your Own Device
C2C	consumer to consumer
CA	certificate authority
CAC	Common Access Card
CAN	computer network attack
CAN-SPAM	Controlling the Assault of Non-Solicited Pornography and Marketing Act
CAP	Certification and Accreditation Professional
CAUCE	Coalition Against Unsolicited Commercial Email

CBA	cost-benefit analysis
CBF	critical business function
CBK	common body of knowledge
CCC	CERT Coordination Center
CCNA	Cisco Certified Network Associate
CDR	call-detail recording
CERT	Computer Emergency Response Team
CFE	Certified Fraud Examiner
C-I-A	confidentiality, integrity, availability
CIPA	Children's Internet Protection Act
CIR	committed information rate
CIRT	computer incident response team
CISA	Certified Information Systems Auditor
CISM	Certified Information Security Manager
CISSP	Certified Information System Security Professional
CMIP	Common Management Information Protocol
CMMI	Capability Maturity Model Integration
CND	computer network defense
CNE	computer network exploitation
COPPA	Children's Online Privacy Protection Act
COS	class of service
CRC	cyclic redundancy check
CSA	Cloud Security Alliance
CSF	critical success factor
CSI	Computer Security Institute
CSP	cloud service provider
CTI	Computer Telephony Integration
CVE	Common Vulnerabilities and Exposures
DAC	discretionary access control
DBMS	database management system

DCS	distributed control system
DDoS	distributed denial of service
DEP	data execution prevention
DES	Data Encryption Standard
DHCPv6	Dynamic Host Configuration Protocol v6 for IPv6
DHS	Department of Homeland Security
DIA	Defense Intelligence Agency
DISA	direct inward system access
DMZ	demilitarized zone
DNS	Domain Name Service OR Domain Name System
DOD	Department of Defense
DoS	denial of service
DPI	deep packet inspection
DR	disaster recovery
DRP	disaster recovery plan
DSL	digital subscriber line
DSS	Digital Signature Standard
DSU	data service unit
EDI	Electronic Data Interchange
EIDE	Enhanced IDE
ELINT	electronic intelligence
EPHI	electronic protected health information
EULA	End-User License Agreement
FACTA	Fair and Accurate Credit Transactions Act
FAR	false acceptance rate
FCC	Federal Communications Commission
FDIC	Federal Deposit Insurance Corporation
FEP	front-end processor
FERPA	Family Educational Rights and Privacy Act
FIPS	Federal Information Processing Standard
FISMA	Federal Information Security Management Act
FRCP	Federal Rules of Civil Procedure
FRR	false rejection rate
FTC	Federal Trade Commission
FTP	File Transfer Protocol
GAAP	generally accepted accounting principles

GIAC	Global Information Assurance Certification
GigE	Gigibit Ethernet LAN
GLBA	Gramm-Leach-Bliley Act
HIDS	host-based intrusion detection system
HIPAA	Health Insurance Portability and Accountability Act
HIPS	host-based intrusion prevention system
HTML	Hypertext Markup Language
HTTP	Hypertext Transfer Protocol
HTTPS	Hypertext Transfer Protocol Secure
HUMINT	human intelligence
IaaS	Infrastructure as a Service
IAB	Internet Activities Board
ICMP	Internet Control Message Protocol
IDEA	International Data Encryption Algorithm
IDPS	intrusion detection and prevention
IDS	intrusion detection system
IEEE	Institute of Electrical and Electronics Engineers
IETF	Internet Engineering Task Force
IGP	interior gateway protocol
IMINT	imagery intelligence
InfoSec	information security
IP	intellectual property OR Internet protocol
IPS	intrusion prevention system
IPSec	Internet Protocol Security
IPv4	Internet Protocol version 4
IPv6	Internet Protocol version 6
IS-IS	intermediate system-to-intermediate system
(ISC)²	International Information System Security Certification Consortium
ISO	International Organization for Standardization
ISP	Internet service provider
ISS	Internet security systems
ITIL	Information Technology Infrastructure Library

ITRC	Identity Theft Resource Center	**PIN**	personal identification number
IVR	interactive voice response	**PKI**	public key infrastructure
L2TP	Layer 2 Tunneling Protocol	**PLC**	programmable logic controller
LAN	local area network	**POAM**	plan of action and milestones access tool
MAC	mandatory access control	**PoE**	power over Ethernet
MAN	metropolitan area network	**POS**	point-of-sale
MAO	maximum acceptable outage	**PPTP**	Point-to-Point Tunneling Protocol
MASINT	measurement and signals intelligence	**PSYOPs**	psychological operations
MD5	Message Digest 5	**RA**	registration authority OR risk assessment
modem	modulator demodulator	**RAID**	redundant array of independent disks
MP-BGP	Multiprotocol Border Gateway Protocol for IPv6	**RAT**	remote access Trojan OR remote
MPLS	multiprotocol label switching	**RFC**	Request for Comments
MSTI	Multiple spanning tree instance	**RIPng**	Routing Information Protocol next generation for IPv6
MSTP	Multiple Spanning Tree Protocol	**RIPv2**	Routing Information Protocol v2 for IPv4
NAC	network access control	**ROI**	return on investment
NAT	network address translation	**RPO**	recovery point objective
NFIC	National Fraud Information Center	**RSA**	Rivest, Shamir, and Adleman (algorithm)
NIC	network interface card	**RSTP**	Rapid Spanning Tree Protocol
NIDS	network intrusion detection system	**RTO**	recovery time objective
NIPS	network intrusion prevention system	**SA**	security association
NIST	National Institute of Standards and Technology	**SaaS**	Software as a Service
NMS	network management system	**SAN**	storage area network
NOC	network operations center	**SANCP**	Security Analyst Network Connection Profiler
NSA	National Security Agency	**SANS**	SysAdmin, Audit, Network, Security
NVD	national vulnerability database	**SAP**	service access point
OPSEC	operations security	**SCADA**	supervisory control and data acquisition
OS	operating system	**SCSI**	small computer system interface
OSI	open system interconnection	**SDSL**	symmetric digital subscriber line
OSINT	open source intelligence	**SET**	secure electronic transaction
OSPFv2	Open Shortest Path First v2 for IPv4	**SGC**	server-gated cryptography
OSPFv3	Open Shortest Path First v3 for IPv6	**SHA**	secure hash algorithm
PaaS	Platform as a Service	**S-HTTP**	secure HTTP
PBX	private branch exchange	**SIEM**	Security Information and Event Management system
PCI	Payment Card Industry	**SIGINT**	signals intelligence
PCI DSS	Payment Card Industry Data Security Standard	**SIP**	Session Initiation Protocol
PGP	Pretty Good Privacy	**SLA**	service level agreement
PII	personally identifiable information	**SLE**	single loss expectancy

B

Standard Acronyms

SMFA	specific management functional area
SNMP	Simple Network Management Protocol
SOX	Sarbanes-Oxley Act of 2002 (also Sarbox)
SPOF	single point of failure
SQL	Structured Query Language
SSA	Social Security Administration
SSCP	Systems Security Certified Practitioner
SSID	service set identifier (name assigned to a Wi-Fi network)
SSL	Secure Sockets Layer
SSL-VPN	Secure Sockets Layer virtual private network
SSO	single system sign-on
STP	shielded twisted pair OR Spanning Tree Protocol
TCP/IP	Transmission Control Protocol/ Internet Protocol
TCSEC	Trusted Computer System Evaluation Criteria
TFA	two-factor authentication
TFTP	Trivial File Transfer Protocol
TGAR	trunk group access restriction
TNI	Trusted Network Interpretation

TPM	technology protection measure OR trusted platform module
UC	unified communications
UDP	User Datagram Protocol
UPS	uninterruptible power supply
USB	universal serial bus
UTP	unshielded twisted pair
VA	vulnerability assessment
VBAC	view-based access control
VLAN	virtual local area network
VoIP	Voice over Internet Protocol
VPN	virtual private network
W3C	World Wide Web Consortium
WAN	wide area network
WAP	wireless access point
WEP	wired equivalent privacy
Wi-Fi	wireless fidelity
WLAN	wireless local area network
WNIC	wireless network interface card
WPA	Wi-Fi Protected Access
WPA2	Wi-Fi Protected Access 2
XML	Extensible Markup Language
XSS	cross-site scripting

Glossary of Key Terms

A

Access control list (ACL) | A list of rules that permit or deny access based on a network address and/or port.

Active defense | A network defense strategy that emphasizes the ability to respond as attacks occur at varying levels of intensity against the right attackers, and using shared intelligence.

Active response | A network defense strategy that uses attacks and probes in response to attacks to attempt to stop an attack.

Activists | Individuals or organizations that seek to change opinions or effect political change through peaceful, nonviolent means.

Administrative privileges | Rights provided to an administrator that give control over the system or network.

Advanced persistent threat (APT) | A sophisticated cyberwarfare threat group characterized by advanced technical tools and a persistent focus on compromising targets thought to be of strategic value.

Allow lists | The process of creating a list of allowed items, typically programs, websites, individuals, or other groups (sometimes called whitelisting).

Analysis and Production | The phase of the intelligence cycle where intelligence analysts transform collected information into intelligence that satisfies the intelligence requests made by decision-makers.

ANY/ANY rule | Firewall rule that allows all traffic.

Anomaly based detection | Relies on a baseline understanding of what a system, network, or device does and monitors for differences from that baseline that may be malicious or unwanted behavior.

Anonymous | A hacktivist organization composed of many individuals. Anonymous has no set leaders, although some members act as leaders for specific actions or subgroups.

Application-aware firewalls | Firewalls that are able to read and analyze application protocols then apply rules based on the content of packets using those protocols.

Application whitelisting | A technological solution that uses known, allowed programs to run on trusted systems. It also prohibits unknown programs not on the whitelist from running.

Asymmetric encryption | An encryption algorithm that requires both a private and a public key, allowing secure key distribution, authentication, and nonrepudiation when combined with cryptographic hashes.

Asymmetric warfare | Warfare fought between opponents with significantly different capabilities or strategies. Guerrilla wars and insurrections are both examples of asymmetric warfare.

Authenticity | Whether a message or data is genuine and can be proven to be from the person it claims to be from.

Authority | The principle of influence that says individuals will often defer to someone perceived to have power granted by law, due to his or her position within an organization or because of his or her social leadership.

Availability | Whether an information system is accessible and usable when needed. A system without the ability to ensure availability may be offline or inaccessible. Availability is part of the C-I-A triad.

B

Backups | Copies of data used to provide disaster recovery and to help ensure the ability to restore systems to working order.

Baiting | A social engineering attack where the attacker creates a flash drive that contains malicious code and leaves it in a parking lot, lobby, or other location where a target is likely to discover it.

Baseline configuration | A defined configuration to which systems are expected to conform. It is often used as part of a security design to ensure that systems comply with required settings and configurations.

Black-hat hackers | Hackers who use their skills with malicious intent, seeking to gain unauthorized access to systems for financial gain, to advance a political agenda, or for other purposes.

Block list | The process of creating a list of prohibited sites, software, users, or other items (sometimes called blacklisting).

Bot | A system that has been compromised by an attacker and is attached to a command-and-control server.

Botnet | A collection of bots under the control of the same attacker that may be leveraged for a common purpose, such as a distributed denial of service attack.

Buffer overflow | A category of exploits that attempt to write to areas of memory beyond that reserved for a particular purpose in the hope that the system will execute the overwritten areas as programs.

C

Careto | The name given to an advanced persistent threat of unknown origin that infected systems from 2007 through its discovery in 2014.

Certificate authority (CA) | The authority responsible for issuing certificates in a public key infrastructure.

Certification | A process used to verify that a system, application, or device meets a known standard. Certifying bodies use a known, documented process to ensure that certified items meet those standards.

Checksums | Mathematical calculations used to verify that data is intact.

C-I-A triad | An information security model describing the relationship between confidentiality, integrity, and availability in relation to the security of systems and data.

Ciphers | Algorithms for encrypting or decrypting data or messages.

Ciphertext | The resulting data when plaintext is encrypted.

Civilian infrastructure | The infrastructure and systems used for civilian purposes. At times, this can include infrastructure used by the military for nonmilitary purposes, such as Facebook or Twitter.

Classification | The process of assigning a file or object security or sensitivity levels, or the levels of security or sensitivity assigned to a file or object.

Clone phishing | A phishing attack that uses a modified copy of a legitimate message in the hopes of prompting the recipient to visit a link or open a file.

Cloud computing | Computing using resources hosted elsewhere, usually in a shared, virtual environment, which may exist in multiple data centers around the world.

Code | A series of substitutions for letters or words.

Code Red | A computer worm that affected more than 350,000 systems running Microsoft Internet Information Server in 2001.

Collection | The phase of the intelligence cycle during which intelligence professionals use assets at their disposal to gather essential elements of information.

Command, Control, Communications, Computers, and Intelligence (C4I) | A U.S. Department of Defense strategy for integration of technology command, control, and information management.

Commitment and consistency | The principle of influence that states that once someone has made a commitment to a particular course of action, his or her future actions will likely be consistent with that committed decision.

Common Criteria | An international standard (ISO/IEC 15408) for computer security certification and testing.

Communications intelligence (COMINT) | The collection of communications between individuals for intelligence purposes. COMINT may include the collection of telephone calls, email messages, and web communications.

Computer network attack (CNA) | One of the core capabilities of offensive information operations and cyberwarfare, consisting of actions taken through the use of computer networks to deny, corrupt, or destroy an adversary's information and/or information systems.

Computer network defense (CND) | Activities designed to protect, monitor, analyze, detect, and respond to unauthorized activity in friendly information systems and networks.

Computer network exploitation (CNE) | Cyberespionage capabilities of the military that include the ability to infiltrate computer systems and steal sensitive information.

Confidentiality | The ability to prevent disclosure of information or data to unauthorized individuals or users. A system without the ability to provide assurance of confidentiality is likely to expose data. Confidentiality is part of the C-I-A triad.

Configuration management systems | Systems that monitor and maintain the configuration of workstations or other devices to a defined standard.

Containerization | Packaging software, data, and any dependencies so they can run on any system.

Control | A nation-state's ability to prevent or allow actions by units or organizations.

Controlled Unclassified Information (CUI) | A program employed by the U.S. government to protect and label sensitive information that is not classified.

Conventional warfare | Warfare between two nation-states, typically following the traditional rules of warfare or treaties like the Geneva Conventions.

Corporations | For-profit businesses that are organized by individuals and officially registered by a government.

Cryptanalysis | The discipline of studying and defeating encryption technology to gain access to the plaintext of encrypted messages.

Cryptocurrency | Electronic currency that relies on cryptographic algorithms to provide proof of payment and proof of value. Bitcoin and Litecoin are examples of cryptocurrencies.

Cryptosystem | The set of algorithms required to perform a specific type of encryption and decryption.

Cyberattacks | Nonkinetic offensive operations that are intended to cause some form of physical or electronic damage.

Cyber domain | The domain of warfare that encompasses all cyberwarfare operations. The cyber domain complements the traditional domains of land, sea, air, and space.

Cyberespionage | Intrusions onto computer systems and networks designed to steal sensitive information that may be used for military, political, or economic gain.

Cyberhygiene | A term used by the U.S. Department of Defense to describe the overall health and maintenance of a computer system, including patching, vulnerability scanning, and other good system administration techniques.

Cyberintelligence | Intelligence activities related to cyberthreats, including identifying and analyzing their existence and capabilities.

Cyber Kill Chain | A model used to describe the process of engaging in a cyberwarfare attack, from conducting reconnaissance through acting on strategic objectives.

Cyberwarfare | Acts of war that include a wide range of activities using information systems as weapons against an opposing force.

Cyberwarriors | Combatants in cyberwar, either formally or informally trained.

D

Darknet | A segment of network space that is unused, but monitored. Traffic sent to darknets can be presumed to be hostile, or at least unwanted because no legitimate systems exist there.

Data Execution Prevention (DEP) | A technology that prevents the execution of code stored in certain parts of memory to protect against malware exploitation of buffer overflows.

Data labeling | The addition of flags or tags to data that provide information about the data, such as its classification, handling requirements, date of creation, or declassification date.

Data Lifecycle Management (DLM) | The use of policy and procedures to manage data throughout its life cycle from creation to deletion.

Declassified | A description of data that has been removed from a classification system.

Decryption | The process of decoding ciphertext, typically using a secret key.

Demilitarized zone (DMZ) | A special-purpose network designed to contain and isolate systems that offer public-facing services from other systems on the network.

Deny list | The process of creating a list of prohibited sites, software, users, or other items (sometimes called blacklisting).

Digital certificates | Electronic documents that are used to connect a public key and an individual, system, or other organization's identity.

Dissemination | The phase of the intelligence cycle that includes the delivery of finished intelligence products to decision-makers and the integration of intelligence information into user processes.

Distributed control systems (DCSs) | Systems that use a combination of sensors and feedback systems to control and adjust processes as they receive feedback.

Distributed denial of service (DDoS) | A network attack conducted by many machines using their combined resources to consume resources on the targeted system or systems.

Doctrine | Military planning documents that provide commanders with a shared philosophy and language for military operations.

Downgraded | Data that have been moved from a higher classification level to a lower classification level, but that have not been declassified.

Drive encryption | Encryption of a full disk or volume (partition) of a drive.

Duqu | A computer worm discovered in 2011 that is believed to be an advanced variant of the Stuxnet worm used for reconnaissance against industrial control systems.

E

Economy of force | The principle of warfare that states commanders should use the minimum amount of force necessary to achieve secondary objectives.

Electronic intelligence (ELINT) | The collection of electronic signals generated by nonhuman communications for intelligence purposes. ELINT includes the electronic emissions generated by radar systems, aircraft, ships, missiles, and other communications systems.

Electronic warfare | Information operations that include all military actions designed to use electromagnetic or directed energy to either control the electromagnetic spectrum or attack the enemy.

Eligible Receiver | The code name for a military cyberwarfare exercise that took place in 1997, raising awareness of the U.S. vulnerability to cyberwarfare attacks.

Encryption | The process of encoding ciphertext using an algorithm and a secret key.

Enterprise | Devices that typically have additional functionality, may have redundant power supplies or other features, and provide more management and configuration options.

Essential elements of information (EEIs) | The specific pieces of information that may be collected to help answer the priority intelligence requirement.

F

File encryption | Encryption of one or more files.

Financial intelligence (FININT) | The collection of information about financial transactions that may be exploited for intelligence purposes.

Firewalls | Network security devices that use a group of rules known as a ruleset to allow or deny traffic from passing through them.

Flame | A computer worm discovered in 2012 that is believed to be a later version of Stuxnet and Duqu used to target Iranian computer systems.

FOXACID | The code name for a program designed to compromise targeted computer systems by tricking them into visiting infected websites.

G

Geneva Conventions | The primary treaty basis for the internationally accepted laws of war.

Geospatial intelligence (GEOINT) | The collection of information gathered through the use of photography, maps, and other information about terrain.

Gray-hat hackers | Hackers who use their skills to advise companies of system vulnerabilities, but perform their actions without permission.

Guerrilla warfare | A type of asymmetric warfare; fought between opponents with significantly different capabilities or strategies.

H

Hacker | An individual who exploits computer security weaknesses for fame, financial gain, or other purposes.

Hacktivism | Political activism conducted via hacking or by using cyberwarfare techniques.

Hacktivist | A cyberactivist who uses hacking for political or social protest and activism. Members of Anonymous, one of the most active hacktivist groups, have also ventured into cybercrime activities.

Hashes | Algorithms that generate a fixed-length output given variable-length input. The fixed-length output is unique for each given set of input data and cannot be reversed to determine the original input.

Heuristics | A behavior-based detection that focuses on how a program acts and what it does to determine if it is a threat.

Honeynets | Networks or groups of systems set up to allow attackers to attack a number of systems, which then capture information about the attacks for defenders to analyze.

Honeypot | A fake or decoy system designed to allow attackers to successfully compromise it

while providing information about the tools and techniques the attackers used.

Host intrusion prevention system (HIPS) | A software-based system that attempts to detect and prevent attacks before they can reach the system they are installed on.

Human intelligence (HUMINT) | The collection of information from interactions among people.

I

Imagery intelligence (IMINT) | The collection of photographic information by aircraft or satellites overflying an area of intelligence interest.

Indicators | Actions taken by friendly forces and publicly available information that reveal critical information to the enemy.

Industrial control system (ICS) | A term that covers DCS, SCADA, and PLC systems used to control and oversee industrial processes and systems.

Industrial espionage | Intelligence activities conducted for business purposes, rather than for national security reasons.

Information assurance | A discipline that covers the protection of information and the management of risks to that information.

Information operations | Actions taken to affect adversary information and information systems while defending your own information and information systems.

Information warfare | Information operations conducted during a time of crisis or conflict to achieve or promote specific objectives over a specific adversary.

Integrity | The ability to prevent unauthorized modification of data, settings, or other elements of a device or system. Integrity is part of the C-I-A triad.

Intelligence | The collection, analysis, and dissemination of information about the capabilities, plans, intentions, and operations of an adversary.

Intelligence cycle | A model used to describe the process of intelligence operations with five phases: Planning and Direction, Collection, Processing and Exploitation, Analysis and Production, and Dissemination.

Intelligence gathering | Information operations actions that include efforts to gather information about an adversary's capabilities, plans, and actions.

Intelligence requirements | General or specific subjects for which a decision-maker has determined that there is a need for the collection of information or the production of intelligence.

International organized crime | Self-perpetuating associations of individuals who operate internationally for the purpose of obtaining power, influence, monetary, and/or commercial gains, wholly or in part by illegal means, while protecting their activities through a pattern of corruption and/or violence.

Internet of Things (IoT) | A term used to refer to the large number of networked devices that can now connect to the Internet.

Intrusion detection systems (IDSs) | Network- or host-based network protection systems that detect and report attacks.

Intrusion prevention systems (IPSs) | Network- or host-based network protection systems that detect and can prevent attacks from passing through them.

Irregular warfare | A form of warfare characterized as a violent struggle among state and nonstate actors for legitimacy and influence over the relevant population(s).

J

Jurisdiction | A nation-state's territory and areas where it is the sovereign power.

Jus ad bellum | Latin for "the right to war"; a set of criteria that defines when war is allowed.

Jus in bello | Latin for "laws of war"; laws of warfare like the Geneva Conventions.

K

Key | The part of an encryption algorithm that controls the output of the algorithm. Keys are kept secret because they can be used to encrypt or decrypt information using the algorithm.

Key distribution | The method by which keys are securely provided to those who need them.

Key pair | A private and a public key that are used by an individual or other entity in asymmetric encryption.

Kill chain | A model describing the process followed during an attack, from identifying appropriate targets to assessing the damage after an attack.

Kinetic warfare | Traditional warfare that is active, such as bombing and killing opposing troops.

L

Legitimacy | The principle of joint operations that states military operations must take place within the confines of appropriate authority.

Liking | The principle of influence that says people are more apt to be influenced by people who they know and appreciate.

Local area networks (LANs) | Networks that connect systems within a limited or nearby area, such as a building or campus.

M

Malware | Malicious software, such as a virus, Trojan, or other software designed to attack or take over a computer or system.

Maneuver | The principle of warfare that states applying power in a flexible manner keeps the adversary off guard.

Mass | The principle of warfare that states military commanders should concentrate their combat forces at the decisive time and location.

Mean time between failure (MTBF) | A measure of the reliability of devices and drives; the average time until the device fails.

Measurement and signature intelligence (MASINT) | The collection of information gathered from unintended electromagnetic emissions generated by a target.

Mercenaries | Combatants who are involved in a conflict for personal gain and who are not members of an involved party's military or other governmental functions.

Message digests | Cryptographic functions that protect the integrity of a message by allowing you to check that the message has not changed.

Metadata | Data about communications other than the actual content of the communication. For example, metadata for a telephone call may include the called number, calling number, and length of the call.

Military deception | Actions designed to mislead adversary forces about the operational capabilities, plans, and actions of friendly forces.

Mission assurance | The combination of risk management, system engineering, design, quality assurance, and management to ensure that systems and networks remain available and usable.

Mission Assurance Categories (MAC) | A U.S. Department of Defense mission assurance categorization scheme ranging from MAC I, which provides extremely high levels of integrity and availability, to MAC III, which matches common industry practices for integrity and availability of data and systems.

Moonlight Maze | The code name for a series of attacks, believed to be of Russian origin, that began targeting U.S. government systems in 1998.

N

Nation-states | Formally recognized countries or nations.

Network address translation (NAT) | A technology that allows one or more addresses to be used by a group of systems inside of a network without exposing their internal addresses to the outside world.

Network admissions control (NAC) | A system that requires systems to authenticate and/or provide proof that they meet required configuration standards before connecting to a network.

Network attacks | Attacks conducted across or via a network.

Network enclave | A separated portion of an internal network. Network enclaves are used to separate sections of the network based on usage, data, or the security requirements of systems or devices on that segment.

New York Times Co. v. United States | U.S. Supreme Court decision that legitimized the constitutional basis for publishing the *Pentagon Papers* under the press's First Amendment protections.

Nongovernmental organizations (NGOs) | Organizations that exist outside of the context of a government or a for-profit corporation.

Nonkinetic warfare | Cyber and electronic warfare.

Nonrepudiation | The ability in a communication to ensure that the sender cannot claim that he or she did not send a message.

Nonstate actors | Individuals or groups that seek to participate in cyberwarfare but do so independently, without the endorsement of a national government.

O

Objective | The principle of warfare that states every military action should have a clearly defined and articulated purpose.

Offensive | The principle of warfare that states military commanders must seize, retain, and exploit the initiative.

Open source intelligence (OSINT) | The collection of publicly available information to satisfy intelligence requirements.

Operation Aurora | A series of cyberattacks launched by Chinese sources against Google and other American companies in 2009.

Operational level | The level of warfare that links strategy and tactics through the planning and execution of operations.

Operations security (OPSEC) | Activities designed to deny an adversary access to information about friendly forces that would reveal capabilities, plans, or actions.

Opportunistic attack | An attack type that uses a brute-force approach against thousands or millions of targets in an attempt to find a handful of vulnerable systems.

Organ | Every nation-state agency, person, or entity that is part of the official government or associated bodies of a nation-state.

Glossary

P

Packet filter firewalls | Firewalls that provide very basic filtering capabilities based on the network address, port, or protocol that traffic is coming from or going to.

Penetration testing | Security testing that involves attacking systems to attempt to gain access or control as part of a test.

Pentagon Papers | The collection of classified U.S. government documents related to the Vietnam War published by the *New York Times* in 1971 as part of a famous leak of embarrassing government records.

Perseverance | The principle of joint operations that states military commanders must take measures to ensure that the commitment exists to achieve the desired end state.

Phishing | A social engineering attack where the attacker sends the victim an electronic message (via email, text, or other means) attempting to solicit sensitive information from the victim.

Pivot | Attackers use a technique known as pivoting when they compromise a system inside of a defensive layer. They attack other systems inside of the segment or security zone they are in, using those systems to attempt to gain more access in other zones or to access additional data and systems.

Plaintext | Unencrypted text or data.

Planning and Direction | The phase of the intelligence cycle that includes the identification of intelligence requirements, the development of an intelligence architecture, the design of a collection plan, and the issuance of collection requests.

Poison Ivy | A remote access Trojan developed in 2005 and still in use today to remotely control systems infiltrated by hackers.

Port scanning | The process of scanning systems to determine what services they are offering on numbered ports.

Pretexting | A social engineering attack where the attacker creates a false set of circumstances and uses them to convince the target to take some form of action.

Priority intelligence requirements (PIRs) | The questions identified by decision-makers as needing critical attention from intelligence operations.

Processing and Exploitation | The phase of the intelligence cycle during which data collected by intelligence assets in the Collection phase is transformed into information useful to intelligence analysts.

Programmable logic controllers (PLCs) | Special-purpose computers designed to handle specialized input and output systems.

Psychological operations (PSYOPs) | Military operations planned to convey selected information and indicators to foreign governments, organizations, groups, and individuals in order to influence their emotions, motives, objective reasoning, and behavior.

Public key encryption | A method of encryption that can provide all four critical requirements: confidentiality, message integrity, authentication, and nonrepudiation. *See* asymmetric encryption.

Public key infrastructure (PKI) | The software, hardware, policies, procedures, and staff needed to create, manage, distribute, and use digital certificates.

Q

Quantum cryptography | Cryptography that uses quantum mechanics to perform cryptographic tasks like encrypting and decrypting data or providing secure key exchange.

R

Ransomware | Malicious computer software that takes over a system, encrypting files with a secret key rendering them inaccessible to the legitimate user until he or she pays a ransom.

Reciprocity | A relationship between two people that involves the exchange of goods or services of approximately equal value.

Redacting | The removal of information from a document or data. Paper records are often redacted using a black marker, whereas digital documents often have words or sections replaced with blanks or black boxes.

Red Cell | The U.S. government's title for opposing forces in exercises, including specialized teams that attack networks and other cyberinfrastructure.

Red team | *See* Red Cell.

Redundancy | Providing multiple copies of data, multiple servers, or otherwise providing multiple versions of a system or data to ensure availability.

Registration authority (RA) | The organization or individual responsible for verifying the identities of entities requesting certificates in a PKI.

Remote Access Trojan (RAT) | Malicious software that allows continued access to a compromised system from a remote location.

Restraint | The principle of joint operations that states that, while conducting military operations, commanders should limit collateral damage and prevent the unnecessary use of force.

Routers | Network devices used to interconnect networks and direct traffic between them.

Rulesets | The list of rules that define what traffic is allowed through or stopped by a firewall or other network security device.

Ruse | A trick or act intended to fool an enemy during wartime.

S

Sanitizing | The removal of all data from a drive or other media.

Script kiddies | Hackers who do not discover vulnerabilities on their own but instead download exploit scripts written by others and run them against target systems without a real understanding of the technical details behind the attack.

Secure Sockets Layer (SSL) | The predecessor to TLS, a security protocol for the creation and use of an encrypted link between a client and a server.

Security | The principle of warfare that states commanders must never allow enemy forces to gain an advantage.

Security information and event management (SIEM) | A network security system that gathers data about attacks as well as security information from logs and other data sources, then correlates that data for security practitioners.

Security Orchestration, Automation, and Response (SOAR) | Relatively new category of security solution that focuses on case management, incident response workflows, and playbook-based responses using alarm and event data.

Self-defending network | A network that modifies itself to protect against attacks as they occur.

Semi-targeted attack | An attack that seeks to infiltrate a specific organization or type of target but not a specific individual.

Senior Suter | The code name for a U.S. Air Force program designed to manipulate enemy air defense systems.

Side-channel attacks | In encryption, attacks that target the physical implementation of encryption, such as measuring the amount of power a cryptographic processor draws to determine what calculation it is performing.

Signals intelligence (SIGINT) | The collection of intelligence information through the interception of communications and other electronic signals.

Signature-based detection | A detection capability that uses details of a file or program to match it to a list of known files. Malware detection systems often use cryptographic hashing and other techniques to perform this function.

Simplicity | The principle of warfare that states military plans should be simple, clear, and concise to reduce confusion and misunderstanding.

Glossary

Simulations | Exercises intended to simulate actual attacks to test processes and procedures.

Social engineering | The art of manipulating human behavior through social influence tactics in order to achieve a desired behavior.

Social proof | The principle of influence that says an individual in a social situation who is unsure how to act will likely follow the example of the crowd and behave in the same way as the people around him or her.

Software testing | Testing via either human or programmatic means that attempts to validate a program's function and to determine if it has any flaws.

Solar Sunrise | The code name for cyberattacks launched in 1998 by three teenagers against computer systems operated by the U.S. government.

Sovereignty | A nation's right to be the final authority over its territory, citizens, and resources.

Spear phishing | A phishing attack that is targeted at a specific individual and uses personal information to add legitimacy to the phishing message.

SQL Slammer | A computer worm that affected more than 75,000 systems running Microsoft SQL Server in 2003.

Stateful packet inspection firewalls | Firewalls that track the state of traffic between systems, allowing ongoing permitted traffic to pass through without additional inspection.

Strategic level | The level of warfare that involves national-level policies and military strategies across an entire theater of operations.

Stuxnet | The computer worm allegedly used by a joint U.S.-Israeli operation to destroy Iranian uranium enrichment centrifuges in 2010.

Supervisory control and data acquisition (SCADA) | Systems used to monitor and control remote equipment.

Surprise | The principle of warfare that states that, when attacking, military forces should strike the enemy at an unexpected time or place.

Symmetric encryption | Encryption in which the same secret key is used for both decryption and encryption.

 T

Tactical level | The level of warfare concerned with the employment and ordered arrangement of forces in relation to each other.

Tallinn Manual | An in-depth analysis of the current international law in the context of cyberwarfare created for the United Nations.

Terrorists | Individuals or organizations that seek to change opinions or effect political change through violent, unlawful means.

Tiger teams | Teams assigned to solve a specific problem. In security testing, a tiger team is assigned to identify and possibly fix flaws and vulnerabilities.

Titan Rain | The code name for the investigation into a series of computer intrusions launched against U.S. government systems from Chinese web servers from 2003 through 2005.

Total warfare | Warfare fought without any restrictions or boundaries. Total warfare is fought entirely to win, and ignores normal conventions regarding targeting of civilians, use of some types of weapons, and other restrictions.

Traditional warfare | A form of warfare consisting of large-scale conflict between nations or groups of nations that is characterized as a violent struggle for domination between nation-states or coalitions and alliances of nation-states.

Transport Layer Security (TLS) | A security protocol for the creation and use of an encrypted link between a client and a server. TLS is most often associated with secure web traffic.

Treachery | Harmful trickery used in wartime that can result in death or injury.

Trusted Platform Module (TPM) | A cryptographic processor that provides encryption capabilities as well as helping to prove a system's identity when decrypting a drive. Also commonly called a TPM chip.

U

Unconventional warfare | Warfare fought primarily on a psychological level, with the goal of changing how an enemy feels about the conflict. Unconventional warfare attempts to persuade the enemy to surrender or negotiate peace.

Unified Threat Management devices | Advanced security devices that add additional features beyond firewall capabilities, including intrusion detection and prevention, anti-malware capabilities, and other features that each device manufacturer decides are important.

Unity of Command | The principle of warfare that states a single commander should be responsible for achieving each military objective.

U.S. Cyber Command (USCYBERCOM) | A major command of the U.S. military organized to operate in the cyber domain.

V

Virtualization | The creation of virtual, rather than physical, environments for servers and devices. Virtualization can allow multiple virtual servers or systems to run on the same hardware that would normally host a single system or device.

Virtual LAN (VLAN) | A virtual local area network, or logically separated network created through software-based tags on a network.

Virtual private network (VPN) | A network constructed through other networks that relies on encryption or encapsulation to keep its content separate.

VLAN hopping attacks | Attacks against VLANs that allow attackers to send traffic to other VLANs.

Vulnerabilities | Flaws in a system, software, network, or device that can leave it vulnerable to attack or exploit.

Vulnerability scanning | A scanning process conducted against systems and devices to identify any vulnerabilities in their services or software.

W

War | Socially sanctioned violence to achieve a political purpose.

War games | Exercises intended to simulate actual attacks to test processes and procedures.

White-hat hackers | Benevolent security practitioners who use their hacking skills for good purposes, seeking to secure the information systems of their employers or consulting clients.

Wide area networks (WANs) | Networks that interconnect over a broad area, such as those across geographic regions.

Window of vulnerability | The time that exists between when a vulnerability is discovered and when the software vendor releases a patch to correct the vulnerability.

Worm | A piece of malicious software that is able to spread between computer systems without any human intervention.

Z

Zero-day attacks | Attacks that occur during the window of vulnerability when no patch is available to successfully defend against the attack.

Zero-day vulnerabilities | Vulnerabilities in a computer system or network that are unknown to anyone other than the attacker, making it extremely difficult to defend against.

Zero trust | A network designed without a presumption of trust and with a presumption that other devices may be malicious or compromised.

Glossary

References

Abdo, Alex. "You May Have 'Nothing to Hide' but You Still Have Something to Fear." American Civil Liberties Union Blog of Rights. August 2, 2013. Retrieved May 17, 2014, from https://www.aclu.org/blog/national-security/you-may-have-nothing-hide-you-still-have-something-fear.

Abrams, Lawrence. "CryptoLocker Ransomware Information Guide and FAQ." December 20, 2014. Retrieved March 30, 2014, from http://www.bleepingcomputer.com/virus-removal/cryptolocker-ransomware-information.

Abrams, Marshall D., and Joe Weiss. "Malicious Control System Cyber Security Attack Case Study—Maroochy Water Services, Australia." August 2008. Retrieved May 1, 2014, from http://csrc.nist.gov/groups/SMA/fisma/ics/documents/Maroochy-Water-Services-Case-Study_briefing.pdf.

Adair, Steven, and Ned Moran. "Cyber Espionage & Strategic Web Compromises—Trusted Websites Serving Dangerous Results." May 15, 2012. Retrieved May 19, 2014, from http://blog.shadowserver.org/2012/05/15/cyber-espionage-strategic-web-compromises-trusted-websites-serving-dangerous-results/.

Arooj, Ahmed. "New Report Sheds Light on the State of Cyber Security and Malware" Digitalinformationworld.com. September 2, 2020. Retrieved October 25, 2020, from https://www.digitalinformationworld.com/2020/09/security-report-sheds-light-on-the-facts-and-analysis-regarding-malware-attacks.html.

Aid, Matthew M. "Inside the NSA's Ultra-Secret China Hacking Group." *Foreign Policy*, June 10, 2013. Retrieved March 10, 2014, from http://www.foreignpolicy.com/articles/2013/06/10/inside_the_nsa_s_ultra_secret_china_hacking_group.

Alexander, Keith B. "Statement of General Keith B. Alexander Commander United States Cyber Command Before the Senate Committee on Armed Services." Defense.gov. March 12, 2013. Retrieved June 1, 2014, from http://www.defense.gov/home/features/2013/0713_cyberdomain/docs/Alexander%20testimony%20March%202013.pdf.

———. "Statement of General Keith B. Alexander Commander United States Cyber Command Before the Senate Committee on Armed Services." Defense Innovation Marketplace. February 27, 2014. Retrieved May 11, 2014, from http://www.defenseinnovationmarketplace.mil/resources/Cyber_Command_Alexander_02-27-14.pdf.

Alford, Lionel D. "Cyber Warfare: The Threat to Weapons Systems." *Weapon Systems Technology Information Analysis Center Quarterly* 9, no. 4. Retrieved April 29, 2014, from https://wstiac.alionscience.com/pdf/WQV9N4_ART01.pdf.

Alperovitch, Dmitri. "Active Defense: Time for a New Security Strategy." February 25, 2013. Retrieved May 12, 2014, from http://www.crowdstrike.com/blog/active-defense-time-new-security-strategy/.

———. "Revealed: Operation Shady RAT." Vers. 1.1. McAfee Web site. August 8, 2011. Retrieved March 23, 2014, from http://www.mcafee.com/us/resources/white-papers/wp-operation-shady-rat.pdf.

Anderson, Nate. "'Operation Payback' Attacks to Go on Until 'We Stop Being Angry.'" Ars Technica Web site. September 30, 2010. Retrieved May 5, 2014, from http://arstechnica .com/tech-policy/2010/09/operation-payback-attacks-continue-until-we-stop-being -angry/.

Anderson, Ross. *Security Engineering: A Guide to Building Dependable Distributed Systems.* 2nd ed. New York: John Wiley & Sons, Inc., 2008.

Applebaum, Jacob, Judith Horchert, and Christian Stocker. "Shopping for Spy Gear: Catalog Advertises NSA Toolbox." *Der Spiegel*, December 29, 2013. Retrieved March 23, 2014, from http://www.spiegel.de/international/world/catalog-reveals-nsa-has-back-doors-for -numerous-devices-a-940994.html.

Applegate, Scott D. "The Principle of Maneuver in Cyberspace Operations." *4th International Conference on Cyber Conflict.* Tallinn: NATO CCD COE Publications, 2012, 183–195.

Archives.gov. "Controlled Unclassified Information (CUI)." National archives website. n.d. Retrieved October 16, 2020, from https://www.archives.gov/cui.

Arquilla, John. Interview with FRONTLINE. March 4, 2003. Transcript retrieved on May 3, 2014, from http://www.pbs.org/wgbh/pages/frontline/shows/cyberwar/interviews/arquilla .html.

Arquilla, John, and David Ronfeldt, "Cyberwar Is Coming!" Spring 1993. Retrieved June 1, 2014, from http://www.rand.org/content/dam/rand/pubs/monograph_reports/MR880/MR880 .ch2.pdf.

Arquilla, John, and David F. Ronfeldt. *Networks and Netwars: The Future of Terror, Crime, and Militancy.* Santa Monica, CA: Rand, 2001.

Bamford, James. *Body of Secrets.* New York: Anchor Books, 2002.

Banks, William. "The Role of Counterterrorism Law in Shaping ad Bellum Norms for Cyber Warfare." *International Law Studies* (U.S. Naval War College) 89 (April 2013): 157–197.

Bencsath, Boldizsar et al. "Duqu: A Stuxnet-like Malware Found in the Wild." Laboratory of Cryptography and System Security (CrySyS). October 14, 2011. Retrieved May 4, 2014, from http://www.crysys.hu/publications/files/bencsathPBF11duqu.pdf.

Bender, Jason M. "The Cyberspace Operations Planner." Small Wars Journal website. November 5, 2013. Retrieved May 18, 2014, from http://smallwarsjournal.com/jrnl/art/the-cyberspace -operations-planner.

Bennett, James T., Ned Moran, and Nart Villeneuve. "Poison Ivy: Assessing Damage and Extracting Intelligence." FireEye Labs blog. August 21, 2013. Retrieved March 26, 2014, from http://www.fireeye.com/resources/pdfs/fireeye-poison-ivy-report.pdf.

Blank, Laurie R. "International Law and Cyber Threats from Non-State Actors." *International Law Studies* 89 (2013): 406–437.

Borger, Julian. "Hacking Trail Leads to Swedish Teen." *The Guardian*, May 10, 2005. Retrieved May 4, 2014, from http://www.theguardian.com/technology/2005/may/11/usnews. internationalnews.

Braun, Stephen. "U.S. Network to Scan Workers with Secret Clearances." The Associated Press, March 10, 2014. Retrieved March 12, 2014, from http://hosted.ap.org/dynamic/stories/U /US_EYES_ON_SPIES.

Brown, Anthony Cave. *Bodyguard of Lies: The Extraordinary True Story Behind D-Day.* New York: Lyons Press, 2007.

Burgess, Ronald L. Remarks to the Association of Former Intelligence Officers. August 12, 2011. Retrieved May 23, 2014, from https://web.archive.org/web/20131004054724/http://www .dia.mil/public-affairs/testimonies/2011-08-12.html.

Butler, Sean C. "Refocusing Cyber Warfare Thought." *Air & Space Power Journal* (2013): 44–57. Retrieved May 24, 2014, from http://www.airpower.maxwell.af.mil/digital/pdf/articles /Jan-Feb-2013/F-Butler.pdf.

Center for Strategic and International Studies Defense-Industrial Initiatives Group. "An Assessment of the National Security Software Industrial Base." Center for Strategic and International Studies Web site. October 19, 2006. Retrieved April 27, 2014, from https://csis .org/files/media/csis/pubs/061019_softwareindustrialbase.pdf.

Central Intelligence Agency. "Centers in the CIA." June 18, 2013. Retrieved May 23, 2014, from https://www.cia.gov/library/publications/additional-publications/the-work-of-a-nation/cia -director-and-principles/centers-in-the-cia.html.

Cialdini, Robert. *Influence: Science and Practice*. 5th ed. Boston: Allyn & Bacon, 2008.

Clapper, James R. "Statement for the Record: Worldwide Threat Assessment of the US Intelligence Community: Senate Select Committee on Intelligence." Office of the Director of National Intelligence. Washington, DC, March 12, 2013. Retrieved March 10, 2014, from http:// www.dni.gov/files/documents/Intelligence%20Reports/2013%20ATA%20SFR%20for%20 SSCI%2012%20Mar%202013.pdf.

———. "Statement for the Record: Worldwide Threat Assessment of the US Intelligence Community: Senate Select Committee on Intelligence." Office of the Director of National Intelligence. Washington, DC, January 29, 2014. Retrieved March 12, 2014, from http:// www.dni.gov/files/documents/Intelligence%20Reports/2014%20WWTA%20%20SFR _SSCI_29_Jan.pdf.

Committee on National Security Systems. "CNSS Instruction 1253." CNSS Web site. March 27, 2014. Retrieved May 13, 2014, from https://www.cnss.gov/CNSS/openDoc .cfm?UoK1JHI3+7FhQcOk1ZPE4A==.

Constantin, Lucian. "Researchers Identify Stuxnet-like Cyberespionage Malware Called 'Flame.'" *PCWorld*, May 28, 2012. Retrieved February 27, 2014, from http://www.pcworld.com /article/256370/researchers_identify_stuxnetlike_cyberespionage_malware_called_flame .html.

"Control System Security Center: Major business activities." Retrieved October 25, 2020, from http://www.css-center.or.jp/en/aboutus/activity.html.

Convertino, Sebastian. "Flying and Fighting in Cyberspace." July 2007. Air University.

Conway, Tom. "DoD Can Use USB Securely." McAfee blogs. March 9, 2010. Retrieved May 27, 2014, from http://blogs.mcafee.com/business/security-connected/dod-can-use-usb -securely.

Cooney, Michael. "Healthcare Industry Group Builds Cybersecurity Threat Center." April 24, 2012. Retrieved June 8, 2014, from http://www.networkworld.com/article/2187972/data -center/healthcare-industry-group-builds-cybersecurity-threat-center.html.

———. "Statement of General Keith B. Alexander Commander United States Cyber Command Before the Senate Committee on Armed Services." Defense Innovation Marketplace. February 27, 2014. Retrieved May 11, 2014, from http://www.defenseinnovationmarketplace.mil /resources/Cyber_Command_Alexander_02-27-14.pdf.

Corbin, Jane. Cyber Attack! Interview with Senator John Kyl. British Broadcasting Company. June 3, 2000. Transcript retrieved on May 3, 2014, from http://news.bbc.co.uk/hi/english/static /audio_video/programmes/panorama/transcripts/transcript_03_07_00.txt.

Crowdstrike. "Crowdstrike Global Threat Report: 2013 Year in Review." Crowdstrike website. January 22, 2014. Retrieved March 19, 2014, from https://s3.amazonaws.com/download .crowdstrike.com/papers/CrowdStrike_Global_Threat_Report_2013.pdf.

Danchev, Dancho. "The Russian Business Network." Dancho Danchev's blog. October 18, 2007. Retrieved April 23, 2014, from http://ddanchev.blogspot.com/2007/10/russian-business -network.html.

Deeks, Ashley. "The Geography of Cyber Conflict: Through a Glass Darkly." *International Law Studies 89* (2013): 1–20.

Defense Information Systems Agency. "CNDSP Subscription Services." disa.mil website. 2014. Retrieved May 13, 2014, from http://www.disa.mil/Services/Information-Assurance/CNDSP.

———. "Continuous Monitoring and Risk Scoring." Retrieved May 22, 2014, from http://www .disa.mil/Services/Information-Assurance/CMRS.

———. "Frequently Asked Questions: DoD Policy Memorandum 'Encryption of Sensitive Unclassified Data at Rest on Mobile Computing Devices and Removable Storage Media.'" DISA Information Assurance Support Environment. March 19, 2008. Retrieved May 27, 2014, from http://iase.disa.mil/policy-guidance/faq_dar_encryption_policy_memo _18mar08 _update-6_final.doc.

———. "Host Based Security System: Components." DISA website. n.d. Retrieved April 29, 2014, from http://www.disa.mil/Services/Information-Assurance/HBSS /Components.

———. "Network Defense." disa.mil website. 2014. Retrieved May 13, 2014, from http://www .disa.mil/Services/Information-Assurance/CNDSP.

Defense Logistics Agency. "Data Loss Prevention Solicitation." September 2011. Retrieved May 24, 2014, from https://www.fbo.gov/?s=opportunity&mode=form&tab=core&id=bd9cb808b5bd 73f37dd74390455560ad&_cview=0.

———. "HBSS Components." DISA Web site. n.d. Retrieved May 27, 2014, from http://www .disa.mil/Services/Information-Assurance/HBSS/Components.

Defense Science Board. "Resilient Military Systems and the Advanced Cyber Threat." ACQ Web. January 2013. Retrieved May 18, 2014, from http://www.acq.osd.mil/dsb/reports /ResilientMilitarySystems.CyberThreat.pdf.

Defense Systems. "Al-Qaeda Reportedly Targeting U.S. Drones." Defense Systems website. September 2013. Retrieved April 27, 2014, from http://defensesystems.com /articles/2013/09/05/counter-drone.aspx.

Dennesen, Kristen, John Feker, Tonya Feyes, and Sean Kern. *Strategic Cyber Intelligence*. Intelligence and National Security Alliance Cyber Intelligence Task Force, March 2014. Retrieved March 27, 2014, from http://www.insaonline.org/i/d/a/Resources/StrategicCyber .aspx.

Department of the Navy. "Computer Network Defense Roadmap." May 2009. Retrieved March 29, 2014, from http://www.doncio.navy.mil/uploads/1019TSI85933.pdf.

DISA Field Security Operations. "Enclave Security Technical Implementation Guide, Version 4, Release 4." DISA Information Assurance Support Environment. January 9, 2014. Retrieved March 18, 2014, from http://iase.disa.mil/stigs/net_perimeter/enclave_dmzs/u_enclave _v4r4_stig.zip.

Dorp, Evan. "CryptoLocker—A New Ransomware Variant." Emsisoft blog. September 10, 2013. Retrieved March 30, 2014, from http://blog.emsisoft.com/2013/09/10/cryptolocker-a-new -ransomware-variant/.

Drogin, Bob. "NSA Blackout Reveals Downside of Secrecy." *Los Angeles Times*, March 13, 2000. Retrieved May 3, 2014, from https://www.fas.org/irp/news/2000/03/e20000313nsa.htm.

Dvorak, Daniel L., ed. "NASA Study on Flight Software Complexity." NASA website. March 5, 2009. Retrieved April 27, 2014, from http://www.nasa.gov/pdf/418878main_FSWC_Final _Report.pdf.

Elkjaer, Bo, and Kenan Seeberg. "Echelon Singles Out the Red Cross." Cryptome. March 8, 2000. Retrieved May 17, 2014, from http://cryptome.org/echelon-red.htm.

"Eligible Receiver." GlobalSecurity.org. Retrieved May 4, 2014, from http://www.globalsecurity.org/military/ops/eligible-receiver.htm.

Fahrenkrug, David T. "Countering the Offensive Advantage in Cyberspace: An Integrated Defensive Strategy." *4th International Conference on Cyber Conflict.* Tallinn: NATO CCD COE Publications, 2012, 197–207.

Federal Bureau of Investigation. "Intelligence Cycle." n.d. Retrieved March 22, 2014, from http://www.fbi.gov/about-us/intelligence/intelligence-cycle.

———. "Safety and Security for the Business Professional Traveling Abroad." n.d. Retrieved May 17, 2014, from http://www.fbi.gov/about-us/investigate/counterintelligence/business-travel-brochure.

———. "U.S. Charges Five Chinese Military Hackers with Cyber Espionage Against U.S. Corporations and a Labor Organization for Commercial Advantage." May 19, 2014. Retrieved June 8, 2014, from http://www.fbi.gov/pittsburgh/press-releases/2014/u.s-charges-five-chinese-military-hackers-with-cyber-espionage-against-u.s-corporations-and-a-labor-organization-for-commercial-advantage/.

Ferrer, Zarestel, and Methusela Cebrian Ferrer. "In-Depth Analysis of Hydraq: The Face of Cyberwar Enemies Unfolds." Computer Associates website. 2010. Retrieved March 23, 2014, from http://www.ca.com/us/~/media/files/securityadvisornews/in-depth_analysis_of_hydraq_final_231538.aspx.

Field Manual 2-0: Intelligence. Washington, DC: Department of the Army, May 17, 2004. Retrieved March 23, 2014, from http://www.globalsecurity.org/intell/library/policy/army/fm/2-0/index.html.

Field Manual 3-05.30: Psychological Operations. Washington, DC: Department of the Army, April 2005. Retrieved March 10, 2014, from http://www.fas.org/irp/doddir/army/fm3-05-30.pdf.

Field Manual 34-52: Intelligence Interrogation. Washington, DC: Department of the Army, September 28, 1992. Retrieved March 23, 2014, from http://www.loc.gov/rr/frd/Military_Law/pdf/intel_interrrogation_sept-1992.pdf.

"Follow the Money: NSA Spies on International Payments." *Der Spiegel,* September 15, 2013. Retrieved March 23, 2014, from http://www.spiegel.de/international/world/spiegel-exclusive-nsa-spies-on-international-bank-transactions-a-922276.html.

Foster, Peter. "'Bogus' AP Tweet About Explosion at the White House Wipes Billions Off US Markets." *The Telegraph,* April 23, 2013. Retrieved May 17, 2014, from http://www.telegraph.co.uk/finance/markets/10013768/Bogus-AP-tweet-about-explosion-at-the-White-House-wipes-billions-off-US-markets.html.

FoxIT. "Black Tulip: Report of the Investigation into the DigiNotar Certificate Authority Breach." August 13, 2012. Retrieved April 6, 2014, from http://www.rijksoverheid.nl/bestanden/documenten-en-publicaties/rapporten/2012/08/13/black-tulip-update/black-tulip-update .pdf.

"Frontline Interview with John Hamre." PBS website. February 18, 2003. Retrieved May 18, 2014, from http://www.pbs.org/wgbh/pages/frontline/shows/cyberwar/interviews/hamre.html.

Fryer-Biggs, Zachary. "DoD's New Cyber Doctrine." *Defense News,* October 13, 2012. Retrieved May 24, 2014, from http://www.defensenews.com/article/20121013/DEFREG02/310130001/DoD-8217-s-New-Cyber-Doctrine.

Fulghum, David A. "Black Surprises." *Aviation Week and Space Technology 162,* no. 7 (February 14, 2005): 68–69.

Gander, Kashmira. Isis and al-Qaeda sending coded messages through eBay, pornography, and Reddit. March 2, 2015. Retrieved November 7, 2020, from https://www.independent .co.uk/news/world/middle-east/isis-and-al-qaeda-sending-coded-messages-through-ebay -pornography-and-reddit-10081123.html.

Gellman, Barton, and Ashkan Soltani. "NSA Infiltrates Links to Yahoo, Google Data Centers Worldwide, Snowden Documents Say." *Washington Post*, October 30, 2013. Retrieved March 9, 2014, from http://www.washingtonpost.com/world/national-security/nsa-infiltrates-links -to-yahoo-google-data-centers-worldwide-snowden-documents-say/2013/10/30/e51d661e -4166-11e3-8b74-d89d714ca4dd_story.html.

Gentry, Craig. "A Fully Homomorphic Encryption Scheme." Doctoral Thesis, Computer Sciences, Palo Alto, CA: Stanford University, 2009.

Gentry, Craig, and Shai Halevi. "Implementing Gentry's Fully-Homomorphic Encryption Scheme." *Advances in Cryptology—EUROCRYPT 2011*, 2011: 129–148.

Gertz, Bill. "Commander: U.S. Military Not Ready for Cyber Warfare." The Washington Free Beacon website. February 27, 2014. Retrieved May 11, 2014, from http://freebeacon.com /national-security/commander-u-s-military-not-ready-for-cyber-warfare/.

Google. "A New Approach to China." Google Official blog website. January 12, 2010. Retrieved March 23, 2014, from http://googleblog.blogspot.com/2010/01/new-approach-to-china .html.

Goodchild, Joan. "The Robin Sage Experiment: Fake Profile Fools Security Pros." NetworkWorld website. July 8, 2010. Retrieved April 7, 2014, from http://www.networkworld.com /news/2010/070810-the-robin-sage-experiment-fake.html.

Gorman, Siobhan, Yochi J. Dreazen, and August Cole. "Insurgents Hack U.S. Drones." *Wall Street Journal*, December 17, 2009. Retrieved March 1, 2014, from http://online.wsj.com/news /articles/SB126102247889095011.

Graham, Bradley. "Hackers Attack Via Chinese Web Sites." *Washington Post*, August 25, 2005. Retrieved May 4, 2014, from http://www.washingtonpost.com/wp-dyn/content/article /2005/08/24/AR2005082402318.html .

Greenberg, Andy. "Hackers Reveal Nasty New Car Attacks—With Me Behind the Wheel." *Forbes*, July 24, 2013. Retrieved June 3, 2014, from http://www.forbes.com/sites/ andygreenberg/2013/07/24/hackers-reveal-nasty-new-car-attacks-with-me-behind-the -wheel-video/.

———. "How the Syrian Electronic Army Hacked Us: A Detailed Timeline." *Forbes*, February 20, 2014. Retrieved March 4, 2014, from http://www.forbes.com/sites /andygreenberg/2014/02/20/how-the-syrian-electronic-army-hacked-us-a-detailed -timeline/.

———. *Sandworm: A New Era of Cyberwar and the Hunt for the Kremlin's Most Dangerous Hackers.* New York: Doubleday, 2019.

Greene, Thomas C. "Chapter One: Kevin Mitnick's Story: Here It Is." *The Register*, January 13, 2003. Retrieved April 19, 2014, from http://www.theregister.co.uk/2003/01/13/chapter _one_kevin_mitnicks_story/.

Greenwald, Glenn. "NSA Collecting Phone Records of Millions of Verizon Customers Daily." *The Guardian*, June 5, 2013. Retrieved March 23, 2014, from http://www.theguardian.com /world/2013/jun/06/nsa-phone-records-verizon-court-order.

Greenwald, Glenn, and Ewan MacAskill. "NSA Prism Program Taps into User Data of Apple, Google and Others." *The Guardian*, June 6, 2013. Retrieved May 17, 2014, from http://www .theguardian.com/world/2013/jun/06/us-tech-giants-nsa-data.

Grow, Brian, and Mark Hosenball. "Special Report: In Cyberspy vs. Cyberspy, China Has the Edge." Reuters. April 14, 2011. Retrieved May 4, 2014, from http://www.reuters.com /article/2011/04/14/us-china-usa-cyberespionage-idUSTRE73D24220110414?page Number=1.

Hagen, Christian, and Jeff Sorenson. "Delivering Military Software Affordably." *Defense AT&L Magazine* (March/April 2013): 30–34.

Halperin, Daniel, et al. "Pacemakers and Implantable Cardiac Defibrillators: Software Radio Attacks and Zero Power Defenses." Retrieved April 28, 2014, from http://www.secure-medicine.org /public/publications/icd-study.pdf.

Hang, Teo Cheng. "Non-Kinetic Warfare: The Reality and the Response." Singapore Ministry of Defense. 2010. Retrieved April 19, 2014, from http://www.mindef.gov.sg/imindef /publications/pointer/journals/2010/v36n1/feature5/_jcr_content/imindefPars/0003 /file.res/Pointer%20V36N1%20inside%2045-57.pdf.

Hansen, Marc, and Robert F. Nesbit. *Report of Defense Science Board Task Force on Defense Software.* Advisory Report, Washington, DC: Defense Science Board, 2000.

Harris, Shane. "Hack Attack." ForeignPolicy.com. March 3, 2014. Retrieved June 17, 2014, from http://www.foreignpolicy.com/articles/2014/03/03/hack_attack.

———. "Inside the FBI's Fight Against Chinese Cyber-Espionage." ForeignPolicy.com. May 27, 2014. Retrieved June 1, 2014, from http://www.foreignpolicy.com/articles/2014/05/27 /exclusive_inside_the_fbi_s_fight_against_chinese_cyber_espionage.

Hayden, Michael. Address to Kennedy Political Union of American University. February 17, 2000. Transcript retrieved on May 3, 2014, from http://www.fas.org/irp/news/2000/02 /dir021700.htm.

Healey, Jason, and Karl Frindal (editors). *A Fierce Domain: Conflict in Cyberspace, 1986 to 2012.* Cyber Conflict Studies Association, June 2013.

"Heartbleed Bug." Heartbleed.com. April 29, 2014. Retrieved May 13, 2014, from http:// heartbleed.com/.

Holmes, James, and John Posner. "Presentation to the House Armed Services Committee Subcommittee on Tactical Air and Land Forces U.S. House of Representatives." March 20, 2012. Retrieved May 23, 2014, from http://armedservices.house.gov/index.cfm/files /serve?File_id=2cb71c16-223f-4421-8272-279666f034ce.

Hoover, J. Nicholas. "Stolen VA Laptop Contains Personal Data." Dark Reading website. May 14, 2010. Retrieved May 27, 2014, from http://www.darkreading.com/risk-management/stolen -va-laptop-contains-personal-data/d/d-id/1089135.

Howard, Michael, and David LeBlanc. *Writing Secure Code.* 2nd ed. Redmond: Microsoft Press, 2002.

HP Security Research. "Companion to HPSR Threat Intelligence Podcast Episode 11." HP Security Research blog. February 2014. Retrieved March 26, 2014, from http://h30499 .www3.hp.com/hpeb/attachments/hpeb/off-by-on-software-security-blog/177/1 /Companion%20to%20HPSR%20Threat%20Intelligence%20Briefing%20Episode%2011 %20Final.pdf.

"H.R. 1960—113th Congress: National Defense Authorization Act for Fiscal Year 2014." U.S. Government Printing Office. July 8, 2013. Retrieved April 14, 2014, http://www.gpo.gov /fdsys/pkg/BILLS-113hr1960pcs/pdf/BILLS-113hr1960pcs.pdf.

Hutchins, Eric M., Michael J. Cloppert, and Rohan M. Amin. "Intelligence-Driven Computer Network Defense Informed by Analysis of Adversary Campaigns and Intrusion Kill Chains." *6th Annual International Conference on Information Warfare & Security.* Washington, DC: Academic Publishing International Limited, 2011.

Hyatt, Kyle. "New Study Shows Just How Bad Vehicle Hacking Has Gotten." CNet Road Show. December 18, 2019. Retrieved October 25, 2020, from https://www.cnet.com/roadshow /news/2019-automotive-cyber-hack-security-study-upstream/.

"Information Operations Roadmap." Washington, DC: Department of Defense, October 30, 2003. Retrieved March 10, 2014, from http://www2.gwu.edu/~nsarchiv/NSAEBB /NSAEBB177/info_ops_roadmap.pdf.

InformationWeek. "Cloud Providers Align With FedRAMP Security Standards." January 21, 2014. Retrieved June 7, 2014, from http://www.informationweek.com/government /cybersecurity/cloud-providers-align-with-fedramp-security-standards/d/d-id/1113499.

Internet Crime Complaint Center. "CryptoLocker Ransomware Encrypts User's Files." October 28, 2013. Retrieved May 17, 2014, from http://www.ic3.gov/media/2013/131028.aspx.

Israeli Defense Forces. "Yesterday the IDF Thwarted Cyber Attack; Today the IDF General Speaks About Future of Cyber Warfare." April 8, 2014. Retrieved June 1, 2014, from http://www .idfblog.com/2014/04/08/future-cyber-warfare-speech-head-idf-telecommunications -branch/.

Jardin, Xeni. "Prominent Tibetan Dissident Blogger Hacked, Impersonated on Skype." Boing Boing. May 28, 2008. Retrieved May 3, 2014, from http://boingboing.net/2008/05/28 /prominent-tibetan-di.html.

Jensen, Eric Talbot. "Cyber Attacks: Proportionality and Precautions in Attack." *International Law Studies 89* (2013): 198–217.

"Joint Publication 2-0: Joint Intelligence." Washington, DC: Joint Chiefs of Staff, October 22, 2013. Retrieved March 22, 2014, from http://www.dtic.mil/doctrine/new_pubs/jp2_0.pdf.

"Joint Publication 3-13: Information Operations (Change 1)." Washington, DC: Joint Chiefs of Staff, November 20, 2014. Retrieved May 6, 2021, from https://www.jcs.mil/Portals/36 /Documents/Doctrine/pubs/jp3_13.pdf.

"Joint Publication 3-13.1: Electronic Warfare." Washington, DC: Joint Chiefs of Staff, January 25, 2007. Retrieved March 10, 2014, from http://www.fas.org/irp/doddir/dod/jp3-13 -1.pdf.

"Joint Publication 3-13.3: Operations Security." Washington, DC: Joint Chiefs of Staff, January 6, 2016. Retrieved May 6, 2021, from https://media.defense.gov/2020/Oct/28/2002524944 /-1/-1/0/JP%203-13.3-OPSEC.PDF.

"Joint Publication 3-60: Joint Targeting." Washington, DC: Joint Chiefs of Staff, January 31, 2013. Retrieved March 30, 2014, from http://www.fas.org/irp//doddir/dod/jp3_60.pdf.

Kahn, David. *The Codebreakers: The Comprehensive History of Secret Communication from Ancient Times to the Internet.* Rev. Sub. ed. New York: Scribner, 1996.

Kahn, David. *Seizing the Enigma: The Race to Break the German U-Boats Codes, 1939–1943.* Boston: Houghton Mifflin, 1991.

Kan, Michael. "China Accuses Cisco of Supporting US Cyberwar Efforts." May 27, 2014. Retrieved June 8, 2014, from http://www.computerworld.com/s/article/9248595/China _accuses_Cisco_of_supporting_US_cyberwar_efforts.

Kaplan, David A. "Suspicions and Spies in Silicon Valley." *Newsweek*, September 17, 2006. Retrieved April 19, 2014, from http://www.newsweek.com/suspicions-and-spies-silicon -valley-109827.

Kelley, Michael. "The Stuxnet Attack on Iran's Nuclear Plant Was 'Far More Dangerous' Than Previously Thought." *Business Insider*, November 20, 2013. Retrieved May 4, 2014, from http://www.businessinsider.com/stuxnet-was-far-more-dangerous-than-previous-thought -2013-11#!IfUEr.

Kissel, Richard, Matthew Scholl, Steven Skolochenko, and Xing Li. "NIST Special Publication 800-88: Guidelines for Media Sanitization." National Institute of Standards and Technology website. September 2006. Retrieved May 27, 2014, from http://csrc.nist.gov/publications /nistpubs/800-88/NISTSP800-88_with-errata.pdf.

Krebs, Brian. "Amnesty International Site Serving Java Exploit." Krebs on Security. December 22, 2011. Retrieved May 17, 2014, from https://krebsonsecurity.com/2011/12/amnesty -international-site-serving-java-exploit/.

———. "The New Normal: 200-400 Gbps DDoS Attacks." Krebs on Security. February 14, 2014. Retrieved March 2, 2014, from http://krebsonsecurity.com/2014/02/the-new-normal -200-400-gbps-ddos-attacks/.

———. "Shadowy Russian Firm Seen as Conduit for Cybercrime." Washingtonpost.com. October 13, 2007. Retrieved April 23, 2014, from http://www.washingtonpost.com/wp-dyn/content /article/2007/10/12/AR2007101202461.html?sid=ST2007101202661.

———. "Stolen Laptop Exposes Personal Data on 207,000 Army Reservists." Krebs on Security. May 13, 2010. Retrieved May 27, 2014, from http://krebsonsecurity.com/2010/05/stolen -laptop-exposes-personal-data-on-207000-army-reservists/.

Krekel, Bryan A., Patton Adams, and George Bakos. *Occupying the Information High Ground: Chinese Capabilities for Computer Network Operations and Cyber Espionage.* Washington, DC: U.S. -China Economic and Security Review Commission, 2012.

"Kremlin Warns of Cyberwar After Report of U.S. Hacking into Russian Power Grid," *New York Times,* June 19, 2019.

Langner, Ralph. "To Kill a Centrifuge: A Technical Analysis of What Stuxnet's Creators Tried to Achieve." The Langner Group website. November 2013. Retrieved April 10, 2014, from http://www.langner.com/en/wp-content/uploads/2013/11/To-kill-a-centrifuge.pdf.

Le, Dong. "China Hopes to Dispel 'Copy Others' Reputation." British Broadcasting Corporation website. January 30, 2014. Retrieved June 3, 2014, from http://www.bbc.com/news /business-25944840.

Leed, Maren. "Offensive Cyber Capabilities at the Operational Level: The Way Ahead." Center for Strategic and International Studies website. September 2013. Retrieved April 10, 2014, from http://csis.org/files/publication/130916_Leed_OffensiveCyberCapabilities_Web.pdf.

Levy, Steven. *Hackers: Heroes of the Computer Revolution.* Garden City, NY: Anchor Press /Doubleday, 1984.

Leyden, John. "Anonymous Joins Forces with Arch-Enemy The Jester Against Norks." *The Register.* April 4, 2014. Retrieved April 6, 2014, from http://www.theregister. co.uk/2013/04/04/anon_nork_cyber_offensive/.

Libicki, Martin C. "Cyberdeterrence and Cyberwar." RAND Corporation. 2009. Retrieved May 8, 2014, from http://www.rand.org/content/dam/rand/pubs/monographs/2009/RAND _MG877.pdf.

Lieber, Francis. "General Orders No. 100." Retrieved March 11, 2014, from http://avalon.law .yale.edu/19th_century/lieber.asp.

Lipsky, Jessica. "China Bets on Homegrown OS." EE Times website. January 31, 2014. Retrieved June 3, 2014, from http://www.eetimes.com/document.asp?doc_id=1320848.

Lynn, William J., III. "Defending a New Domain: The Pentagon's Cyberstrategy." *Foreign Affairs Magazine* (September/October 2010). Retrieved April 29, 2014, from http://www .foreignaffairs.com/articles/66552/william-j-lynn-iii/defending-a-new-domain.

Mandiant. "APT1: Exposing One of China's Cyber Espionage Units." Mandiant Intelligence Center. February 18, 2013. Retrieved March 2, 2014, from http://intelreport.mandiant.com /Mandiant_APT1_Report.pdf.

Manning, David. "The Secret Downing Street Memo." *The Sunday Times,* May 1, 2005. Retrieved March 10, 2014, from http://web.archive.org/web/20110723222004/http://www .timesonline.co.uk/tol/news/uk/article387374.ece.

Markoff, John, and Thom Shanker. "Halted '03 Iraq Plan Illustrates U.S. Fear of Cyberwar Risk." *New York Times*, August 1, 2009. Retrieved March 13, 2014, from http://www.nytimes .com/2009/08/02/us/politics/02cyber.html.

Marlatt, Greta. "Information Warfare and Information Operations (IW/IO): A Bibliography." Naval Postgraduate School, January 2008. Retrieved March 10, 2014, from http://edocs .nps.edu/npspubs/scholarly/biblio/Jan08-IWall_biblio.pdf.

Maurer, Kevin. "'Psychological Operations' Are Now 'Military Information Support Operations.'" The Associated Press, July 2, 2010. Retrieved May 23, 2014, from http://publicintelligence .net/psychological-operations-are-now-military-information-support-operations/.

Maurer, Tim. "Cyber Norm Emergence at the United Nations—An Analysis of the UN's Activities Regarding Cyber-Security." Discussion paper 2011–11, Cambridge, MA: Belfer Center for Science and International Affair, Harvard Kennedy School, September 2011.

Melzer, Nils. "Cyberwarfare and International Law." United Nations Institute for Disarmament Research. 2011. Retrieved April 20, 2014, from http://www.unidir.org/files/publications /pdfs/cyberwarfare-and-international-law-382.pdf.

Menga, Rich. "How Long Does Backup Media Last?" March 2009. Retrieved May 26, 2014, from http://www.pcmech.com/article/how-long-does-backup-media-last/.

Moore, David, and Colleen Shannon. "The Spread of the Code-Red Worm (CRv2)." The Cooperative Association for Internet Data Analysis. 2001. Retrieved May 3, 2014, from http://www.caida.org/research/security/code-red/coderedv2_analysis.xml.

Muncaster, Phil. "Honker Union Sniffs 270 Hacktivism Targets." *The Register*. September 18, 2013. Retrieved May 3, 2014, from http://www.theregister.co.uk/2013/09/18/honker _union_270_japan_targets_manchurian_incident/.

Mulrine, Anna. "Welcome to CyberCity." *Air Force Magazine*, June 2013. Retrieved June 2, 2014, from http://www.airforcemag.com/MagazineArchive/Pages/2013/june%20 2013/0613cybercity.aspx.

Nakashima, Ellen. "Chinese Hackers Who Breached Google Gained Access to Sensitive Data, U.S. Officials Say." *Washington Post*, May 20, 2013. Retrieved March 23, 2014, from http://www .washingtonpost.com/world/national-security/chinese-hackers-who-breached-google -gained-access-to-sensitive-data-us-officials-say/2013/05/20/51330428-be34-11e2-89c9 -3be8095fe767_story.html.

Nakashima, Ellen et al. "U.S., Israel Developed Flame Computer Virus to Slow Iranian Nuclear Efforts, Officials Say." *Washington Post*, June 19, 2012. Retrieved May 4, 2014, from http:// www.washingtonpost.com/world/national-security/us-israel-developed-computer-virus-to -slow-iranian-nuclear-efforts-officials-say/2012/06/19/gJQA6xBPoV_story.html.

National Archives. "Frequently Asked Questions (FAQs) About Optical Storage Media: Storing Records on CDs and DVDs." n.d. Retrieved May 24, 2014, from http://www.archives.gov /records-mgmt/initiatives/temp-opmedia-faq.html.

National Geospatial-Intelligence Agency. "About NGA." Retrieved May 23, 2014, from https:// www1.nga.mil/About/Pages/default.aspx.

National Institute of Standards and Technology. "NIST Guide for Applying the Risk Management Framework to Federal Information Systems." February 2010. Retrieved May 11, 2014, from http://csrc.nist.gov/publications/nistpubs/800-37-rev1/sp800-37-rev1 -final.pdf.

———. "Special Publication 800-60, Revision 1, Guide for Mapping Types of Information and Information Systems to Security Categories." August 2008. Retrieved May 24, 2014, from http://csrc.nist.gov/publications/nistpubs/800-60-rev1/SP800-60_Vol1-Rev1.pdf.

National Security Agency. "IAD's Top 10 Information Assurance Mitigation Strategies." National Security Agency website. November 2013. Retrieved March 16, 2014, from http://www.nsa.gov/ia/_files/factsheets/I43V_Slick_Sheets/Slicksheet_Top10IAMitigationStrategies_Web.pdf.

———. "Manageable Network Plan." April 2012. Retrieved May 23, 2014, from http://iase.disa.mil/cgs/downloads/implementation/Manageable_Network_Plan.pdf.

———. "Securing Data and Handling Spillage Events." October 2012. Retrieved May 22, 2014, from http://www.nsa.gov/ia/_files/factsheets/Final_Data_Spill.pdf.

Neil Jr. "Spy Agency Taps into Undersea Cable." May 23, 2001. Retrieved May 18, 2014, from http://www.zdnet.com/news/spy-agency-taps-into-undersea-cable/115877.

Ney, Paul. "DOD General Counsel Remarks at U.S. Cyber Command Legal Conference." Retrieved March 2, 2020, from https://www.defense.gov/Newsroom/Speeches/Speech/Article/2099378/dod-general-counsel-remarks-at-us-cyber-command-legal-conference/.

"NSA Spied on Vatican and Top Catholic Cardinals, Says Italian Report." *Huffington Post*, October 31, 2013. Retrieved May 17, 2014, from http://www.huffingtonpost.com/2013/10/30/nsa-vatican_n_4177882.html.

Obama, Barack H. "Executive Order 13526 - Classified National Security Information." Whitehouse.gov website. December 29, 2009. Retrieved May 11, 2014, from http://www.whitehouse.gov/the-press-office/executive-order-classified-national-security-information.

———. "Executive Order 13556—Controlled Unclassified Information." Whitehouse.gov website. November 4, 2010. Retrieved May 11, 2014, from http://www.whitehouse.gov/the-press-office/2010/11/04/executive-order-controlled-unclassified-information.

———. "Guide to Recordkeeping in the Army." August 2008. Retrieved May 26, 2014, from http://www.apd.army.mil/jw2/xmldemo/p25_403/main.asp.

———. *National Security Strategy*. Washington, DC: The White House, May 2010. Retrieved March 10, 2014, from http://www.whitehouse.gov/sites/default/files/rss_viewer/national_security_strategy.pdf.

O'Connor, T. J. "An Analysis of Jester's QR Code Attack. (Guest Diary)." Infosec Handlers Diary blog. March 3, 2012. Retrieved April 6, 2014, from http://isc.sans.edu/diary/An+Analysis+of+Jester+s+QR+Code+Attack+Guest+Diary+/12760.

———. "The Jester Dynamic: A Lesson in Asymmetric Unmanaged Cyber Warfare." SANS Institute Reading Room. December 30, 2011. Retrieved April 6, 2014, from http://www.sans.org/reading-room/whitepapers/attacking/jester-dynamic-lesson-asymmetric -unmanaged-cyber-warfare-33889.

Office of the Director of National Intelligence. "Data Gathering." Intelligence.gov. Retrieved March 6, 2014, from http://www.intelligence.gov/mission/data-gathering.html.

Office of the Director of National Intelligence: Office of General Counsel. *Intelligence Community Legal Reference Book—2012*. dni.gov, 2012. Retrieved March 7, 2014, from http://www.dni.gov/files/documents/IC_Legal_Ref_2012.pdf.

Office of the National Counterintelligence Executive. "Annual Report to Congress on Foreign Economic Collection and Industrial Espionage." February 2004. Retrieved on May 17, 2014, from http://www.ncix.gov/publications/reports/fecie_all/fecie_2003/fecie_2003.pdf.

———. "Foreign Spies Stealing U.S. Economic Secrets in Cyberspace: A Report to Congress on Foreign Economic Collection and Industrial Espionage, 2009–2011." Office of the National Counterintelligence Executive, October 2011. Retrieved March 9, 2014, from http://www.ncix.gov/publications/reports/fecie_all/Foreign_Economic_Collection_2011.pdf.

Office of the Secretary of Defense. "Report of the Office of the Secretary of Defense Vietnam Task Force." January 15, 1969. Retrieved May 17, 2014, from http://www.archives.gov/research/pentagon-papers/.

Olson, Parmy. *We Are Anonymous: Inside the Hacker World of LulzSec, Anonymous, and the Global Cyber Insurgency*. New York: Little, Brown and Company, 2012.

Panda Security. "Annual Report PandaLabs 2013 Summary." Retrieved June 7, 2014, from http://press.pandasecurity.com/wp-content/uploads/2010/05/PandaLabs-Annual-Report_2013.pdf.

Parrish, Karen. "Lynn: Cyber Strategy's Thrust Is Defensive." July 14, 2011. American Forces Press Service. Retrieved June 1, 2014, from http://www.defense.gov/news/newsarticle.aspx?id=64682.

Perlroth, Nicole. "Cyberattack on Saudi Firm Disquiets U.S." *New York Times*, October 24, 2012: A1.

Perrone, Jane. "The Echelon Spy Network." *The Guardian*, May 29, 2001. Retrieved March 23, 2014, from http://www.theguardian.com/world/2001/may/29/qanda.janeperrone.

Pham, Sherisse. "How Much Has the US Lost from China's IP Theft?" CNN Business, March 23, 2018. Retrieved October 25, 2020, from https://money.cnn.com/2018/03/23/technology/china-us-trump-tariffs-ip-theft/index.html#:~:text=Total%20theft%20of%20US%20trade,the%20most%20of%20that%20theft.

Power, Richard. "The Solar Sunrise Case: Mak, Stimpy and Analyzer Give the DoD a Run for Its Money." InformIT. October 30, 2000. Retrieved May 3, 2014, from http://www.informit.com/articles/article.aspx?p=19603&seqNum=4.

Reichert, Corinne. "Payroll Data for 29,000 Facebook Employees Stolen." C|Net, December 13, 2019. Retrieved October 29, 2020, from https://www.cnet.com/news/payroll-data-of-29000-facebook-employees-reportedly-stolen/.

Reuters. "Aramco Says Cyberattack Was Aimed at Production." *New York Times*, December 10, 2012: B2.

Riley, Michael. "Obama Invokes Cold-War Security Powers to Unmask Chinese Telecom Spyware." Bloomberg website. November 30, 2011. Retrieved May 18, 2014, from http://www.bloomberg.com/news/2011-11-30/obama-invokes-cold-war-security-powers-to-unmask-chinese-telecom-spyware.html.

Roculan, Jensenne et al. "SQLExp SQL Server Worm Analysis." Symantec DeepSight Threat Management System Threat Analysis. January 28, 2003. Retrieved May 4, 2014, from http://securityresponse.symantec.com/avcenter/Analysis-SQLExp.pdf.

RSA FraudAction Research Labs. "Anatomy of an Attack." April 2011. Retrieved March 29, 2014, from https://blogs.rsa.com/anatomy-of-an-attack/.

Rumsfeld, Donald. "Annual Report to Congress: The Military Power of the People's Republic of China 2005." Office of the Secretary of Defense. July 2005. Retrieved May 4, 2014, from http://www.defense.gov/news/Jul2005/d20050719china.pdf.

Ryan, Thomas. "Getting in Bed with Robin Sage." Blackhat website. July 2010. Retrieved April 7, 2014, from http://media.blackhat.com/bh-us-10/whitepapers/Ryan/BlackHat-USA-2010-Ryan-Getting-In-Bed-With-Robin-Sage-v1.0.pdf.

Sanger, David E. "Obama Order Sped Up Wave of Cyberattacks Against Iran." *New York Times*, June 1, 2012. Retrieved May 23, 2014, from http://www.nytimes.com/2012/06/01/world/middleeast/obama-ordered-wave-of-cyberattacks-against-iran.html?pagewanted=all&_r=0.

———. "Syria War Stirs New U.S. Debate on Cyberattacks." *New York Times*, February 24, 2014.

———. Interview by Terry Gross. "While Warning of Chinese Cyberthreat, U.S. Launches Its Own Attack." *Fresh Air*. National Public Radio. April 2, 2014.

Sanger, David E., and Thom Shanker. "NSA Devises Radio Pathway into Computers." *New York Times*, January 14, 2014. Retrieved March 23, 2014, from http://www.nytimes .com/2014/01/15/us/nsa-effort-pries-open-computers-not-connected-to-internet.html?hp& _r=0.

Satter, Raphael. "U.S. General: We Hacked the Enemy in Afghanistan." *USA Today*, August 24, 2012.

Schafer, Sarah. "With Capital in Panic, Pizza Deliveries Soar." *Washington Post*, December 19, 1998. Retrieved March 10, 2014, from http://www.washingtonpost.com/wp-srv/politics /special/clinton/stories/pizza121998.htm.

Schmitt, Michael N., ed. *Tallinn Manual on the International Law Applicable to Cyber Warfare*. New York: Cambridge University Press, 2013.

Schneier, Bruce. "How the NSA Attacks Tor/Firefox Users with QUANTUM and FOXACID." Schneier on Security. October 7, 2013. Retrieved March 12, 2014, from https://www .schneier.com/blog/archives/2013/10/how_the_nsa_att.html.

———. "The NSA's New Risk Analysis." Schneier on Security. October 9, 2013. Retrieved May 4, 2014, from https://www.schneier.com/blog/archives/2013/10/the_nsas_new_ri.html.

Schwartau, Winn. *Information Warfare: Chaos on the Electronic Superhighway*. Berkeley, CA: Publishers Group West, 1994.

Secretary of the Air Force. "Legal Reviews of Weapons and Cyber Capabilities." Federation of American Scientists. July 27, 2011. Retrieved April 10, 2014, from http://static.e-publishing .af.mil/production/1/af_a3_5/publication/afi51-402/afi51-402.pdf.

Shakarian, Paulo. "The 2008 Russian Cyber Campaign Against Georgia." *Military Review* (2011): 63–68.

Shimomura, Tsutomo, and John Markoff. *Takedown: The Pursuit and Capture of Kevin Mitnick, America's Most Wanted Computer Outlaw—By the Man Who Did It*. New York: Hyperion, 1996.

Sieberg, Daniel. "Report: Hacker Infiltrated Government Computers." CNN.com. May 10, 2005. Retrieved April 23, 2014, from http://www.cnn.com/2005/TECH/05/10/govt.computer .hacker/.

Singer, Abe. "Tempting Fate." ;login (February 2005).

Singh, Simon. *The Code Book*. New York: Doubleday, 1999.

Soumenkov, Igor. "The Mystery of the Duqu Framework Solved." SecureList. March 14, 2012. Retrieved June 7, 2014, from https://www.securelist.com/en/blog/677/The_mystery_of _Duqu_Framework_solved.

Speigel Staff. "Inside TAO: Documents Reveal Top NSA Hacking Unit." Speigel Online: International website. December 29, 2013. Retrieved May 5, 2014, from http://www .spiegel.de/international/world/the-nsa-uses-powerful-toolbox-in-effort-to-spy-on-global -networks-a-940969.html.

Stamos, Alex. "Aurora Response Recommendations." iSEC Partners. February 17, 2010. Retrieved May 4, 2014, from https://www.isecpartners.com/media/10932/isec_aurora _response_recommendations.pdf.

Stewart, Joe. "Operation Aurora: Clues in the Code." Dell SecureWorks Research blog. January 19, 2010. Retrieved May 3, 2014, from http://www.secureworks.com/resources/blog /research/research-20913/.

Stoll, Clifford. "Stalking the Wily Hacker." *Communications of the ACM 31*, no. 5 (May 1988): 484–500.

Tenable Network Security, Inc. "Tenable Delivers Best-of-Breed Configuration Compliance and Vulnerability Managment for U.S. Department of Defense." Tenable website. 2013. Retrieved April 29, 2014, from http://www.tenable.com/sites/drupal.dmz.tenablesecurity.com/files/case-studies/ACAS_CS_(EN)_v3_web.pdf.

Traynor, Ian. "Russia Accused of Unleasing Cyberwar to Disable Estonia." *The Guardian*, May 16, 2007. Retrieved May 17, 2014, from http://www.theguardian.com/world/2007/may/17/topstories3.russia.

Tzu, Sun. *The Art of War*. Calgary, Alberta: Theophania Publishing, 2011.

"U.K. 'Spied on UN's Kofi Annan.'" British Broadcasting Company. February 26, 2004. Retrieved May 17, 2014, from http://news.bbc.co.uk/2/hi/uk_news/politics/3488548.stm.

United Nations. "Report of the Group of Governmental Experts on Developments in the Field of Information and Telecommunications in the Context of International Security." The United Nations website. June 24, 2013. Retrieved March 12, 2014, from http://www.un.org/ga/search/view_doc.asp?symbol=A/68/98.

United Nations. "Report of the Group of Governmental Experts on Developments in the Field of Information . "Responsibility of States for Internationally Wrongful Acts. The United Nations International Law Commission. Retrieved April 22, 2014, from http://legal.un.org/ilc/texts/instruments/english/draft%20articles/9_6_2001.pdf.

United Nations Office on Drugs and Crime. "Comprehensive Study on Cybercrime." United Nations Office on Drugs and Crime website. February 2013. Retrieved April 23, 2014, from http://www.unodc.org/documents/organized-crime/UNODC_CCPCJ_EG.4_2013/CYBERCRIME_STUDY_210213.pdf.

United States Air Force. "Air Force Doctrine Document 3-12, Cyberspace Operations." July 15, 2010. Retrieved May 18, 2014, from http://www.fas.org/irp/doddir/usaf/afdd3-12.pdf.

United States Army. "Army Doctrine Publication 3-0, Unified Land Operations." October 2011. Retrieved May 24, 2014, from http://usarmy.vo.llnwd.net/e2/rv5_downloads/info/references/ADP_3-0_ULO_Oct_2011_APD.pdf.

United States Army. "Army Field Manual 3-0, Operations." February 27, 2008. Retrieved May 24, 2014, from http://www.fas.org/irp/doddir/army/fm3-0.pdf.

United States Cyber Command. *Achieve and Maintain Cyberspace Superiority.* Retrieved May 5, 2021, from https://www.cybercom.mil/Portals/56/Documents/USCYBERCOM%20Vision%20April%202018.pdf.

United States Department of Defense. *CAC Security*. March 30, 2014. Retrieved March 30, 2014, from http://cac.mil/common-access-card/cac-security/.

———. "Compendium of Key Joint Doctrine Publications." Defense Technical Information Center. November 8, 2010. Retrieved May 24, 2014, from http://www.dtic.mil/doctrine/new_pubs/jp1_02.pdf.

———. "Department of Defense Cybersecurity Culture and Compliance Initiative (DC3I)". dod.defense.gov website. Retrieved November 1, 2020, from https://dod.defense.gov/Portals/1/Documents/pubs/OSD011517-15-RES-Final.pdf.

———. "Department of Defense Information Enterprise Strategic Plan 2010–2012." April 2010. Retrieved May 23, 2014, from http://dodcio.defense.gov/Portals/0/Documents/ISE/DoDIESP-r16.pdf.

———. "Department of Defense Instruction 8500.01: Cybersecurity." Defense Technical Information Center Online. March 14, 2014. Retrieved May 18, 2014, from http://www.dtic.mil/whs/directives/corres/pdf/850001_2014.pdf.

———. "Department of Defense Instruction 8510.01, Risk Management Framework (RMF) for DoD Information Technology (IT)." July 28, 2017. Executive Services Directorate website.

Retrieved October 30, 2020, from https://www.esd.whs.mil/Portals/54/Documents/DD/issuances/dodi/851001p.pdf?ver=2019-02-26-101520-300.

———. "Department of Defense Strategy for Operating in Cyberspace." U.S. Department of Defense website. July 2011. Retrieved April 27, 2014, from http://www.defense.gov/home/features/2011/0411_cyberstrategy/docs/DoD_Strategy_for_Operating_in_Cyberspace_July_2011.pdf.

———. "DoD Cybersecurity Discipline Implementation Plan" dodcio.defense.gov website. October 2015. Retrieved November 1, 2020, from https://dodcio.defense.gov/Portals/0/Documents/Cyber/CyberDis-ImpPlan.pdf.

———. "DOD Directive 2311.01E: Law of War Program." Defense Technical Information Center Online. May 9, 2006. Retrieved February 27, 2014, from http://www.dtic.mil/whs/directives/corres/pdf/231101e.pdf.

———. "DoD Instruction 3020.45 Mission Assurance (MA) Construct". Executive Services Directorate website. August 14, 2018. Retrieved October 30, 2020, from https://www.esd.whs.mil/Portals/54/Documents/DD/issuances/dodi/302045p.pdf?ver=2018-08-14-081232-450.

———. "DoD Instruction 5000.02 Operation of the Adaptive Acquisition Framework." Executive Services Directorate. January 23, 2020. Retrieved November 1, 2020, from https://www.esd.whs.mil/Portals/54/Documents/DD/issuances/dodi/500002p.pdf?ver=2019-05-01-151755-110.

———. "Joint Publication 1: Doctrine for the Armed Forces of the United States." Defense Technical Information Center. March 25, 2013. Retrieved May 24, 2014, from http://www.dtic.mil/doctrine/new_pubs/jp1.pdf.

———. "Joint Publication 1-02: Department of Defense Dictionary of Military and Associated Terms." Defense Technical Information Center. November 8, 2010. Retrieved March 17, 2014, from http://www.dtic.mil/doctrine/new_pubs/jp1_02.pdf.

———. "Recommended Practice: Improving Industrial Control Systems Cybersecurity with Defense-In-Depth Strategies." October 2009. Retrieved March 21, 2014, from https://ics-cert.us-cert.gov/sites/default/files/recommended_practices/Defense_in_Depth_Oct09.pdf.

United States Department of Defense CIO. "Encryption of Sensitive Unclassified Data at Rest on Mobile Computing Devices and Removable Storage Media." DISA Information Assurance Support Environment. July 3, 2007. Retrieved May 27, 2014, from http://iase.disa.mil/policy-guidance/dod-dar-tpm-decree07-03-07.pdf.

United States Department of Justice. "Overview of the Law Enforcement Strategy to Combat International Organized Crime." April 2008. Retrieved May 17, 2014, from http://www.justice.gov/criminal/icitap/pr/2008/04-23-08combat-intl-crime-overview.pdf.

United States Government Accountability Office. "GAO-11-421: Defense Department Cyber Efforts—More Detailed Guidance Needed to Ensure Military Services Develop Appropriate Cyberspace Capabilities." General Accountability Office website. May 2011. Retrieved May 18, 2014, from http://www.gao.gov/assets/320/318604.pdf.

———. "Future Warfare: Army Is Preparing for Cyber and Electronic Warfare Threats, but Needs to Fully Assess the Staffing, Equipping, and Training of New Organizations." August 2019. Retrieved May 5, 2021 from https://www.gao.gov/assets/710/700940.pdf.

United States Supreme Court. *New York Times Co. v. United States.* June 30, 1971. Retrieved May 17, 2014, from http://www.gwu.edu/~nsarchiv/NSAEBB/NSAEBB48/decision.pdf.

"Unveiling 'Careto'—The Masked APT." Kaspersky Lab. February 2014. Retrieved May 4, 2014, from http://www.securelist.com/en/downloads/vlpdfs/unveilingthemask_v1.0.pdf.

Verton, Dan. E-mail Correspondence. InfoSec News mailing list. October 12, 2001. Retrieved May 4, 2014, from http://seclists.org/isn/2001/Oct/88.

Vise, David A. *The Bureau and the Mole: The Unmasking of Robert Philip Hanssen, the Most Dangerous Double Agent in FBI History*. New York: Grove Publishers, 2001.

Vistica, Gregory. "Inside the Secret Cyberwar: Facing Unseen Enemies, the Feds Try to Stay a Step Ahead." *Newsweek*, February 21, 2000: 48.

———. "We're in the Middle of a Cyberwar." *Newsweek*, September 20, 1999: 50.

"W32.Duqu: The Precursor to the Next Stuxnet." Symantec Security Response. November 23, 2011. Retrieved May 4, 2014, from http://www.symantec.com/content/en/us/enterprise /media/security_response/whitepapers/w32_duqu_the_precursor_to_the_next_stuxnet.pdf.

Waterman, Shaun. "Fictitious Femme Fatale Fooled Cybersecurity." *Washington Times*, July 18, 2010. Retrieved April 7, 2014, from http://www.washingtontimes.com/news/2010/jul/18 /fictitious-femme-fatale-fooled-cybersecurity/.

Webster, William H. *A Review of FBI Security Programs*. Commission for the Review of FBI Security Programs, U.S. Department of Justice, March 2002. Retrieved March 23, 2014, from http://www.fas.org/irp/agency/doj/fbi/websterreport.html.

Welsh, William. "Cyber Warriors: The Next Generation." Defense Systems website. January 23, 2014. Retrieved May 18, 2014, from http://defensesystems.com/Articles/2014/01/23/Next -generation-cyber-warriors.aspx.

Winkler, J. R., C. J. O'Shea, and M. C. Stokrp. "Information Warfare, INFOSEC, and Dynamic Information Defense." PRC, Inc. 1996. Retrieved March 29, 2014, from http://csrc.nist.gov /nissc/1996/papers/NISSC96/paper016/jrwink.pdf.

Zetter, Kim. "'The Analyzer' Gets Time Served for Million-Dollar Bank Heist." Wired Threat Level. July 5, 2012. Retrieved May 3, 2014, from http://www.wired.com/2012/07/tenenbaum -sentenced/.

———. "Obama: NSA Must Reveal Bugs Like Heartbleed, Unless They Help the NSA." Wired Threat Level. April 15, 2014. Retrieved May 18, 2014, from http://www.wired .com/2014/04/obama-zero-day/.

Index

Index